**ELEMENTS OF
SPACE TECHNOLOGY**
FOR AEROSPACE ENGINEERS

ELEMENTS OF SPACE TECHNOLOGY
FOR AEROSPACE ENGINEERS

RUDOLF X. MEYER

Department of Mechanical and Aerospace Engineering
University of California
Los Angeles, California

ACADEMIC PRESS

San Diego London Boston
New York Sydney Tokyo Toronto

This book is printed on acid-free paper. ∞

Copyright © 1999 by Academic Press

All rights reserved.
No part of this publication may be reproduced or
transmitted in any form or by any means, electronic
or mechanical, including photocopy, recording, or
any information storage and retrieval system, without
permission in writing from the publisher.

ACADEMIC PRESS
525 B. St. Suite 1900, San Diego, California 92101-4495, USA
http://www.apnet.com

Academic Press
24-28 Oval Road, London NW1 7DX, UK
http://www.hbuk.co.uk/ap/

Library of Congress Cataloging-in-Publication Number
Meyer, Rudolf X.
    Elements of space technology for aerospace engineers / Rudolf X. Meyer.
      p.   cm.
    Includes bibliographical references and index.
    ISBN 0-12-492940-0
    1. Space vehicles—Design and construction. 2. Aerospace engineering. 3. Astronautics. I. Title. II. Title: Elements of space technology
  TL795 .M48   1999
  629.4—ddc21
                                                       98-52665
                                                          CIP

Printed in the United States of America
99 00 01 02 03 IP 9 8 7 6 5 4 3 2 1

CONTENTS

Preface		ix
CHAPTER 1.	REFERENCE FRAMES AND TIME	1
	1.1 Reference Frames	1
	1.2 Motion in Accelerated Reference Frames	7
	1.3* Example: The Yo-Yo Despin Mechanism	9
	1.4 Euler Angles and Transformations of Coordinates	12
	1.5 Time Intervals and Epoch	14
CHAPTER 2.	FORCES AND MOMENTS	21
	2.1 Gravity	21
	2.2 Thrust	31
	2.3 Aerodynamic Forces and Moments	37
	2.4* Free Molecule Flow	40
	2.5* Solar Radiation Pressure	47
	2.6 Atmospheric Entry	49
CHAPTER 3.	ORBITS AND TRAJECTORIES IN AN INVERSE SQUARE FIELD	59
	3.1 Kepler Orbits and Trajectories	60
	3.2 Position as a Function of Time	65
	3.3* D'Alembert and Fourier–Bessel Series	66
	3.4 Orbital Elements	68
	3.5* Spacecraft Visibility above the Horizon	71
	3.6* Satellite Observations and the f and g Series	73
	3.7 Special Orbits	75
	3.8* Perturbations by Other Astronomical Bodies	81
	3.9* Planetary Flyby and Gravity Assist	85
	3.10* Relativistic Effects	91
CHAPTER 4.	CHEMICAL ROCKET PROPULSION	97
	4.1 Configurations of Liquid-Propellant Chemical Rocket Motors	99
	4.2 Configurations of Solid-Propellant Motors	101

v

4.3*	Rocket Stages	103
4.4	Idealized Model of Chemical Rocket Motors	111
4.5	Ideal Thrust	116
4.6	Rocket Motor Operation in the Atmosphere	117
4.7*	Two- and Three-Dimensional Effects	122
4.8	Critique of the Ideal Model	126
4.9*	Elements of Chemical Kinetics	127
4.10*	Chemical Kinetics Applications to Rocket Motors	135
4.11	Liquid Propellants	139
4.12*	Propellant Tanks	143
4.13*	Propellant Feed Systems of Launch Vehicles	147
4.14	Thrust Chambers of Liquid-Propellant Motors	155
4.15*	Pogo Instability and Prevention	158
4.16*	Thrust Vector Control	160
4.17*	Engine Control and Operations	162
4.18	Liquid-Propellant Motors and Thrusters on Spacecraft	165
4.19	Components of Solid-Propellant Rocket Motors	170
4.20	Hybrid-Propellant Rocket Motors	175

CHAPTER 5. ORBITAL MANEUVERS — 181

5.1	Minimum Energy Paths	181
5.2*	Lambert's Theorem	184
5.3	Maneuvers with Impulsive Thrust	188
5.4	Hohmann Transfers	192
5.5*	Other Transfer Trajectories	194
5.6	On-Orbit Drift	196
5.7	Launch Windows	197
5.8*	Injection Errors and Their Corrections	199
5.9*	On-Orbit Phase Changes	202
5.10*	Rendezvous Maneuvers	205
5.11*	Gravity Turn	209

CHAPTER 6. ATTITUDE CONTROL — 215

6.1	Principal Axes and Moments of Inertia of Spacecraft	218
6.2*	The Euler Equations for Time-Dependent Moments of Inertia	223
6.3	The Torque-Free Spinning Body	225
6.4	Attitude Control Sensors	230
6.5	Attitude Control Actuators	240
6.6	Spin-Stabilized Vehicles	251
6.7*	Gravity Gradient Stabilization	262

CHAPTER 7. SPACECRAFT THERMAL DESIGN — 269

7.1*	Fundamentals of Thermal Radiation	269
7.2	Spacecraft Surface Materials	278

7.3	Model of a Spacecraft as an Isothermal Sphere	281
7.4	Earth Thermal Radiation and Albedo	284
7.5*	Diurnal and Annual Variations of Solar Heating	285
7.6*	Thermal Blankets	286
7.7*	Thermal Conduction	289
7.8	Lumped Parameter Model of a Spacecraft	290
7.9	Thermal Control Devices	299

Appendices   307

Index   325

PREFACE

This book is intended as a first introduction to space technology for aerospace engineering students, at either the senior or graduate level. It is hoped also to be useful to professional engineers who wish to become more familiar with the aerospace aspects of space systems.

It would have been tempting to include all major disciplines that are represented in space systems engineering, including electronics, space communications, and even the civil engineering of launch site construction. That such a broad, interdisciplinary approach would have merit is suggested by the fact that quite often the effort (and cost) in developing a new spacecraft and its ground support is about equally divided between the mechanical and electronic systems. However, the division of most engineering curricula into separate aerospace and electronic engineering disciplines would have made such a collection quite impractical as a textbook. The concentration here is therefore on the mechanical ("aerospace") aspects, omitting the equally important electronic and communications disciplines. Nevertheless, in instances in which the disciplines clearly overlap, as in the thermal control of electronic components, an effort has been made to broaden the scope sufficiently to make the student aware of the problems and solutions that are addressed by the electronics and communications engineers.

Most of the material in this book can be covered by graduate students in one semester and similarly by seniors at American universities. Several parts, particularly in Chapter 4 on the chemistry of propellants are suited for reading by the student, without need for class presentation. Major parts have been taught by the author to engineering students at the University of California, Los Angeles.

The book has been arranged such that, at the choice of the instructor, entire sections can be omitted without losing continuity. These sections, which sometimes contain more advanced topics, are indicated in the Table of Contents by an asterisk. The problems at the end of each chapter that relate to these sections are indicated in the same manner.

Chapters 1 and 2 are preliminary to the discussion of orbits and trajectories in Chapters 3 and 5. Some of this material may already be familiar to the reader but may be used for review or may be skipped. Classical orbital mechanics is treated relatively briefly, forgoing the elegance gained by Hamiltonian mechanics. The analysis of spacecraft orbits, in good part borrowed from classical celestial mechanics, has become the specialty of numerical analysts. Instead, more importance in this book has been placed on the trajectories encountered in rendezvous and docking maneuvers and on launch trajectories with thrust and aerodynamic drag.

For lack of space in the typical aerospace curriculum, more specialized topics, such as spacecraft testing methods or reliability analyses, had to be omitted or are mentioned only very briefly. Similarly, topics that are usually dealt with elsewhere in the curriculum are introduced only in instances in which their application to space technology is unique in some sense. In spite of their importance, the design and analysis of launch vehicle and spacecraft structures are therefore hardly touched upon in this book because the principles involved are very similar to those of aircraft structures, a topic that usually precedes this in the aerospace curriculum. Similarly, in the chapter on attitude control, the emphasis is on sensors and actuators, not on control theory, in which specialized courses are offered at most institutions. The author may also have to apologize for not having included a chapter on space mission design. The excuse is that at present mission designs and their future capabilities change so rapidly that it seemed better to limit the lecture course to the fundamentals that are likely to have more permanent validity.

Because the aim is to emphasize the more fundamental aspects, the treatment must necessarily be in good part theoretical. However, the author hopes that he has resisted successfully the temptation to make the mathematical arguments appear to be more sophisticated than warranted by the subject matter. Where it seemed desirable, some brief mathematical comments are made. In spite of this emphasis on the scientific and engineering foundations and on analysis, an effort has been made to include the more practical aspects encountered in the design of space vehicles and their components whenever these topics were felt to be more than ephemeral, soon to be replaced by more advanced designs. Emphasis has also been placed on providing a "feel for numbers" so that the reader may develop a sense of what is numerically important and what is not.

Each chapter is followed by some exercises, varying from the simple to the more difficult. It is the author's contention, however, that such exercises, which can be stated in a few lines, are less valuable to the engineering student than assigned projects, be they mission analyses or hardware designs. To support such projects, space was found to include at least some of the more important technical data in an appendix.

Some technicalities:

Throughout this text SI units are used. Only in a very few instances, and only when sanctioned by almost universal usage, will other units, such as electron volts for energy, appear. ESA and now also NASA publications have been using exclusively the SI units, with only an occasional and parenthetical mention of English units when dictated by very long usage, and as still occur in major data collections.

For the convenience of the reader, the principal symbols used are summarized at the end of each chapter, often with the number of the defining equation.

Many of the subscripts are used uniformly throughout this book. They are, with their meaning:

$( )_a$	ambient	$( )_h$	sun; solar
$( )_{av}$	average	$( )_{pl}$	planet
$( )_{el}$	electric	$( )_{pr}$	propellant
$( )_{ex}$	nozzle exit	$( )_s$	spacecraft
$( )_g$	earth; also gravitational	$( )_{si}$	sidereal
$( )_G$	Greenwich		

To help distinguish the forest from the trees, equation numbers that are printed in bold font indicate that the equation is either important, likely to be referred to in a later chapter, or else the final result of a preceding development. Simple corollaries or special cases of a preceding equation are given the same number, but with a prime.

As would seem appropriate for a textbook of this type, journal papers or monographs are referenced only when the text or illustrations are so close to the original that it would be unprofessional not to refer to the author. Instead, references are mainly to books, monographs, or data collections, where the reader may find additional material or different presentations.

It remains for me to thank my colleagues and friends, far too many to mention by name here, at *The Aerospace Corporation* and at the *University of California* who have inspired much of this work. Space technology is a national and international enterprise with many workers. To them go my thanks.

Ad astra pontem fecerunt fabri, in majorem Dei gloriam.

1

Reference Frames and Time

This chapter includes a discussion of the reference frames commonly used in space vehicle dynamics. They are employed to describe and analyze orbits and trajectories and also to describe the orientation in space ("attitude") of the vehicle. Euler angles are introduced to characterize the rotation of reference frames.

Because of the importance of relative motion in such problems as fluid sloshing in the vehicle's propellant tanks or the effect of vehicle motion on gyroscopes, the transformation equations for velocities and accelerations are rederived. Although the reader will already be familiar with this topic and other parts of this chapter, they are included because later chapters will refer to them. A second purpose is to introduce notation that will remain uniform for the remainder of this book.

The chapter concludes with a discussion of the various time measurement systems that are in use in the design of space missions.

1.1 Reference Frames

A quantitative description of the motion of space vehicles, be they launch vehicles, upper stage vehicles, artificial satellites, or deep-space probes, requires one or several *reference frames* with respect to which the vehicle's position, velocity, and acceleration as functions of time are defined. Reference frames are also needed to describe the vehicle's attitude relative, for instance, to points on the earth, to the sun, or to guide stars. Similarly, the time derivatives of the vehicle's attitude, particularly angular rates and angular accelerations, are often needed. Implied in the use of reference frames is also the existence of a precisely defined, "uniformly" evolving time.

Easily defined reference frames are those that are rigidly attached to the earth, to some other astronomical body, or to the structure—assumed rigid—of the space vehicle itself. This latter frame is useful, for instance, in studying the dynamic effects on the vehicle resulting from the deployment of solar panels, antennas, or scientific probes. Although easily defined, these reference frames will generally be accelerating; that is, they may be rotating or may also accelerate by parallel translation. Table 1.1 contains approximate values of the accelerations of several of such reference frames, expressed by their ratios to the standard gravity $g_0 = 9.80665$ m/s^2.

The first few accelerations listed in Table 1.1 are large and must be taken into account, together with other residual accelerations of comparable magnitude, in the planning of typical space missions. Particularly sensitive to small residual accelerations are missions in which a satellite is required to

Table 1.1 Magnitudes of Various Residual Accelerations

Acceleration at the equator due to the planet's rotation	
Earth	$3.44 \; 10^{-3} \; g_0$
Mars	$1.74 \; 10^{-3} \; g_0$
Acceleration on the earth's surface	
Due to earth–moon mutual attraction, with moon at zenith	$1.13 \; 10^{-7} \; g_0$
Due to sun–earth mutual attraction, with sun at zenith	$0.53 \; 10^{-7} \; g_0$
Acceleration (at the solar system location) due to rotation about the galactic center	$2.24 \; 10^{-11} \; g_0$

execute, during its lifetime, a large number of orbits about a planet or moon. Unless corrected, small errors will tend to accumulate and will lead to large deviations from the intended path. On the other hand, the effect produced by the rotation of the solar system about the galactic center as a part of a spiral arm of our galaxy is sufficiently small to be negligible for present purposes.

Newton's second law refers only to *inertial reference frames*, that is, nonaccelerated frames. This then raises two questions: How can an inertial frame be found? And how is one to relate motions that are best described relative to an accelerating frame—such as a frame attached to a spinning spacecraft—to the same motion relative to an inertial frame?

An important postulate of Newtonian mechanics asserts that once an inertial frame has been established, all other frames that move relative to this frame with constant velocity and no rotation are also inertial frames ("Galilean invariance") [1, 2].

Attempts to define inertial frames and time as those for which the laws of Newtonian mechanics are valid are in essence circulatory. Definitions that attach inertial frames to the "fixed stars," taking advantage of the near-zero apparent rotation of distant stars, leave open the question of acceleration by parallel displacement.

The meaning of an inertial reference frame (often referred to as "inertial space") in Newtonian mechanics can be made clear as a limiting case in the theory of general relativity. Newtonian mechanics is a model of space and time that postulates a homogeneous, isotropic space and a uniform time proceeding at a uniform rate. It is a sufficiently accurate representation for nearly all purposes that are of interest to space technology engineers.

Confining oneself to classical, Newtonian mechanics, one has at least the heuristic fact that no contradiction has ever been found by postulating the existence of an inertial space. In particular, no such contradictions have been found by observations of spacecraft paths. The assumption that a frame of reference that is nonrotating with respect to the most distant stars, and is not accelerated with respect to the center of our galaxy, is for all purposes a more than sufficient approximation to an ideal inertial space in Newtonian mechanics. Similarly, no contradiction has been found by assuming that the time in Newtonian mechanics is the same as that measured by the most reproducible timepieces, particularly by atomic clocks.

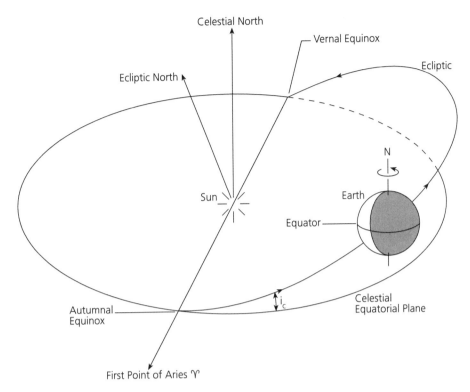

Figure 1.1 Ecliptic and celestial equatorial planes.

1.1.1 The Ecliptic and Celestial Equatorial Planes

The ecliptic (the path of the earth about the sun) is said to be in the *ecliptic plane*, which therefore contains the sun's and the earth's mass centers (Fig. 1.1). One also speaks of the *celestial equatorial plane*, which is the plane parallel to the earth's equatorial plane and through the sun's mass center. The ecliptic and celestial equatorial planes intersect, as illustrated in the figure, in a line referred to as the *equinox line* (because, when the earth on its annual path crosses this line, day and night have equal length). The crossing points are called the *vernal equinox* (the earth is at this point on about the 21st of March) and the *autumnal equinox* (about the 22nd of September). As discussed in Chapter 7, the vernal and autumnal equinox points play an important role in the operation of geosynchronous and other spacecraft. During two periods each year, centered around these points, and centered around local midnight, the sun will be eclipsed for these satellites. Special operational procedures are then needed to compensate for the lack of solar radiation.

The direction of the equinox line that points from the earth at the vernal equinox toward the sun is referred to, for historical reasons, as the *first point of Aries* and is often designated by the symbol ♈. (As the name suggests, the line was found by early astronomers some 2500 years ago to point toward the constellation Aries. But since it moves by about 0.8′ per year, it is presently in Pisces, moving into Aquarius.)

The ecliptic plane is inclined to the equatorial plane at an angle i_c called the *obliquity of the ecliptic*. At present it is 23.44°. (Primarily as a consequence of the gravitational attraction by the sun and the moon on the earth's equatorial bulge, the earth's spin axis, and hence also the equatorial plane, precesses relative to the most distant stars with a period of 25,920 years. The obliquity fluctuates between 21.5° and 24.5° with a period of about 41,000 years and is presently decreasing by about 0.5″ per year. The eccentricity of the earth's orbit about the sun also has a small fluctuation, with a period of about 100,000 years. All these effects, however, are too small to be significant for most space operations.)

1.1.2 Reference Frames and Coordinate Axes

The following reference frames and coordinates are in frequent use [3]. They are arranged in order of increasing closeness to ideal inertial frames.

(1a) *Earth-fixed geocentric equatorial frame*: As shown in Fig. 1.2, this is the frame of reference conventionally used in cartography. Longitude is measured eastward from the Greenwich meridian (the former site of the Royal Observatory in Greenwich, a suburb of London). The latitude is measured positive northward, negative southward, starting at the equator. These two spherical coordinates determine the location, for instance, of a ground station. They are also useful in stating the instantaneous location of a near-earth space vehicle, by radially projecting its location in space on the earth's surface (a

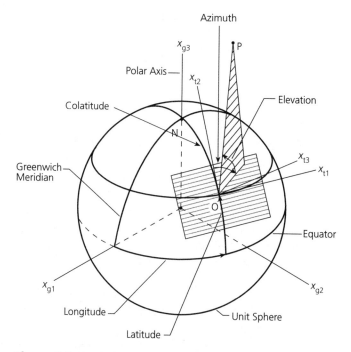

Figure 1.2 Earth-fixed geocentric equatorial (x_{g1}, x_{g2}, x_{g3}) and topocentric (x_{t1}, x_{t2}, x_{t3}) reference frames. P, space vehicle location; O, ground station.

spherical approximation to the geoid is assumed here). The curve obtained by this projection of a space vehicle moving relative to the earth is often referred to as the vehicle's *ground track*.

The same system of spherical coordinates is also useful for other planets and for moons. The reference frames are then referred to as "planetocentric" and "selenocentric," respectively. Because longitude and latitude are angles independent of the radius of the astronomical body, it is convenient to think of them as arc lengths on the "unit sphere" (the sphere of radius one).

(1b) *Topocentric frame*: The line of sight to the instantaneous position of a space vehicle is given by its azimuth and by its elevation (Fig. 1.2). The azimuth angle is measured in the local horizon plane—that is, the plane tangent to the (spherical) earth at the observer's location—and starts at local north, positive for the direction toward local east. The elevation angle is zero at the horizon and positive for lines of sights above it.

With the possible exceptions of short-range sounding rockets or tactical missiles, the two reference systems (1a) and (1b) listed here are hardly ever suitable in space technology as approximations to an inertial frame. They are needed, however, for specifying space vehicle locations in coordinates that are directly observable. Coordinate transformations, to be discussed later, will therefore often be required to convert from one system to another.

(2) *Geocentric equatorial reference frame*: In contrast to frame (1a), this frame does *not* co-rotate with the earth. The origin of the spherical coordinate system is in the earth's mass center (or, correspondingly, in the mass center of other planets or moons). As shown in Fig. 1.3, the longitude is again

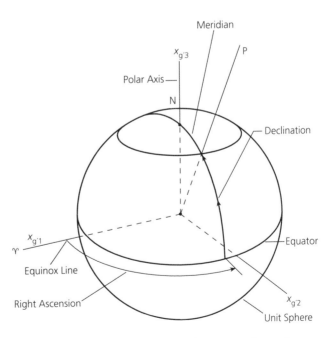

Figure 1.3 Geocentric, equatorial, nonrotating reference frame ($x_{g'1}, x_{g'2}, x_{g'3}$). For many space missions this frame is a sufficient approximation to the ideal inertial reference frame. P, location of space vehicle.

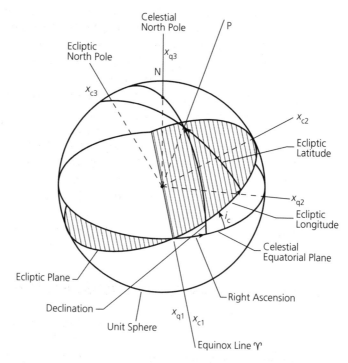

Figure 1.4 Heliocentric equatorial (x_{q1}, x_{q2}, x_{q3}) and ecliptic (x_{c1}, x_{c2}, x_{c3}) reference frames. P, location of space vehicle.

taken positive eastward along the equator, but starts at the first point of Aries (equinox line). It is referred to as the *right ascension* of a point. The latitude is now referred to as the *declination*. This reference frame, which is very nearly nonrotating, is a suitable approximation to an inertial frame for the analysis of near-earth space vehicle trajectories and for their attitude control.

(3) *Heliocentric equatorial reference frame*: This reference frame, illustrated in Fig. 1.4, is the same as the one in item (2), except that the origin of the coordinate system is now in the mass center of the sun and the celestial equatorial plane now replaces the usual equatorial plane. This frame is an even closer approximation to an inertial frame than is the geocentric equatorial reference frame.

(4) *Heliocentric ecliptic reference frame*: Again, as is also illustrated in Fig. 1.4, the origin of the spherical coordinate system is in the sun's mass center. The ecliptic plane now replaces the celestial equatorial plane. Longitude and latitude are now referred to as the *ecliptic longitude* and *ecliptic latitude*, respectively. Again, the ecliptic longitude is measured eastward from the first point of Aries. The ecliptic latitude is taken positive extending toward the ecliptic north pole, negative when pointing toward the ecliptic south pole. This reference frame is most often used as a suitable approximation to an ideal inertial frame when analyzing space missions to the planets.

(5) *Barycentric reference frame*: This frame, which is a still closer approximation to an ideal inertial frame, has its coordinate origin in the center of mass of the solar system. It accounts, therefore, for the acceleration of the sun by

the planets, principally by Jupiter, and to a lesser extent by Saturn. Because of the large mass of the sun relative to the planets (Sun/Jupiter mass ratio approximately 1047), the barycenter stays at all times close to the sun, at a distance from the sun's center that is comparable to the solar radius. A nearly perfect inertial frame can then be constructed by using in place of the ecliptic plane the *Laplace invariable plane*, which is the plane through the barycenter and perpendicular to the orbital angular momentum of all the masses in the solar system. The inclination of this plane is intermediate between the orbital planes of Jupiter and Saturn. Its inclination relative to the ecliptic plane is 1.65°. Although it is important in astronomy when calculating planetary positions over long time spans, either forward or backward in time, for the much shorter times of flights of spacecraft within the solar system, the slight deviation of the nonrotating heliocentric ecliptic reference frame from an ideal inertial frame can almost always be neglected.

1.2 Motion in Accelerated Reference Frames

The need to consider two or more reference frames that are in relative motion to each other arises frequently in all branches of dynamics. It is also important in space technology. An example is the relation between an inertial frame and a reference frame attached to a spin-stabilized spacecraft. A second example, in which both frames are accelerated, is the relation between an earth-fixed reference frame and the spacecraft-fixed frame. In analyzing, for instance, the effect of the sloshing motion of propellant in a vehicle tank, it will be advantageous to formulate the equations of motion of the fluid and the boundary conditions in a reference frame that is rigidly attached to the propellant tank, hence to the vehicle, yet taking into account the space vehicle's own translation and rotation.

We consider here velocities and accelerations, hence derivatives of the position vector of a point with respect to *time*. Of interest here are reference spaces that are in motion relative to each other.

As illustrated in Fig. 1.5, the reference spaces are indicated by their Cartesian coordinate axes. The coordinate frame (x_1, x_2, x_3) translates and rotates relative to the (X_1, X_2, X_3) frame. The first frame's motion relative to the second one is determined by the time derivative of the position vector $\mathbf{R}_0(t)$ and by the relative angular velocity vector $\boldsymbol{\omega}(t)$. The position vector $\mathbf{r}(t)$ defines the point P in the (x_1, x_2, x_3) coordinate frame. In applications to space technology, this point is often the instantaneous center of mass of a space vehicle. (The fact that the mass itself may be a function of time as a consequence of expenditure by the vehicle of propellant is immaterial for what follows.) Time derivatives in the space spanned by X_1, X_2, X_3 will be designated by d/dt, in the space spanned by x_1, x_2, x_3 by the dot notation (˙).

Let $\mathbf{u}_i$, $i = 1, 2, 3$ designate the orthonormal base vectors of the (x_1, x_2, x_3) coordinate frame and $\mathbf{b}(t)$ an arbitrary vector with components $b_i(t)$ in this frame. Hence

$$\mathbf{b} = \sum_i b_i \mathbf{u}_i$$

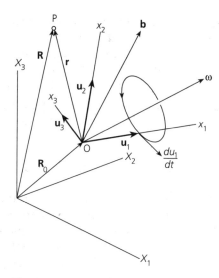

Figure 1.5 Illustrates derivation of Eqs. (1.1) to (1.3).

By definition $\dot{\mathbf{u}}_i = 0$, whereas

$$d\mathbf{u}_i/dt = \omega \times \mathbf{u}_i$$

so that

$$\dot{\mathbf{b}} = \sum_i \dot{b}_i \mathbf{u}_i = \sum_i (db_i/dt)\mathbf{u}_i$$

(b_i being a scalar, there is of course no distinction between the two derivatives). From

$$d\mathbf{b}/dt = \sum_i [(db_i/dt)\mathbf{u}_i + b_i(\omega \times \mathbf{u}_i)]$$

then follows

$$d\mathbf{b}/dt = \dot{\mathbf{b}} + \omega \times \mathbf{b} \tag{1.1}$$

This result is basic for relating the two time derivatives $d\mathbf{b}/dt$ and $\dot{\mathbf{b}}$ of an arbitrary vector $\mathbf{b}$.

The transformation equation for the *velocity* of a moving point with position vectors $\mathbf{r}(t)$ in the (x_1, x_2, x_3) coordinate frame and $\mathbf{R}(t)$ in the (X_1, X_2, X_3) coordinate frame is obtained by taking $\mathbf{r}$ for $\mathbf{b}$. Hence, with $\mathbf{R} = \mathbf{R}_0 + \mathbf{r}$,

$$d\mathbf{R}/dt = d\mathbf{R}_0/dt + \dot{\mathbf{r}} + \omega \times \mathbf{r} \tag{1.2}$$

Similarly, the transformation for the *acceleration* of the moving point is obtained by differentiation of (1.1) and identifying $\mathbf{b}$ with $\mathbf{r}$ and then with $\omega \times \mathbf{r}$. Thus the following result for transforming accelerations between two different reference spaces is obtained:

$$d^2\mathbf{R}/dt^2 = d^2\mathbf{R}_0/dt^2 + \ddot{\mathbf{r}} + \omega \times (\omega \times \mathbf{r}) + 2\omega \times \dot{\mathbf{r}} + \dot{\omega} \times \mathbf{r} \tag{1.3}$$

In the last term it is immaterial whether the time derivative of the relative angular velocity is taken in one or the other of the two spaces, since by (1.1), identifying **b** with ω, $d\omega/dt = \dot{\omega}$.

Equations (1.1) to (1.3) apply to any two reference spaces, because these relations are purely kinematic. If, however, as happens frequently in applications, the (X_1, X_2, X_3) coordinate frame defines an *inertial space*, where therefore Newton's second law in its familiar form (where the acceleration is taken relative to inertial space) holds, it then follows from (1.3) that

$$m\ddot{\mathbf{r}} = \mathbf{F} - m[d^2\mathbf{R}_0/dt^2 + \omega \times (\omega \times \mathbf{r}) + 2\omega \times \dot{\mathbf{r}} + \dot{\omega} \times \mathbf{r}] \quad (1.4)$$

where m is the mass concentrated at the point P and **F** the force acting on it. The second term on the right comes from the acceleration of the origin of the (x_1, x_2, x_3) coordinate frame, the third term is the centrifugal force, and the fourth is the Coriolis force. (No particular name has been given to the last term.) The last four terms jointly are referred to as *inertial forces*. They must be added to the force **F** to complete all forces acting on m in a noninertial space. In the next section, we discuss an application of (1.4) to space technology.

1.3 Example: The Yo-Yo Despin Mechanism

Both in orbit and during orbit insertion, spacecraft and upper stage vehicles are frequently provided with spin to ensure attitude stability. At other times it may be necessary to terminate the spin permanently or at least temporarily, for example, to reorient the vehicle prior to a propellant burn. Despinning can be accomplished by firing small retro-rockets mounted on the vehicle such as to produce a torque counter to the direction of spin. Alternatively, despin can be obtained by initiating the so-called yo-yo mechanism discussed in the following.

This type of despin is frequently used in the case of cylindrically shaped vehicles. It has the advantage of great simplicity and reliability. As indicated in Fig. 1.6, one or several weights are attached to cables that are initially wrapped around the vehicle, typically in the plane through the vehicle's center of mass and normal to the spin axis (which is also one of the principal axes of inertia). Initially, the weights are attached to the body of the vehicle. Despin is initiated by releasing the weights pyrotechnically. As the cables now unwrap, the weights swing out on a path that is an involute of a circle when viewed in a reference frame corotating with the vehicle. The sense of the initial wraparound is such that the pull of the cables reduces the vehicle's angular momentum. When the cables are fully extended, a split hinge releases them and the weights.

The following analysis applies to a system with two symmetrically arranged weights, each of mass m. (X_1, X_2) designates an inertial reference frame. The frame (x_1, x_2) with the corresponding unit base vectors $\mathbf{u}_1$ and $\mathbf{u}_2$ is moving with the instantaneous location of the tangent point O of the cable. The angular velocity of the vehicle is designated by $\Omega(t)$, and $\psi(t)$ is the angle between the weight attachment/release point and the cable's

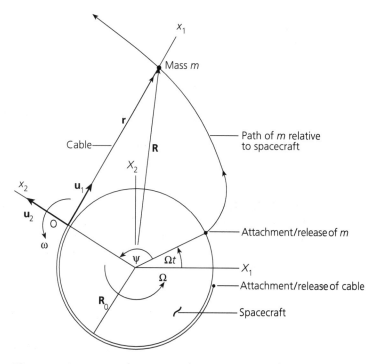

Figure 1.6 Schematic of despin yo-yo mechanism.

tangent point. Hence the angular velocity of the (x_1, x_2) reference frame is

$$\omega = \Omega + \dot{\psi} \tag{1.5}$$

The free length of the cable is $R_0\psi$, where R_0 is the radius of the spacecraft. The position vector for the point mass m in the (x_1, x_2) frame is $\mathbf{r} = R_0\psi\mathbf{u}_1$, and $\mathbf{R} = R_0(\psi\mathbf{u}_1 + \mathbf{u}_2)$ in the (X_1, X_2) system. From (1.2) then follows

$$\begin{aligned}
\frac{d\mathbf{R}}{dt} &= R_0\left(\frac{d\psi}{dt}\mathbf{u}_1 + \psi\frac{d\mathbf{u}_1}{dt} + \frac{d\mathbf{u}_2}{dt}\right) \\
&= R_0(\dot{\psi}\mathbf{u}_1 + \psi\omega\mathbf{u}_2 - \omega\mathbf{u}_1) \\
&= R_0(-\Omega\mathbf{u}_1 + \psi\omega\mathbf{u}_2)
\end{aligned} \tag{1.6}$$

The *total angular momentum* $\mathbf{L}$ of the vehicle and the two masses is in the direction of the X_3 axis and is given by

$$\mathbf{L} = I_3\Omega + 2m(\mathbf{R} \times d\mathbf{R}/dt) \tag{1.7}$$

where I_3 is the vehicle's moment of inertia about the spin axis. Hence

$$L = I_3\Omega + 2mR_0^2(\psi^2\omega + \Omega)$$

With the initial condition $\psi = 0$ at $t = 0$, it follows from the constancy of $\mathbf{L}$ that

$$k(\Omega_0 - \Omega) = \psi^2\omega \tag{1.8}$$

where

$$\Omega_0 = \Omega(t=0), \qquad k = I_3/(2mR_0^2) + 1 \qquad (1.9)$$

The *total kinetic energy* T of the system is

$$T = \tfrac{1}{2}(I_3\Omega^2 + 2m(d\mathrm{R}/dt)^2) = \tfrac{1}{2}I_3\Omega^2 + mR_0^2(\Omega^2 + \psi^2\omega^2) \qquad (1.10)$$

From the constancy of T follows

$$k(\Omega_0^2 - \Omega^2) = \psi^2\omega^2 \qquad (1.11)$$

and dividing by (1.8), $\Omega_0 + \Omega = \omega = \Omega + \dot{\psi}$. Therefore $\dot{\psi} = \Omega_0 = $ const, so that $\psi = \Omega_0 t$. From (1.8) and (1.5), $k(\Omega_0 - \Omega) = \Omega_0^2(\Omega_0 + \Omega)t^2$. Solving for Ω then gives Ω between $t = 0$ and the final time $t_f = l/(R_0\Omega_0)$ where l is the length of each cable. The final result, therefore, for the vehicle angular velocity at some time t is

$$\Omega = \Omega_0 \frac{k - \Omega_0^2 t^2}{k + \Omega_0^2 t^2}, \qquad 0 \le t \le \frac{l}{R_0\Omega_0} \qquad \mathbf{(1.12)}$$

Solving instead for l, one obtains an expression for the line length required to despin the vehicle from an initial angular velocity Ω_0 to the final velocity Ω_f:

$$l = R_0\left(k\frac{\Omega_0 - \Omega_f}{\Omega_0 + \Omega_f}\right)^{1/2} \qquad (1.13)$$

If a complete despin to $\Omega_f = 0$ is required, then $l = R_0 k^{1/2}$, that is,

$$l = \left(\frac{I_3}{2m} + R_0^2\right)^{1/2}, \qquad \Omega_f = 0 \qquad \mathbf{(1.14)}$$

This result is remarkable because it shows that the cable length for complete despin is *independent* of the initial angular velocity of the vehicle. Deviations from the assumed initial velocity therefore do not propagate as errors affecting the intended final zero rate of spin.

As an example, for a fairly typical upper stage vehicle, with $I_3 = 1000$ kg m^2, $R_0 = 1.00$ m, and $m = 1.00$ kg, the required length of each of two cables is 22.4 m, corresponding to about 3.5 turns around the vehicle drum.

The preceding expressions used for the angular momentum and kinetic energy assume that the vehicle is a single rigid mass. Liquid propellants in the spacecraft, however, have their own dynamics, and their rate of rotation during the spin-down may be different from that of the vehicle. Nevertheless, these effects can be minimized by subdividing the propellant stores into several, eccentrically arranged tanks. When a new vehicle is being developed, it is customary to check the analytical results by an actual spin-down test on a spin table. A disadvantage of the yo-yo system is the fact that the free-flying weights and cables could potentially cause a collision with other spacecraft in a similar orbit. This possibility cannot be entirely dismissed

1.4 Euler Angles and Transformations of Coordinates

To describe the orientation of a three-dimensional Cartesian coordinate system with respect to another such system, in general three angles are needed. These then can serve as *generalized coordinates*. Different choices are possible, but choices usually most convenient in applications to spinning rigid bodies, and therefore also to space vehicles, are the *Euler angles* ϕ, θ, ψ.

To define them, we consider the Cartesian coordinate systems (X_1, X_2, X_3) and (x_1, x_2, x_3) as illustrated in Fig. 1.7. Three rotations through angles ϕ, θ, ψ (in this order) will rotate the X_1, X_2, X_3 axes first into the intermediate x_1', x_2', x_3' axes, these then into the x_1'', x_2'', x_3'' axes and these finally into the x_1, x_2, x_3 axes. (Caution is needed because in some texts on rigid body dynamics, the definitions of ϕ and ψ are opposite to those used most often in space technology and also used here.) The ϕ is referred to as the *precession*

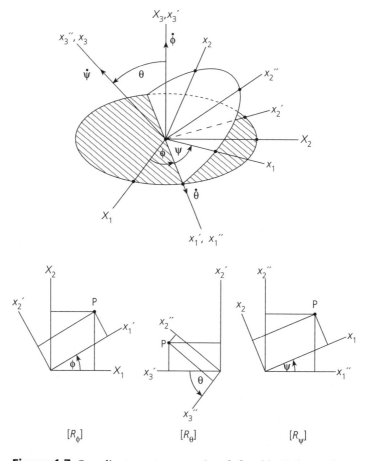

Figure 1.7 Coordinate system rotation defined by Euler angles.

1.4 Euler Angles and Transformations of Coordinates

angle, θ as the *nutation angle*, and ψ as the *spin angle*. The first rotation, R_ϕ, is through the angle ϕ, the second, R_θ, through the angle θ, the third, R_ψ, through ψ.

In matrix notation, these transformations about single axes are

$$\begin{bmatrix} x'_1 \\ x'_2 \\ x'_3 \end{bmatrix} = [R_\phi] \begin{bmatrix} X_1 \\ X_2 \\ X_3 \end{bmatrix} \quad \text{where } [R_\phi] = \begin{bmatrix} \cos\phi & \sin\phi & 0 \\ -\sin\phi & \cos\phi & 0 \\ 0 & 0 & 1 \end{bmatrix} \quad (1.15a)$$

$$\begin{bmatrix} x''_1 \\ x''_2 \\ x''_3 \end{bmatrix} = [R_\theta] \begin{bmatrix} x'_1 \\ x'_2 \\ x'_3 \end{bmatrix} \quad \text{where } [R_\theta] = \begin{bmatrix} 1 & 0 & 0 \\ 0 & \cos\theta & \sin\theta \\ 0 & -\sin\theta & \cos\theta \end{bmatrix} \quad (1.15b)$$

$$\begin{bmatrix} x_1 \\ x_2 \\ x_3 \end{bmatrix} = [R_\psi] \begin{bmatrix} x''_1 \\ x''_2 \\ x''_3 \end{bmatrix} \quad \text{where } [R_\psi] = \begin{bmatrix} \cos\psi & \sin\psi & 0 \\ -\sin\psi & \cos\psi & 0 \\ 0 & 0 & 1 \end{bmatrix} \quad (1.15c)$$

as is easily verified by the geometric relations read from the figure.

The general transformation of the coordinates of a point P, or, what comes to the same, of the components of a vector, is obtained by the matrix product $[R_\psi][R_\theta][R_\phi]$ so that

$$\begin{bmatrix} x_1 \\ x_2 \\ x_3 \end{bmatrix} = [R] \begin{bmatrix} X_1 \\ X_2 \\ X_3 \end{bmatrix} \quad \text{where } [R] = [R_\psi][R_\theta][R_\phi] \quad (1.16)$$

Calculation of this product results in

$$[R] = \begin{bmatrix} c\phi c\psi - s\phi c\theta s\psi & s\phi c\psi + c\theta s\psi & s\theta s\psi \\ -c\phi s\psi - s\phi c\theta c\psi & -s\phi s\psi + c\phi c\theta c\psi & s\theta c\psi \\ s\phi s\theta & -c\phi s\theta & c\theta \end{bmatrix} \quad (1.17)$$

where s stands for the sine and c for the cosine. Although complicated, this matrix plays an important part in much of the software needed for the attitude control of space vehicles.

All matrices (with the exception of (1.19)) in this section belong to the class of orthogonal matrices. Mathematical properties of orthogonal matrices that play a role here may be summarized as follows:

(1) Real valued, nonsingular matrices $[A]$ are orthogonal if and only if $[A^T][A] = [A][A^T] = [I]$ (where $[\ ]^T$ indicates the transpose and $[I]$ the identity matrix). It follows from this that $[A]^{-1} = [A]^T$. If $[A]$ is orthogonal, so are $[A]^T$ and $[A]^{-1}$. If $[A]$ and $[B]$ are orthogonal, the same is the case for the matrix product $[A][B]$.
(2) The determinant of an orthogonal matrix equals either $+1$ or -1. If $+1$, the orthogonal matrices effect the transformation of a right-handed (left-handed) orthogonal coordinate system into another right-handed (left-handed) orthogonal coordinate system. Products of orthogonal matrices

with determinant +1 are again of the same type. Examples are the rotation matrices $[R_\phi]$, $[R_\theta]$, $[R_\psi]$, and $[R]$ in this section.

The transformation inverse to (1.16) is therefore

$$\begin{bmatrix} X_1 \\ X_2 \\ X_3 \end{bmatrix} = [R^T] \begin{bmatrix} x_1 \\ x_2 \\ x_3 \end{bmatrix} \quad \text{where} \quad [R^T] = [R_\phi^T][R_\theta^T][R_\psi^T] \quad (1.18)$$

hence simply obtained by taking the transpose of $[R]$.

If the coordinate system (x_1, x_2, x_3) rotates with respect to the (X_1, X_2, X_3) system with the angular velocity $\omega(t)$, this can be expressed either by the temporal derivatives $\dot\phi$, $\dot\theta$, $\dot\psi$ of the Euler angles or by the components of ω in one or the other of the two coordinate systems.

Most useful in applications to spacecraft is to express ω in terms of its components in the (x_1, x_2, x_3) system, which here is assumed to be rigidly attached to the spacecraft. From the geometrical relations shown in Fig. 1.7 and the transformation equations, one finds for these components

$$\begin{bmatrix} \omega_1 \\ \omega_2 \\ \omega_3 \end{bmatrix} = \begin{bmatrix} \sin\theta\sin\psi & \cos\psi & 0 \\ \sin\theta\cos\psi & -\sin\psi & 0 \\ \cos\theta & 0 & 1 \end{bmatrix} \begin{bmatrix} \dot\phi \\ \dot\theta \\ \dot\psi \end{bmatrix} \quad (1.19)$$

The matrix associated with this transformation is generally **not** orthogonal.

A difficulty arises in the use of Euler angles when θ is approximately zero. In this case, the two planes shown in Fig. 1.7 nearly coincide, with the consequence that the nodal line (the x_1' and x_1'' axis) is now poorly defined. In computations performed by a spacecraft computer, this will be evidenced by the subtraction of pairs of large, nearly equal numbers, leading to large errors.

During normal operation of the spacecraft, it will usually be possible to choose the orientation of the axes such that θ always remains fairly large. This, however, will not solve the problem that may arise, for instance, when a spacecraft may be tumbling out of control and must be brought back to a stable attitude. In this case, all possible values of the angles may occur. This problem of indeterminacy of the nodal line can be avoided by introducing more than three, hence interdependent angles, for example, by using two Euler angle systems with their x_3 axes at 90° to each other. Guidance computer software then can provide for switching from one to the other subprogram depending on the magnitude of θ.

1.5 Time Intervals and Epoch

For precise definitions of time, one needs to distinguish between *time intervals* between two events and the time of a single event as conventionally found from a calendar and a clock. This latter time is technically referred to as the *epoch* of the event.

A highly precise definition and standard for time intervals is needed, for instance, for radar measurements that support spacecraft navigation or altimetry. Thus, to determine the distance of a spacecraft from a ground

1.5 Time Intervals and Epoch

station by a one-way microwave signal propagating at the speed of light, the precision of the time interval measurement must be of the order of 3 ns if a position accuracy of 1 m is to be achieved.

Examples of the epoch of an event are the time at which a rendezvous of two space vehicles occurs and the time of a launch. Because, historically, the accuracy of time measurements has greatly improved with time, ever more precise definitions of time have been introduced. Many of the older definitions are still in use, however, in part because important astronomical data are stated in these terms [4].

An ideal system of timekeeping would be one in which time elapses perfectly uniformly, in the sense that the laws of physics are invariant to the time when the physical event takes place.

1.5.1 International Atomic Time (TAI)

Since 1967, the accepted standard unit for *time intervals* has been the **Système International (SI) second**. It is defined as equal to 9,192,631,770 periods of a certain atomic resonance frequency, in the microwave range, of cesium 133. A large number of atomic standard clocks exist worldwide. To achieve great accuracy, measurements from these clocks are compared and combined into an international standard.

There also exist other types of atomic clocks, as indicated in Fig. 1.8 from Ref. 5. Comparison of such clocks shows that they tend to have slightly different drift rates, some short term and some longer term. For an averaging time of 1000 seconds the standard deviation of cesium or rubidium clocks

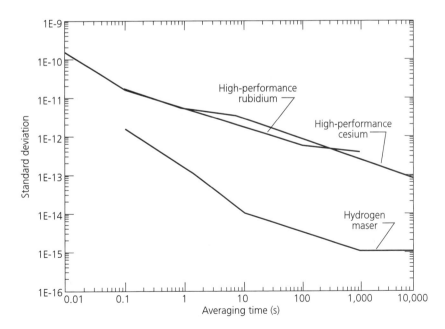

Figure 1.8 Performance of atomic time standards. From Suder, J., Ref. 5, in Pisacane V. L. and Moore, R. C., eds., "Fundamentals of Space Systems." Copyright ©1994 by Oxford University Press, Inc.. Used by permission of Oxford University Press.

is a few times 10^{-13}, corresponding to an error of about 1 second in 100,000 years [5]. Accuracy of this order is sufficient to observe, for instance, the slight irregularities in the rate of rotation of the earth such as caused by the tides, ocean currents, or variations of the mass of the polar ice caps. Hydrogen masers are capable of still higher precision but are relatively bulky.

The *epoch* has been chosen so that it is equal to UT1 on 0 hour on 1 January 1958. (UT1 is one of a family of the so-called Universal Time (UT) standards, discussed in the following.)

Atomic clocks on satellites are being used for worldwide, highly precise position determination and navigation. Thus the United States' Global Positioning System (GPS), which is based on 24 satellites in three different orbits, uses atomic clocks in each of the spacecraft and also in the ground-based control and command center. Using portable receivers, positions on the ground can be determined to an accuracy of about 10 m. For purposes of surveying, centimeter accuracies of *relative* positions of two points can be obtained by taking advantage of a comparison of the carrier frequency phases received by the two points and by long integration times.

1.5.2 Universal Time (UT) and Greenwich Mean Solar Time (GMST)

In astrodynamics, for both historical and practical reasons, **Universal Time** is most often used. Closely related to it by a mathematical formula is the **Greenwich Mean Solar Time**. These systems of timekeeping are derived from observations of the sun's crossing the Greenwich meridian at noon.

A **solar day** is the time interval between two successive crossings by the sun of the observer's meridian. Primarily because of the eccentricity of the earth's orbit about the sun, the length of the solar day is not constant (the earth is closest to the sun in early January, most distant in early July). Solar days therefore must be averaged to serve as a basis for modern timekeeping.

This averaging is accomplished by introducing a fictitious mean sun that moves uniformly along the celestial equator. Time measured in this way eliminates the irregularities caused by the eccentricity of the earth's orbit and by the lack of coincidence of the equatorial and ecliptic planes. It retains, however, the small irregularities caused by variations of the rate of rotation of the earth and of the location of the earth's poles. The resulting system of time is referred to as **Universal Time**. The term **Greenwich Mean Solar Time** is frequently used as synonymous with Universal Time, although, in principle, there is a small difference that manifests itself over centuries, a difference that is unimportant in space technology, considering the much shorter time spans of space missions.

Except for the addition or subtraction of an integer number of hours, **Standard Time** or **Zone Time**, used in everyday life, is the same as GMST.

There are several versions of the system referred to as **Universal Time**. To some extent, they depend on the earth's rate of rotation and on pole wandering. The latest and most precise standard is referred to as the **Coordinated Universal Time** (UTC). It is now based on TAI but differs from it by an integer number of seconds, called leap seconds. Such leap seconds are introduced into UTC on 1 January or 1 July whenever needed to maintain consistency within 0.9 second between the atomic time reckoning and a system that is based on astronomical measurements.

1.5.3 Sidereal Time

Similarly to UT and GMST, **sidereal time** (Latin *sidus*: star) is based on astronomical determinations. It uses the vernal equinox as a reference point. A **sidereal day** is therefore the time interval between two successive passages of the vernal equinox across the observer's meridian. Sidereal time is internally consistent with the geocentric equatorial reference frame discussed in Sect. 1.1.2. Because of variations in the rate of rotation of the earth, sidereal time is not exactly uniform. However, for much of space technology, its precision is entirely sufficient.

Sidereal time and still other time systems referring to the solar system are often convenient for purposes of analysis. In applications to space technology, the results of such calculations are then usually translated into either TAI or GMST.

Important in space technology is the difference in length between a **sidereal day** and a **solar day**, as illustrated in Fig. 1.9. The direction of motion of the earth on its orbit is the same as that of the earth's diurnal rotation. Both are counterclockwise when viewed from the north. The earth moves each day through a heliocentric angle of 360° divided by 365.2 mean solar days, which amounts to approximately 1 degree per day. As it takes the earth about 4 minutes to rotate through this angle, the solar day is about 4 minutes longer than the sidereal day. A more accurate calculation that takes into account the obliquity of the ecliptic plane shows that the length of the sidereal day is 23 h 56 m 4.1 s.

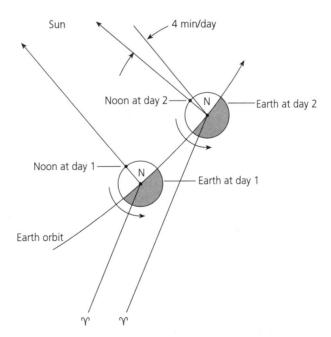

Figure 1.9 Schematic illustrating the difference between sidereal and solar day.

Reference 5 contains a more thorough, yet concise discussion of the various systems of timekeeping. A more extensive treatment may be found in Ref. 3. Practical details may be found in the *Explanatory Supplement to the Astronomical Almanac* [4].

Nomenclature

F	external force
$g_0 = 9.80665$ m/s^2	standard gravitational acceleration
i_c	obliquity (inclination) of the ecliptic (Figs. 1.1, 1.4)
I_3	moment of inertia about x_3 axis
k	constant [(Eq. 1.9)]
l	length of cable
L	angular momentum
m	mass
r, R	position vectors (Fig. 1.5)
t	time
T	kinetic energy
u$_i$	orthonormal base vectors (Fig. 1.5)
ϕ, θ, ψ	Euler angles (in Sect. 1.3, ψ denotes the angle shown in Fig. 1.6)
ω, Ω	angular velocities

Problems

(1) At some specified time the position of a spacecraft is at a right ascension of 30.00° and a declination of 60.00°. The obliquity (inclination) of the ecliptic is 23.45°.
   (a) Find the ecliptic longitude and latitude of the position by using spherical trigonometry.
   (b) Obtain the same result by using the transformation of coordinate systems based on Euler angles.

(2) Let $\omega(t)$ be the angular velocity of the orthonormal coordinate frame (x_1, x_2, x_3) relative to the orthonormal coordinate frame (X_1, X_2, X_3). Let $\omega_1, \omega_2, \omega_3$ be the x_1, x_2, x_3 components of ω. (In many applications to the control of the attitude of space vehicles, the first coordinate frame would be fixed to the vehicle, the second frame would be an inertial frame. However, the formulation of this problem is purely kinematic, hence independent of the assumption of an inertial frame.)
   (a) Defining the Euler angles as shown in Fig. 1.7, show that for $\theta \neq 0$

$$\begin{bmatrix} \dot{\phi} \\ \dot{\theta} \\ \dot{\psi} \end{bmatrix} = \begin{bmatrix} \sin\psi/\sin\theta & \cos\psi/\sin\theta & 0 \\ \cos\psi & -\sin\psi & 0 \\ -\sin\psi/\tan\theta & -\cos\psi/\tan\theta & 1 \end{bmatrix} \begin{bmatrix} \omega_1 \\ \omega_2 \\ \omega_3 \end{bmatrix}$$

Note that this matrix is **not** orthogonal.
   (b) Show that this matrix is, as expected, the inverse of the matrix in Eq. (1.19).

References

1. Goldstein, H., "Classical Mechanics," Addison-Wesley Publishing Company, Reading, MA, 1950.
2. Whittaker, E. T., "Analytical Mechanics," Cambridge University Press, London, 1937.
3. Kovalevsky, J., Mueller, I. I., Kolaczek, J., eds., "Reference Frames in Astronomy and Geophysics," Astrophysics and Space Science Library, Vol. 154, Kluwer Academic Publishers, Dordrecht, 1989.
4. Seidelmann, P. K., ed., "Explanatory Supplement to the Astronomical Almanac," University Science Books, Mill Valley, CA, 1992.
5. Pisacane, V. L. and Moore, R. C., eds., "Fundamentals of Space Systems," Chapter 3 by Black, H. D. and Pisacane, V. L. Oxford University Press, New York, 1994.

2

Forces and Moments

The principal forces that act on space vehicles are *gravity*, *thrust*, and, within planetary atmospheres, also *aerodynamic drag and lift*.

Thrust is derived from rocket motors and is an essential feature of launch vehicles and upper stage vehicles. On a much smaller scale, thrust is also used on artificial satellites and other spacecraft for position and attitude control.

Aerodynamic forces are important for launch vehicles. They are also important for space vehicles that operate in a planetary atmosphere. This is particularly so in the case of reentry vehicles, that is, recoverable vehicles that reenter the earth's atmosphere on their return from a space mission. Aerodynamic forces—although often minute—may also result from the interaction of a spacecraft with the tenuous gas of the outer atmosphere. In the case of orbiting spacecraft, after repeated revolutions, the effects on position due to these forces tend to accumulate with time. They can then become highly significant in disturbing the orbit.

The very small force that results from the *solar radiation pressure* can also play a role in spacecraft orbits. Solar sailing, which would use solar radiation pressure for thrust, is a distinct future possibility. Principally because of the difficult problem of deploying the very large sails required, solar sailing has not as yet been put into practice.

Still other forces that are important in space vehicle engineering are *mechanical reaction forces*. An example is the force that results from the action of springs that are used to push apart vehicle stages after stage separation. Other examples are provided by the interaction of launch vehicles with their support during launch. For these forces Newton's third law holds.

Associated with these forces are the corresponding moments about some reference point. Most often, in the case of space vehicles, the center of mass is chosen as this reference point. As a consequence of the expenditure of propellant, the center of mass can shift relative to the vehicle's body. When this occurs, it is necessary to treat the vehicle as a system of variable mass and variable moments of inertia. The theory of dynamics of a space vehicle is therefore generally more complicated than the classical theory of rigid bodies.

2.1 Gravity

According to Newton's *law of universal gravitation*, the gravitational force $\mathbf{F}_g$ exerted by a point mass m_1 on a point mass m_2 is

$$\mathbf{F}_g = -\frac{Gm_1 m_2}{r^3}\mathbf{r} \tag{2.1}$$

where **r** is the radius vector that extends from the first to the second point mass. The constant $G = 6.6732 \; 10^{-11}$ m³/(kg s²) is known as the **universal gravitational constant**.

If instead of a point mass, m_1 is the mass of an astronomical body, not necessarily spherically symmetric, then (2.1) is still valid asymptotically at distances large compared with the dimensions of the body. As discussed in Section 2.1.1, it also remains valid even at arbitrary distances, if the body's mass distribution is spherically symmetric.

It is preferable to introduce in place of G and of m_1 their product, $\mu = Gm_1$, the so-called **gravitation parameter** of the astronomical body. The reason for this preference is that μ can be determined by astronomical and satellite measurements much more precisely than can G or m_1 separately. In applications to astrodynamics, it is only their product that matters.

Some examples of the numerical values of the gravitation parameter are:

Sun: $\mu = 1.32712 \; 10^{11}$ km³/s²
Earth: $= 3.986006 \; 10^5$ km³/s² (GEM-L2 geopotential model, 1983)
Moon: $= 4.90265 \; 10^3$ km³/s²

If one takes for m_2 a unit mass, then (2.1) defines a force field—strictly speaking an acceleration, say, **g(r)**—at points given by the position vector **r**. As is seen from the form of (2.1), this field is conservative and can therefore be expressed by the gradient of a potential, say, $\Phi_g(\mathbf{r})$, so that

$$\mathbf{g} = -\text{grad} \; \Phi_g \tag{2.2}$$

(This definition of the potential with a negative sign is conventional in physics and in space technology, but usually not in astronomy.) It follows from (2.1) that if m_1 is a point mass,

$$\Phi_g = -\frac{\mu}{r} \tag{2.3a}$$

If m_1 is an extended mass, such as that of an astronomical body, one obtains by superposition of the effects of the elements of m_1,

$$\Phi_g = -\frac{\mu}{m_1} \int_{m_1} \frac{dm_1}{|\mathbf{r} - \mathbf{r}'|} \tag{2.3b}$$

where **r** and **r'** are as defined in Fig. 2.1.

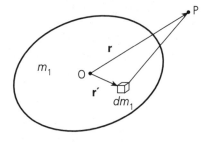

Figure 2.1 Illustration of Eq. (2.3b). O, center of mass.

$\Phi_g(\mathbf{r})$ is seen to have the dimension of an energy per unit mass and is equal to the gravitational potential energy per unit mass at $\mathbf{r}$. The arbitrary constant that can be added to Φ_g conventionally is chosen such that the potential energy is zero at an infinite distance from m_1. Hence it is negative at other distances.

Deviations from an exact spherical mass distribution of an astronomical body result in a gravitational potential that is no longer perfectly symmetric. This can result in important long-term effects on orbiting spacecraft, particularly those that orbit at low altitudes. Conversely, observations of the orbits of near-earth artificial satellites provide the most precise method for determining the exact figure and mass distribution of the earth, of other planets, and of the moon.

2.1.1 Spherically Symmetric Mass Distribution

If the density of an attracting body is such that it depends only on the radius, the expression for the gravitational potential is particularly simple. The result is applicable as an approximation to the earth, for instance, because the density, although much higher in the nickel–iron core than in the mantle, depends only weakly on latitude and longitude.

The result, which was already known to Newton, can be obtained by integration over the elements of mass composing the body. The same result can also be obtained by the use of Gauss' integral theorem, a theorem that proves useful in other problems related to gravitational fields. For this reason, we use it here.

Let S be a spherical surface of radius r, concentric with and outside the body. Also let $\mathbf{n}$ be the inward unit normal to S and $d\mathbf{g}$ the gravitational acceleration, caused by an element dm of the body, at the element dS of S. The distance between dS and dm is designated by r'. Hence from (2.1),

$$\mathbf{n} \cdot d\mathbf{g} = G\, dm \cos\theta / r'^2$$

where θ is the angle between the inward normal and the line from dS to dm. Integration over S results in

$$\int_S \mathbf{n} \cdot d\mathbf{g}\, dS = G\, dm \int_S \frac{\cos\theta}{r'^2} dS = G\, dm \int_S d\Omega = 4\pi G\, dm$$

where $d\Omega$ is the solid angle subtended from dm to dS. The result is seen to be independent of the location of dm within the body.

Summing over all elements dm, we obtain the acceleration $\mathbf{g}$ at a point on S. By symmetry, $\mathbf{g}$ is normal to S and has constant magnitude on S. Therefore $r^2 g = Gm_1 = \mu$ so that if r_0 is the radius of the body

$$\mathbf{g} = -\mu \mathbf{r}/r^3, \qquad r \geq r_0 \tag{2.4}$$

a result that is identical with what follows from (2.1) for a point mass.

To summarize: for a spherically symmetric body, its gravitational field outside it is the same as if the total mass were concentrated at the center. This field, which falls off with distance as the inverse square, is therefore an example of a *central, inverse-square field*.

2.1.2 Gravity Gradient Effect

If, over the space occupied by a space vehicle, the external gravitational field were perfectly uniform, the resultant of the forces on the various parts of the vehicle would be simply the force obtained by placing the entire mass at the vehicle's center of mass.

In most applications, the assumption that the gravitational field is *uniform* over the extent of the space vehicle is satisfied to very high precision. Exceptions, referred to as **gravity gradient effects**, occur only when the dimensions of the vehicle are relatively large or when the time over which dynamical effects resulting from the nonuniformity can accumulate is very long.

One can distinguish two such effects: (1) The resultant of the gravity forces that act on different parts of the space vehicle is not exactly the force obtained from the gravity at the location of the center of mass. (2) The gravity forces produce a torque about the center of mass.

Applications to space technology in which the first effect is important occur in the maneuvering of two space vehicles before docking, but hardly ever in the dynamics of a *single* vehicle. An exception may occur when an instrumented capsule is connected to the spacecraft by a long **tether**, an arrangement that is useful for *in situ* studies of the upper atmosphere.

The second effect, although much too small to be significant for launch vehicles, upperstage vehicles, or most spacecraft, can be important as a perturbation of the attitude (spatial orientation) of very large space structures. Another application occurs in the **gravity gradient stabilization** of near-earth satellites by means of a long beam that is attached to the spacecraft.

In what follows, we consider as an example a near-earth large structure, such as a truss (Fig. 2.2), which may serve as a component of a space station. The gravitational field is assumed to be a central, inverse-square field.

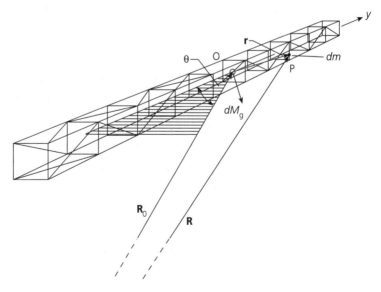

Figure 2.2 Gravity gradient effect. Illustration for Eqs. (2.5) to (2.9).

We designate by $\mathbf{R}_0$ the radius vector from the center of the earth to the center of mass O of the structure and by $\mathbf{R}$ the corresponding vector to an element of mass dm at P. The moment about O induced by the gravitational forces is

$$\mathbf{M}_g = -\mu \int_m \mathbf{r} \times (\mathbf{R}/R^3)\, dm \qquad (2.5)$$

where μ is the earth's gravitational parameter, m the mass of the structure, and $\mathbf{r} = \mathbf{R} - \mathbf{R}_0$. Since $r \ll R_0$, it will be sufficient to consider the terms of lowest order in powers of r/R_0. From the figure, $R^2 = R_0^2 + r^2 + 2\mathbf{R}_0 \cdot \mathbf{r}$. Using the binomial theorem to expand the cubic term, R^3, in (2.5), one finds to the lowest significant order

$$\mathbf{M}_g = -\mu \int_m \mathbf{r} \times (\mathbf{R}_0 + \mathbf{r}) R_0^{-3} \left(1 + 2\mathbf{R}_0 \cdot \mathbf{r}/R_0^2 + \mathbf{r}^2/R_0^2\right)^{-3/2} dm$$

$$= -\frac{3\mu}{R_0^5} \mathbf{R}_0 \times \int_m (\mathbf{R}_0 \cdot \mathbf{r})\mathbf{r}\, dm \qquad (2.6)$$

The term linear in $\mathbf{r}$ vanishes in the integration, since, by the definition of the center of mass, $\int_m \mathbf{r}\, dm = 0$.

The integral can be expressed in terms of the moments of inertia of the structure. For this purpose it is convenient to introduce a Cartesian coordinate system with axes that coincide with the principal axes of inertia. The components of $\mathbf{r}$ and $\mathbf{R}_0$ will be designated by (x_1, x_2, x_3) and (X_{01}, X_{02}, X_{03}) respectively. From the definition of the moment of inertia I_1, when expressed by the principal axes components of the position vector $\mathbf{r}$,

$$I_1 = \int_m (x_2^2 + x_3^2)\, dm, \qquad \int_m x_1 x_2\, dm = 0, \qquad \int_m x_1 x_3\, dm = 0$$

Corresponding expressions for I_2 and I_3 are obtained by cyclic interchange of the indices of the components. By multiplying out the scalar product in (2.6), one obtains for the first component of $(\mathbf{R}_0 \cdot \mathbf{r})\mathbf{r}$

$$(X_{01} x_1 + X_{02} x_2 + X_{03} x_3)\, x_1$$

and correspondingly, by cyclic interchange, for the other components. Therefore

$$\left(\int_m (\mathbf{R}_0 \cdot \mathbf{r})\mathbf{r}\, dm\right)_1 = X_{01} \int_m x_1^2\, dm = \tfrac{1}{2} X_{01}(I_2 + I_3 - I_1)$$

and correspondingly for the other components.

This result can be summarized by introducing the Cartesian tensor $\mathbf{J}$ represented by the matrix

$$\begin{bmatrix} \tfrac{1}{2}(I_2 + I_3 - I_1) & 0 & 0 \\ 0 & \tfrac{1}{2}(I_3 + I_1 - I_2) & 0 \\ 0 & 0 & \tfrac{1}{2}(I_1 + I_2 - I_3) \end{bmatrix} \qquad (2.7)$$

which, when postmultiplied with the vector $\mathbf{R}_0 = (X_{01}, X_{02}, X_{03})^\mathrm{T}$, shows that

$$\int_m (\mathbf{R}_0 \cdot \mathbf{r})\mathbf{r}\, dm = \mathbf{J} \cdot \mathbf{R}_0 \qquad (2.8)$$

The final expression obtained for the **gravity gradient moment**, to the lowest significant order, is therefore

$$\mathbf{M}_g = -\frac{3\mu}{R_0^5} \mathbf{R}_0 \times (\mathbf{J} \cdot \mathbf{R}_0) \qquad (2.9)$$

The following example is of interest because it illustrates the order of magnitude of the effect. To simplify the problem, we approximate the truss in Fig. 2.2 by a thin rod of length $2l$ and uniform mass ϱ per unit length. The moment of inertia, transverse to the rod and about the center of mass, is then $I = (2/3)\varrho l^3 = (1/3)ml^2$ where m is the mass. Let θ designate the angle between the rod and the local vertical. Therefore

$$X_{01} = R_0 \sin\theta, \qquad X_{02} = 0, \qquad X_{03} = -R_0 \cos\theta$$

Evaluation of the expression $\mathbf{R}_0 \times (\mathbf{J} \cdot \mathbf{R}_0)$ in (2.9) then gives the magnitude M_g of the gravitational torque as

$$M_g = \frac{\mu m l^2}{2R_0^3} \sin(2\theta) \qquad (2.10)$$

The torque is seen to have a maximum at $\theta = 45°$. For $\mu = 3.986\,10^5 \text{ km}^3/\text{s}^2$ (earth), $R_0 = 10{,}000$ km (i.e., about 1.57 times the earth's radius), $m = 100$ kg, $l = 10$ m, the maximum torque affecting the rod becomes $1.99\,10^{-3}$ Nm.

2.1.3 The Earth's Gravitational Field

To predict the paths of near-earth orbiting spacecraft, a more exact knowledge of the gravitational acceleration than that afforded by (2.4) is needed. This can be obtained by a combination of geodesic and gravimetric measurements on the surface of the earth, which can then be used to infer the gravitational field everywhere externally to the earth. This is made possible by the application of a well-known theorem in **potential theory**, to be discussed later in this section.

With the advent of space technology, observations of the paths of near-earth satellites have been complementing the earth-bound measurements. Similarly, spacecraft are being used to determine the gravitational fields of other planets or of the moon.

The earth's gravitational field lacks exact spherical symmetry for two reasons: (1) the surface is not exactly spherical; (2) the mass density in the interior is not exactly symmetrically distributed. The presence of these asymmetries has important effects on the orbits of near-earth satellites.

The largest contribution to the asymmetry is caused by the *equatorial bulge*, which is the result of the centrifugal force produced by the earth's rate of spin. Thus, whereas the polar radius (the radius taken along the earth's axis) is approximately 6357 km, the mean equatorial radius is 6378 km, larger by about 21 km.

A useful concept, first introduced by Gauss, for describing the figure of the earth is the **geoid**. This is defined as the surface that would result if the earth were entirely covered by an ocean. Adding to the gravitational force the (much smaller) centrifugal force caused by the earth's rotation, the combined force — for reasons of equilibrium — must be perpendicular to the surface. The geoid is therefore an equipotential surface.

An approximation to the geoid is the **reference ellipsoid** [1], which is usually taken to be the rotationally symmetric ellipsoid that best approximates the geoid, with its axis of symmetry along the polar axis. This reference ellipsoid, however, does not take into account, for instance, the variations of the force field with geographic longitude, nor such anomalous effects as the large gravitational perturbations over oceanic islands. (This latter effect, however, is not significant for calculations of the paths of satellites, because, being highly localized, it falls off rapidly with altitude and produces only a short-term interaction on passing spacecraft.)

More important for predicting the paths of near-earth satellites and their slow drifts out of the original orbital plane is the gravitational influence of the moon and the sun. This topic is beyond the scope of the present section. The determination of the path of near-earth satellites, given the perturbations of the gravitational fields of the earth, moon, and sun, is treated extensively in Ref. 2. It is not considered in this book.

A representation of the earth's gravitational field, more exact than is provided by the reference ellipsoid, will occupy the remainder of this section.

It follows from the form of Newton's universal law of gravitation that the divergence of the force field of a point mass is zero. By superposition of the effects of point masses, this result can be extended to the field exterior to a distributed mass, as is the case for the gravitational field exterior to the earth. Its potential Φ_g therefore satisfies Laplace's equation

$$\nabla^2 \Phi_g = 0 \qquad (2.11)$$

We note first some mathematical preliminaries [3, 4]:

If $\nabla^2 \Phi = 0$ and Φ is prescribed on a closed surface S, then Φ is determined everywhere outside S (the Dirichlet problem). Similarly, if the component normal to S of grad Φ is prescribed on S, then Φ is determined everywhere outside S (the Neuman problem).

In terms of spherical coordinates, Laplace's equation takes the form

$$\frac{1}{\varrho^2} \frac{\partial}{\partial \varrho} \left(\varrho^2 \frac{\partial \Phi}{\partial \varrho} \right) + \frac{1}{\varrho^2 \sin \theta} \frac{\partial}{\partial \theta} \left(\sin \theta \frac{\partial \Phi}{\partial \theta} \right) + \frac{1}{\varrho^2 \sin^2 \theta} \frac{\partial^2 \Phi}{\partial \lambda^2} = 0 \qquad (2.12)$$

where ϱ is the distance from the origin, θ the colatitude (in applications to geodesy, the angle extending south from the north pole, as indicated in Fig. 1.2) and the latitude λ. In these coordinates, Laplace's equation is separable. If one lets

$$\Phi = R(\varrho)\Theta(\theta)\Lambda(\lambda) \qquad (2.13)$$

then the following three equations are obtained ($m, n =$ constants):

$$d^2\Lambda/d\lambda^2 + m^2\Lambda = 0 \tag{2.14a}$$

$$\frac{1}{\sin\theta}\frac{d}{d\theta}\left(\sin\theta\frac{d\Theta}{d\theta}\right) + \left(n(n+1) - \frac{m^2}{\sin^2\theta}\right)\Theta = 0 \tag{2.14b}$$

$$\frac{1}{\varrho^2}\frac{d}{d\varrho}\left(\varrho^2\frac{dR}{d\varrho}\right) - \frac{n(n+1)}{\varrho^2}R = 0 \tag{2.14c}$$

The fundamental solutions of (2.14a) are $\cos(m\lambda)$ and $\sin(m\lambda)$. The requirement of periodicity and continuity of Λ requires that m be an integer. Without loss of generality, we can take $m \geq 0$.

Imposing the condition that Θ be finite in the range $0 \leq \theta \leq \pi$, the fundamental solutions of (2.14b) are the Legendre functions of the first kind $P_n^m(\cos\theta)$ with integer n and $0 \leq m \leq n$. If $m = 0$, the Legendre functions are polynomials, usually written as P_n, omitting the superscript. The Legendre polynomials can be defined by successive differentiation as follows, with $x = \cos\theta$:

$$P_n(x) = \frac{1}{2^n n!}\frac{d^n}{dx^n}(x^2 - 1)^n \tag{2.15a}$$

From this the Legendre functions follow by

$$P_n^m(x) = (1 - x^2)^{m/2}\frac{d^m P_n(x)}{dx^m}, \qquad 0 \leq m \leq n \tag{2.15b}$$

The fundamental solutions of (2.14c) that fall off to zero as the distance from the coordinate origin becomes infinite are $1/\varrho^{n+1}$ (end of mathematical preliminaries).

One needs to add to the gravitational potential Φ_g of **g** the potential Φ_f from the inertial forces that arise from the fact that earth-fixed coordinates are used rather than inertial coordinates. At the equator, the centripetal acceleration is approximately 0.0339 m/s², hence 0.00344 times normal gravity. At general points in near-earth space, as is easily seen, this additional potential is

$$\Phi_f = -\frac{\Omega^2 \varrho^2 \sin^2\theta}{2} \tag{2.16}$$

with $\Omega = 7.292115\ 10^{-5}$ rad/s the rate of rotation of the earth.

The symbol usually given to the combined potentials is U, so that

$$U = \Phi_g + \Phi_f \tag{2.17}$$

The final result for U is obtained from (2.13), the superposition of the fundamental solutions of (2.14a, b, c), and from (2.16). By reason of symmetry, some of the terms in the superposition, referred to as "inadmissible terms," will vanish identically. Thus it can be shown that as a consequence of placing the origin of the spherical coordinate system at the center of mass of the earth, the terms with $n = 1$, that is, with P_1 and P_1^1, vanish. An additional symmetry comes about because the earth's spin axis, for dynamical reasons, coincides with high precision with the axis of maximum moment of inertia, hence with a principal axis. Although there is no exact rotational symmetry about this axis, it nevertheless means that two of the three products of inertia are zero, with the consequence that the term with P_2^1 vanishes. Omitting the inadmissible terms in the summation, the final result for U can be written

$$U = -\frac{\mu}{\varrho}\left\{1 + \sum_{n=2}^{\infty}\left(\frac{a_e}{\varrho}\right)^n J_n P_n(\cos\theta) + \sum_{n=2}^{\infty}\sum_{m=1}^{n}\left(\frac{a_e}{\varrho}\right)^n P_n^m(\cos\theta)\right.$$
$$\left. \times \left(C_n^m \cos m\lambda + S_n^m \sin m\lambda\right)\right\} - \tfrac{1}{2}\Omega^2 \varrho^2 \sin^2\theta \qquad (2.18)$$

(with the proviso that in the double summation the term with P_2^1 also vanishes). The factor used to nondimensionalize the distance from the mass center is the mean equatorial radius $a_e = 6378.140$ km.

The first term in the main bracket comes from P_0 and represents the gravitational potential of a spherically symmetric earth, hence is identical to the potential of a point mass. This term and the terms in the first sum are independent of the longitude. For this reason they are referred to as **zonal spherical harmonics**. The terms in the double sum are referred to as **tesseral spherical harmonics** (Latin *tessera*: a piece of mosaic).

Examples of both types are illustrated in Fig. 2.3, which shows the regions in which the spherical harmonics have positive versus negative values.

The last term in (2.18) represents the potential of the centripetal acceleration.

The constants J_n, C_n^m, and S_n^m represent the mass distribution in the earth and are found by combining gravimetric, geodesic, and satellite measurements. The coefficient with the largest absolute value by far is J_2, which derives from the equatorial bulge or flattening of the poles. Approximate values of the more important coefficients are listed in Table 2.1.

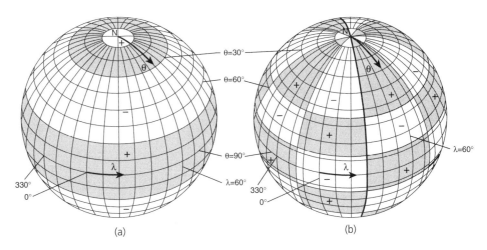

Figure 2.3 Examples of spherical harmonics: (a) zonal harmonic $P_4(\cos\theta)$; (b) tesseral harmonic $P_8^4(\cos\theta)(C_8^4 \cos 4\lambda + S_8^4 \sin 4\lambda)$.

CHAPTER 2 *Forces and Moments*

Table 2.1 Spherical harmonics coefficients for the earth

			Tesseral harmonics	
Zonal harmonics	n	m	C_n^m	S_n^m
$J_2 = -1082.70 \; 10^{-6}$	2	2	$1.57 \; 10^{-6}$	$-0.897 \; 10^{-6}$
$J_3 = 2.56 \; 10^{-6}$	3	1	$2.10 \; 10^{-6}$	$0.16 \; 10^{-6}$
$J_4 = 1.58 \; 10^{-6}$	3	2	$0.25 \; 10^{-6}$	$-0.27 \; 10^{-6}$
$J_5 = 0.15 \; 10^{-6}$	3	3	$0.077 \; 10^{-6}$	$0.173 \; 10^{-6}$
$J_6 = -0.59 \; 10^{-6}$	4	1	$-0.58 \; 10^{-6}$	$-0.46 \; 10^{-6}$
$J_7 = 0.44 \; 10^{-6}$	4	2	$0.074 \; 10^{-6}$	$0.16 \; 10^{-6}$
	4	3	$0.053 \; 10^{-6}$	$0.004 \; 10^{-6}$
	4	4	$-0.0065 \; 10^{-6}$	$0.0023 \; 10^{-6}$

Adapted from Ref. 2.

Different sign conventions for the coefficients are sometimes used, often depending on whether the author's interest is in geodesics or in orbital mechanics. Depending on these conventions, the algebraic signs preceding the summation symbols in (2.18) may differ.

The zonal harmonics have a greater effect on near-earth satellites than the tesseral ones. The former are not only larger but also produce cumulative effects on the satellite orbits, whereas the latter cause oscillatory perturbations that tend to more nearly average out over repeated orbits.

Let

$$\mathbf{u} = -\operatorname{grad} U \tag{2.19}$$

$\mathbf{u}(\varrho, \theta, \lambda)$ is therefore the vector sum of the gravitational and centripetal accelerations. Its radial, meridional (in the direction of increasing colatitude, hence positive from north to south), and longitudinal coordinates are

$$u_\varrho = -\frac{\partial U}{\partial \varrho}, \qquad u_\theta = -\frac{1}{\varrho}\frac{\partial U}{\partial \theta}, \qquad u_\lambda = -\frac{1}{\varrho \sin\theta}\frac{\partial U}{\partial \lambda} \tag{2.20}$$

hence are obtained by differentiation of (2.18).

A numerical example, summarized in Table 2.2, gives some insight into the relative magnitudes of the terms. The table applies to the point where a spacecraft that orbits the earth at an altitude of 300 km crosses 45° north latitude. The effect of the equatorial bulge or flattening of the earth on the meridional gravitational component is evident. It is also of interest to note that the

Table 2.2 Acceleration corrections for the nonspherical earth for $\theta = 45°$ and 300 km altitude

Component	Gravitational acceleration	Centripetal acceleration
ϱ	$+0.678 \; 10^{-3} g_0$	$-1.728 \; 10^{-3} g_0$
θ	$+1.36 \; 10^{-3} g_0$	$+1.728 \; 10^{-3} g_0$
λ	$\mathcal{O}(10^{-6}) g_0$	0

centripetal acceleration is of the same order of magnitude as the radial and meridional gravitational corrections to the spherically symmetric earth.

2.2 Thrust

In defining the thrust of a rocket motor as a force acting on the vehicle, one needs a definition that characterizes, to the extent possible, the motor performance in isolation from the other parts of the vehicle.

When operating in the *vacuum* of space, the thrust is independent of the exterior of the vehicle. However, when operating in the *lower atmosphere* there will unavoidably be some interaction between the rocket plume and the air flow at the base of the vehicle. As is discussed further in Chap. 4, the gas flow in the nozzle and at its exit surface may then substantially deviate from the nominal velocity and pressure distribution present when operating in vacuum. Depending on the design of the rocket motor, flow separation and oblique shocks in the nozzle may occur, further altering the interaction of the rocket plume with the external air flow. One needs to distinguish therefore between the vacuum, or nominal, thrust of a rocket motor and the thrust at various altitudes in the atmosphere. An example of this variation of the thrust with altitude is shown in Fig. 2.4.

The thrust of a rocket motor can be measured by the reaction force on load cells on a test stand. Small and medium-size motors are often tested in either a horizontal or vertical position in a low-pressure chamber. Large pumps, typically steam ejectors, are needed to compensated for the flow of the rocket gas entering the test chamber.

Very large motors are usually tested in a vertical position, with the exhaust directed downward. They must be tested at ambient atmospheric pressure because otherwise the size and power requirement of the pumps would be so large as to be impractical. The corrections needed for determining the vacuum thrust must then be inferred from flight tests of similar

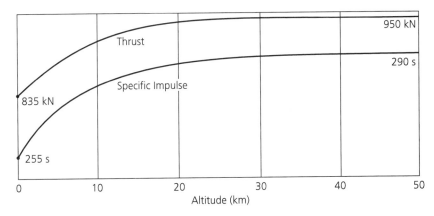

Figure 2.4 Altitude performance of the H-1 liquid-propellant rocket engine. From Ref. 5, Huzel D. K. et al., "Modern Engineering for the Design of Liquid Propellant Rocket Engines." Courtesy of the Rocketdyne Division of Rockwell International. Copyright © 1992, AIAA—reprinted with permission.

motors or must be estimated by analysis, often supported by tests at a reduced scale.

In principle, the gas velocity and thermodynamic variables of state would not have to be exactly the same in a static test firing as compared with actual flight on an accelerating, and possibly spinning, vehicle. However, because the acceleration of the gas in the nozzle is many orders of magnitude larger than the vehicle's acceleration, the difference arising from this cause between static tests and flight is negligible.

In some solid-propellant motors, some slag may be retained in the motor case. Because the amount of slag may differ in flight from that measured in static test firings [6], the mass flow rates at the nozzle exit may differ slightly in the two modes of operation. However, in the majority of cases this effect can be neglected.

If $d\mathbf{A}$ designates the *outward*-directed unit vector, normal to the exit surface A_{ex} of the rocket motor nozzle, $\mathbf{u}$ the gas velocity relative to the vehicle, and ϱ the gas density, the mass flow rate through the nozzle is given by

$$\tilde{m} = \int_{A_{ex}} \varrho \mathbf{u} \cdot d\mathbf{A} \tag{2.21}$$

The change of momentum exiting the nozzle per unit time, relative to the vehicle, is

$$\int_{A_{ex}} \varrho \mathbf{u} (\mathbf{u} \cdot d\mathbf{A}) \tag{2.22}$$

If $m = m(t)$ is the instantaneous mass of the vehicle, $\mathbf{V} = \mathbf{V}(t)$ the vehicle's center of mass velocity relative to inertial space, p the gas pressure at the nozzle exit plane, and p_a the ambient (i.e., atmospheric pressure; $p_a = 0$ in the vacuum of space), then conservation of momentum, applied to a control volume enclosing the vehicle and moving with it, results in the equation of motion

$$d(m\mathbf{V})/dt + \int_{A_{ex}} (\mathbf{V} + \mathbf{u}) \varrho \mathbf{u} \cdot d\mathbf{A} = - \int_{A_{ex}} (p - p_a) \, d\mathbf{A} \tag{2.23}$$

When the same motor is test fired on a test stand, since now $\mathbf{V} = 0$,

$$\int_{A_{ex}} \mathbf{u} \varrho \mathbf{u} \cdot d\mathbf{A} = - \int_{A_{ex}} (p - p_a) \, d\mathbf{A} + \mathbf{T}$$

where $\mathbf{T}$ is the reaction force exerted by the test stand on the vehicle. The measured thrust, $\mathbf{F}_t$, is therefore $-\mathbf{T}$, so that

$$\mathbf{F}_t = - \left[\int_{A_{ex}} \mathbf{u} \varrho \mathbf{u} \cdot d\mathbf{A} + \int_{A_{ex}} (p - p_a) \, d\mathbf{A} \right] \tag{2.24}$$

The first term on the right is often referred to as the **velocity thrust**. The second term represents the part of the thrust that can be ascribed to the pressure difference at the nozzle exit. It is referred to as the **pressure thrust**.

When integrated over the entire vehicle surface, the force produced by the ambient pressure is, of course, zero; hence the term with p_a merely accounts for the fact that at the nozzle exit this pressure is replaced by the pressure of the propellant gas.

In practice, the pressure thrust at *high altitude* or in vacuum is no more than a few percent of the velocity thrust. Nevertheless, in most cases it still needs to be considered because of the high sensitivity of the mass fraction available for payload as a function of the ratio of thrust to total propellant weight. At *low altitude* in the earth's atmosphere, depending on the nozzle design, the pressure thrust is often negative, that is, in a direction opposite to the much larger velocity thrust.

In place of computing the time rate of change of momentum and the pressure at the nozzle exit, the same result for the thrust could also have been obtained by evaluating all pressure force components parallel to the thrust axis that act on the interior surfaces of the motor case and nozzle. This equality follows directly from the momentum equation for a control volume bounded by the interior surfaces and by the nozzle exit surface.

With this definition of the thrust, the momentum equation (2.23) becomes

$$d(m\mathbf{V})/dt + \mathbf{V} \int_{A_{ex}} \varrho \mathbf{u} \cdot d\mathbf{A} - \mathbf{F}_t = 0$$

Noting that

$$\tilde{m} = -dm/dt \qquad (2.25)$$

and using (2.21), one obtains as the final expression of the equation of motion

$$m\, d\mathbf{V}/dt = \mathbf{F}_t \qquad (2.26)$$

The form of this equation is not quite as obvious as it might appear. It is important to note that the first term is $m\, dV/dt$, not $d(mV)/dt$. It may therefore be of historical interest to mention here that Newton expressed his second law, albeit for *constant* mass, by the time derivative of the *momentum* (his "vis motionis"), rather than of the velocity as is done in most elementary textbooks today. Perhaps ironically, Newton's formulation turns out to be correct relativistically (for constant rest mass), whereas the elementary textbook formulation would not hold.

2.2.1 Specific Impulse

An important performance parameter of rocket motors is the **specific impulse**, I_{sp}, defined by the equation

$$F_t = g_0 I_{sp} \tilde{m} \qquad (2.27)$$

where $g_0 = 9.80665 \text{ m/s}^2$ is the internationally agreed standard gravitational acceleration. The higher the specific impulse, the lower the rate of propellant consumption for a given thrust. As a consequence of the lower propellant mass needed for a given space mission, a high specific impulse will therefore result in a higher ratio of payload to propellant mass.

For mainly historical reasons, the specific impulse is defined by the propellant *weight* flow rate $g_0 \tilde{m}$, rather than by the *mass* flow rate $\tilde{m}$ as would have been more appropriate. Numerically, I_{sp} is therefore expressed in seconds.

The specific impulse that can be achieved in practice depends in part on the design of the rocket motor, particularly on the expansion ratio of the nozzle. But principally it depends on the chemical reaction energy of the propellant, or — in the case of nuclear–thermal motors — on the reactor temperature. It depends relatively little on the size of the motor. Typical performance figures will be discussed in Chap. 4.

2.2.2 Tsiolkovsky's Rocket Equation

In the absence of gravity and of all other forces except thrust, particularly simple conclusions can be drawn from (2.25) and (2.26). If, in addition, it is assumed that the thrust is in a direction tangential to the rocket's path, the motion becomes rectilinear and the equation of motion can be readily integrated. It follows that

$$\int_{V_0}^{V_1} d\mathbf{V} = -g_0 I_{sp} \int_{m_0}^{m_1} \frac{dm}{m}$$

where m_0 and m_1 are the initial and final mass, respectively, and similarly for the velocities. Carrying out the integration and solving for the mass ratio yields

$$\frac{m}{m_0} = \exp\left(-\frac{V_1 - V_0}{g_0 I_{sp}}\right) \qquad (2.28)$$

In applications, the mass m_1 would typically be the *empty mass*, that is, the mass remaining after all the usable propellant has been consumed. It is also of interest to note that (2.28) holds independently of the time history of the mass flow rate and therefore of the thrust, as long as I_{sp} is constant.

Equation (2.28), in essentially this form, was first stated by Tsiolkovsky (Russian mathematics teacher; first publication on space travel 1895). Although actual rocket trajectories are almost always more complicated than a simple rectilinear path in a gravity-free space, the equation is important because it points out particularly clearly the need for a high specific impulse in space missions. These almost always require a large velocity increment. Because of its exponential dependence, the mass ratio, unless the specific impulse is high, can become very small, resulting in an unacceptably small payload mass for a given initial mass.

2.2.3 Example: The Sounding Rocket

Sounding rockets are comparatively small vehicles, used most often for scientific purposes, particularly for the study of the earth's upper atmosphere and ionosphere. Sounding rockets therefore do not need to reach orbit. They are often launched along a near-vertical trajectory.

We consider here a single-stage rocket, launched vertically. The mass flow rate $\bar{m}$ is assumed constant until the time when all propellant is exhausted (or until the propellant flow is deliberately terminated so as to control more precisely the final height attained). For simplicity of the calculation, the specific impulse I_{sp} will be assumed to be independent of the altitude $h(t)$ and the aerodynamic drag will be neglected. The altitude reached by the rocket is assumed to be sufficiently small compared with the earth's radius that the gravitational acceleration can be assumed constant and equal to the standard value g_0. The guidance algorithm is programmed to maintain, in an inertial reference, the thrust axis along the local vertical.

The origin of time will be taken at the time of launch. The time at thrust termination will be designated by t_1. The mass of the vehicle is $m(t)$, with m_0 at launch and m_1 at, and after, thrust termination. During the powered phase, from (2.27) and (2.24),

$$m(t)\, dV/dt = g_0 I_{sp} \tilde{m} - g_0 m(t) \qquad (0 \le t \le t_1) \qquad (2.29)$$

where

$$t_1 = \frac{m_0 - m_1}{\tilde{m}} = \left(1 - \frac{m_1}{m_0}\right) \frac{I_{sp}}{\beta_0} \qquad (2.30)$$

and where

$$\beta_0 = \frac{F_t}{g_0 m_0}, \qquad \beta_0 > 1 \qquad (2.30')$$

is the ratio of thrust to initial weight.

Integrated, with $V = 0$ at $t = 0$ and $m(t) = m_0 - \tilde{m}t$, the velocity of the rocket becomes

$$V = g_0 I_{sp} \ln\left(\frac{1}{1 - \beta_0 t/I_{sp}}\right) - g_0 t \qquad (0 \le t \le t_1) \qquad (2.31)$$

The velocity is seen to reach a maximum

$$V_1 = g_0 I_{sp} \ln(m_0/m_1) - g_0(1 - m_1/m_0) I_{sp}/\beta_0 \qquad (2.31')$$

at the time t_1 of thrust termination.

The second term on the right in (2.31') reflects the fact that the maximum velocity achievable for a given mass ratio and specific impulse becomes smaller for smaller values of the ratio of thrust to initial weight. In the limit as $\beta_0 \to \infty$, that is, if an infinite thrust could be applied instantaneously at the time of launch, the second term would vanish. Comparison with the general case then suggests referring to

$$g_0(1 - m_1/m_0) I_{sp}/\beta_0$$

as the **gravity loss** term. This loss becomes important for values of β_0 close to 1. Especially in the case of large vehicles, because of the physical limitations imposed on the size of the rocket motor and on the propellant feed rate, the designer is often forced to accept a substantial loss of this type.

Integrating a second time, with $h = 0$ at $t = 0$, results in

$$h = g_0 I_{sp} \left[\frac{I_{sp}}{\beta_0} - \left(\frac{I_{sp}}{\beta_0} - t\right)\left(1 - \ln\left(1 - \frac{\beta_0}{I_{sp}}t\right)\right)\right] - \frac{g_0 t^2}{2} \qquad (0 \le t \le t_1)$$

$$(2.32)$$

for the height as function of time during the powered phase. At thrust termination, when $t = t_1$, the height becomes

$$h_1 = \frac{g_0 I_{sp}^2}{\beta_0^2}\left[\beta_0 - \beta_0\frac{m_1}{m_0}\left(1 + \ln\frac{m_0}{m_1}\right) - \frac{1}{2}\left(1 - \frac{m_1}{m_0}\right)^2\right] \quad (2.32')$$

Beyond this height, the rocket is coasting, with a velocity

$$V = V_1 - g_0(t - t_1) \quad (t \geq t_1) \quad (2.33)$$

At peak height, $V = 0$. Hence, from (2.31'), the time after the launch needed to reach peak height is

$$t_2 = I_{sp} \ln(m_0/m_1) \quad (2.34)$$

provided, of course, that $F_t > g_0 m_0$.

To obtain the peak height, h_2, we observe that from (2.33) integration from t_1 to t_2 and substitution of V_1 and t_2 gives

$$h_2 = h_1 + \frac{g_0 I_{sp}^2}{2\beta_0^2}\left[1 - \frac{m_1}{m_0} - \beta_0 \ln\frac{m_0}{m_1}\right]^2 \quad (2.35)$$

where h_1 is obtained from (2.32').

Figure 2.5 illustrates the rocket's velocity as a function of time for three values of the ratio of thrust to initial mass. The *ascending* branches in the figure correspond to the powered phase, as given by (2.31), and are *independent of the mass ratio*. The *descending* branch in the nondimensional representation in the figure has a slope of –1 and corresponds to the coasting phase. It is *independent of the ratio of thrust to initial weight* but depends on the mass ratio.

For a numerical example, we assume a specific impulse of 250 s (which is representative for a sounding rocket solid-propellant motor, with altitude-

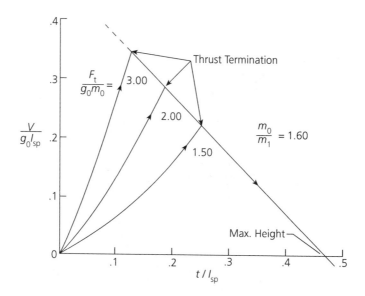

Figure 2.5 Velocity versus time of a vertically ascending sounding rocket, for a mass ratio of 1.60.

averaged performance), a mass ratio of 1.60, a thrust of 10,000 N, and an initial mass of 500 kg. From the relations derived before this, one then finds that the time to burnout (thrust termination) is 46 s, at which time the altitude is 14.1 km and the velocity 701 m/s. Peak height is reached at 117 s after launch, at a height of 39.2 km.

2.3 Aerodynamic Forces and Moments

Aerodynamic forces and moments need to be considered for the **ascent** of launch vehicles in the earth's or other planetary atmospheres. To avoid large structural bending moments—for which they are not designed—major launch vehicles are almost always launched vertically. Starting from low speed, where the fluid mechanics of incompressible fluids applies, they reach hypersonic speeds in the upper atmosphere, where the path may then deviate from the vertical.

The drag experienced by these vehicles is much smaller than the thrust. Preflight predictions of the drag do not require the same accuracy as is needed in the performance prediction of aircraft. Also, the aerodynamic shape of launch vehicles does not need to conform to the exacting demand for low drag contours familiar from aircraft designs. More important, in the case of large launch vehicles, is the reduction of the structural weight by choosing, for instance, simple conical and cylindrical shapes without much fairing or other drag reduction features.

Smaller vehicles may be air launched. They are then usually designed for aerodynamic lift, hence must sustain the ensuing bending moments.

Extremely demanding is the aerodynamic design of vehicles designed for **reentry** into the earth's atmosphere or, more generally, for **aerobraking** in a planetary atmosphere. In addition to lift, drag, and aerodynamic moments, protection of the vehicle's structure against excessive heating then becomes of paramount importance.

Some space missions postulate spacecraft that are designed to lose speed by repeatedly entering and then exiting the atmosphere of planets. They may do this by a sequence of skipping in and out of the atmosphere, followed by a segment of an orbit in space. This technique was first demonstrated in the Venusian atmosphere by the United States Magellan Venus Orbiter. This spacecraft had neither an aerodynamic shroud nor special heat protection, but it could be successfully controlled by its thrusters to avoid aerodynamic instabilities and overheating.

Aerobraking of a spacecraft may also be used for reducing speed prior to the launch of a scientific probe designed to descend to the planet's surface or for changing the orbital parameters of a spacecraft without intent of landing.

2.3.1 Ascent of Launch Vehicles

The reader is likely to be aware of the existence of a very large body of aerodynamic literature, both theoretical and experimental. Much of this American and European literature is contained in monographs published by NASA, ESA (the European Space Agency) and AGARD (the North Atlantic

Treaty Organization), as well as in textbooks, for example, Ref. 7. In the following we confine ourself to a brief, qualitative discussion.

Launch vehicles usually have a roughly cylindrical, elongated shape. Lift and drag coefficients for such shapes have been measured in wind tunnels. Because of the inevitable, practical limitations of wind tunnels, the Reynolds and Mach numbers at which such tests can be conducted are much smaller than what would be required for a complete aerodynamic characterization of the ascent of launch vehicles. Extrapolations based on theoretical considerations must therefore be made. But because the increase in propellant consumption caused by aerodynamic drag is relatively small, great accuracy, comparable to that needed for aircraft, is not required.

There also exist analytic solutions for high-speed, inviscid flows about slender, cylindrical bodies [8]. They can provide approximate information on the lift, aerodynamic moment, and supersonic wave drag.

At the cost of extensive programming and calculating on high-speed computers, more realistic results can be obtained by numerical methods ("computational aerodynamics"). These method allow an accurate and detailed representation of the vehicle's outer surface and of the resulting potential flow. More subject to errors is the modeling of turbulent boundary layers, of the laminar-to-turbulent transition, and of flow separation.

To illustrate, Fig. 2.6 from Ref. 9 shows the pressure distribution at transonic speed ($M = 1.1$) for the Space Shuttle Orbiter in its plane of symmetry on the top side. The pressure, p, is expressed by the "pressure coefficient,"

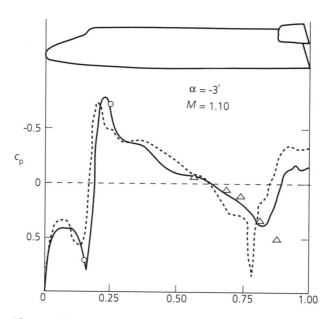

Figure 2.6 Pressure coefficient, Space Shuttle Orbiter upper side. Solid and dashed lines, computed; ($\triangle$) wind tunnel data; ($\circ$) flight data. Angle of attack, $-3°$; Mach number, 1.10. (From Ref. 9 Martin, F. W. and Slotnick, J. P., "Flow Computation for the Space Shuttle in Ascent Mode", in Progress in Astronautics and Aeronautics, Seabass, A. R., ed. Copyright © 1990, AIAA—reprinted with permission.

that is, by the quantity $(p - p_a)/[(\varrho_a/2)V^2]$ shown as a function of the body length (p_a = ambient atmospheric pressure, ϱ_a = ambient density, and V = flight velocity). The solid and dashed lines are the result of two different numerical calculations. The triangles represent wind tunnel data. Actual flight data are indicated by the two circles.

The importance of the lift and of the moment derives largely from the need to consider the effect of **upper atmosphere winds**. Moderate winds at the launch pad are less dangerous to the integrity of the vehicle than those at higher altitude. Not only can the wind speeds be much higher, but also the vehicle velocity is high. Wind shear, in particular, necessitates rapid and forceful steering of the vehicle, imposing large bending moments on the structure. For this reason, prior to the launch of a major vehicle, the wind velocity above the launch site is measured. The required data can be obtained by radar and by balloons.

Because the vehicle's velocity is still low, the aerodynamic loads in the absence of winds are relatively small at low altitude. At high altitude, they are also small because of the low density. Between these extremes there is a density and velocity condition where the loads reach a maximum.

A suitable measure for the severity of these loads is the stagnation pressure. Figure 2.7 shows the ratio of stagnation pressure to sea level ambient

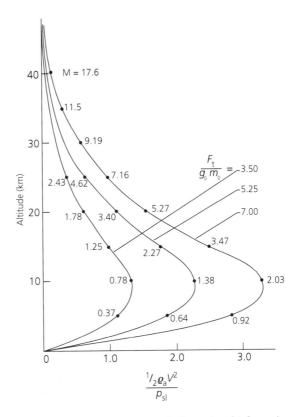

Figure 2.7 Vertical ascent of a launch vehicle: ratio of stagnation pressure to sea level pressure, p_{sl}, as a function of altitude, for three values of thrust to gross mass. I_{sp}, 350 s; M, Mach number.

pressure as a function of altitude. The curves are for three values of the thrust relative to the launch vehicle's initial weight. A specific impulse of 350 s is assumed. Also indicated are the Mach numbers.

As this figure shows, the peak of the stagnation pressure occurs around 10 km altitude, almost independent of the thrust-to-weight ratio. At this altitude, strong **buffeting** may occur, particularly when there is flow separation from parts of the vehicle that are not shaped aerodynamically.

Buffeting can be dangerous because of the vibration loads imposed on the vehicle. For this reason, on some vehicles with engines that can be throttled, the thrust is programmed so that it is decreased in passing the region of high stagnation pressure, before increasing again as the vehicle reaches higher altitudes.

2.4 Free Molecule Flow

With increasing altitude above the earth's or planetary surface, the aerodynamic characteristics — particularly lift, drag, and convective heat transfer — undergo fundamental changes. At one extreme there is classical aerodynamics, which is based on continuum mechanics. In space technology, this is applicable primarily to the takeoff and flight of launch vehicles in the lower atmosphere. At the other extreme is the regime of **free molecule flow**, applicable in particular to satellites in low earth orbits.

Classical aerodynamics applies when the mean free path between successive collisions of the gas molecules is very small in comparison with the thickness of the boundary layer (hence, *a fortiori* when compared with the body dimensions). On the contrary, in free molecule flows the mean free path is much larger than the principal dimensions of the body that is immersed in the flow. For instance, at an altitude of 300 km in the earth's atmosphere, where a number of low-earth-orbiting satellites operate, the length of the mean free path is about 900 m.

There exists a continuity of different flow regimes from the continuum mechanics of classical aerodynamics to the free molecule flow. The two intermediate regimes that are distinguished are referred to as "slip flow," typical for slightly rarefied gas flows, and "transition flow," typical for moderately rarefied gas flows. The theoretical treatment of these intermediate flow regimes is difficult and incomplete and will not be considered here.

Historically, free molecule flows were first studied in laboratory systems [10, 11]. The theoretical and experimental results obtained have since been applied to the study of the hypersonic flight of space vehicles in the rarefied upper atmospheres, in particular of Earth and of Mars.

Other than depending on the nature of the gas or gas mixture and on the density, the mean free path is also somewhat, but only weakly, dependent on the energy of collision of the molecules. Experimental data are usually stated for 760 mm Hg pressure and 15°C temperature and are expressed in terms of the effective collision diameter, D, of the molecules. The conversion of these data to the conditions of a rarefied atmosphere by the ideal gas law is straightforward. If N is the number of molecules per unit volume and if the molecules are assumed to be elastic spheres with a diameter D and a

Table 2.3 Molecular collision diameters and mean free paths at normal temperature and pressure

Gas	Molecular weight	Effective diameter (cm)	Mean free path at 760 mm Hg (cm)
O_2	32.00	3.61×10^{-8}	6.79×10^{-6}
N_2	28.02	3.75×10^{-8}	6.28×10^{-6}
CO_2	44.00	4.59×10^{-8}	4.19×10^{-6}
He	4.002	2.18×10^{-8}	18.62×10^{-6}

Maxwellian velocity distribution, it can be shown that the average length, λ, of the free paths is given by

$$\lambda = \frac{1}{\sqrt{2}\pi N D^2} \quad (2.36)$$

The ratio of the mean free path to the characteristic vehicle dimension, l (such as the length or effective diameter of a reentry probe or spacecraft), is known as the Knudsen number $K = \lambda/l$.

In Table 2.3 the effective diameters for intermolecular collisions and the mean free paths at normal temperature and pressure are listed for several gases. Carbon dioxide is the principal constituent of the atmosphere of Mars. Helium becomes important in the earth's atmosphere above 400 km.

Currently available data on properties such as density and temperature in the upper atmosphere (the thermosphere and the exosphere) of the earth are subject to considerable uncertainties. To some extent, the data depend on the type of experimental method that is being used.

There are strong variations that depend on the solar flux in the extreme ultraviolet and hence are synchronous with the 11-year solar cycle.

In addition to latitude and seasonal variations, there are also variations with local time: on the sunlit side of the earth, the atmosphere is heated and rises, with the consequence that the density in the upper atmosphere increases. The opposite applies to the night side. For low-earth-orbiting satellites these 12-hour variations tend to average out. They are important, however, for predicting the trajectories of reentering spacecraft and for estimating the effects of aerobraking maneuvers in the upper atmosphere.

In the United States, several models for the properties of the earth's upper atmosphere are in current use. Among them is the **U.S. Standard Atmosphere 1976** and the newer **MSIS-86 Model** (Mass Spectrometer Incoherent Scatter model). The latter is based on *in situ* data from satellites and rocket probes and also on measurements from incoherent scatter stations on the ground.

Figure 2.8 indicates average number densities (molecules per unit volume) of the electrically neutral species in the earth's thermosphere [12]. The species, roughly in order of their importance, are atomic oxygen, helium, atomic hydrogen, molecular nitrogen and oxygen, and argon. More recently, atomic nitrogen has also been included in the data.

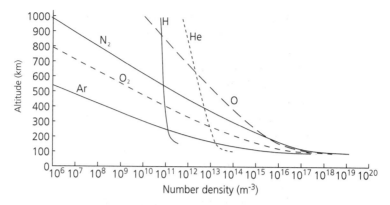

Figure 2.8 Average number densities of the electrically neutral species in the earth's upper atmosphere. (From Ref. 12.) Copyright © 1995 Princeton University Press. By permission.

Figure 2.9 [13] shows averaged data for the combined (mass-weighted) density of these species and for the temperature at 400 km altitude. The data are shown as functions of the solar activity and include three different models.

Making use of a result of the kinetic theory of gases, the mean free path can be related to the kinematic viscosity and therefore to the Reynolds number. This makes it possible to express the Knudsen number in a form that is particularly convenient for applications in fluid mechanics. The result is

$$K = 1.26\sqrt{\gamma}\,M/\text{Re} \tag{2.37}$$

where M is the Mach number, Re the Reynolds number, and γ the ratio of the specific heats. Here, K and Re are based on the same characteristic body length l. Free molecule flow is usually defined as the region for which $M/\text{Re} > 3$.

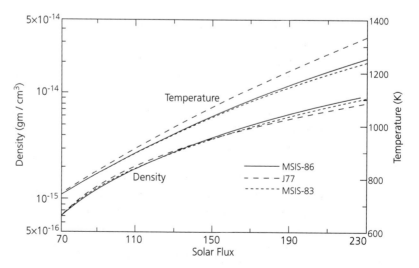

Figure 2.9 Average density and temperature of the electrically neutral earth atmosphere at 400 km altitude, as functions of the solar activity index. For three different models. (From Ref. 13.)

If, as will now be assumed, the mean free path is large compared with the characteristic dimensions of the spacecraft, the paths of the gas molecules prior to their impact on a spacecraft surface are no longer influenced by the presence of the body. Neither shock fronts nor boundary layers are formed.

In applications to orbiting spacecraft in the upper atmosphere, the relative velocity of spacecraft and atmosphere is hypersonic. In this case, the thermal motion of the incident molecules can be neglected. Also, it will be assumed that these molecules reach the surface without having first been reflected from another spacecraft surface. The incident molecules therefore travel on parallel, rectilinear paths, directly incident on the surface.

In turn, the molecules are reflected from the surface. This process can be described approximately as being a mixture of diffuse and specular reflections. The reflected molecules will collide with the incoming stream, as well as among themselves, but only at a large distance from the spacecraft so that they do not react back on the surface.

The *diffusely* reflected molecules, through their multiple interactions with the surface before being reemitted, are at least partially accommodated to the temperature of the surface. Their velocity distribution is approximately in accordance with the Maxwell–Boltzmann distribution

$$f(v)\,dv = 4\pi \left(\frac{m}{2\pi kT}\right)^{3/2} v^2 \exp\left(-\frac{mv^2}{2kT}\right) dv \qquad (2.38)$$

where $f(v)\,dv$ is the probability that a molecule has a velocity between v and $v + dv$ and m is the mass of the molecule, k the Boltzmann constant $= 1.3806\ 10^{-23}$ kg m^2/(s^{2}K), and T the temperature. It follows by integrating (2.38) that the average velocity, $\bar{v}$, and the root mean square velocity, $\bar{\bar{v}}$, are

$$\bar{v} = \left(\frac{8kT}{\pi m}\right)^{1/2}, \qquad \bar{\bar{v}} = \left(\frac{3kT}{m}\right)^{1/2} \qquad (2.39)$$

The average translational kinetic energy per molecule due to the thermal motion is therefore $\frac{3}{2}kT$.

Figure 2.10 shows schematically the incident flow of molecules on a surface element and the reflected flow. For hypersonic flight, the thermal motion of the incident molecules can be neglected. The angle of incidence is designated by ϕ_i, the incident speed by V, and the temperature of the surface (wall) by T_w. The reflected flow can be approximately represented by the sum of two parts: a *specular* reflection at speed V and angle $-\phi_i$, and a *diffuse* reflection with a Maxwellian velocity distribution at temperature T_r into the half-space above the surface element. The specular part represents elastic collisions with the surface, whereas the diffuse part represents multiple interactions with surface atoms, sufficient to establish approximately a thermal equilibrium among the molecules before they scatter in random directions.

The combination of incident and reflected flows imparts momentum and energy to the surface.

The number, say I, of molecules of a specified kind (for instance, atomic oxygen) striking the surface and being reflected, per unit time and unit area

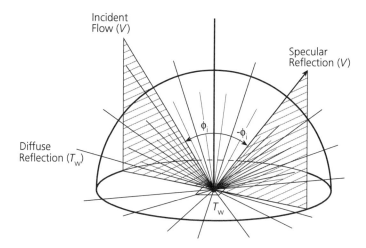

Figure 2.10 Reflection of a free molecule flow from a surface.

of the surface, is given by

$$I = NV \cos \phi_i \qquad (2.40)$$

The fraction of the incident molecules reflected *diffusely* is usually given the symbol σ. Depending on the nature of the surface and of the incident molecules, σ typically ranges from 0.8 to 1. The recoil of these molecules produces a force normal to the surface but, by symmetry, not a tangential force.

The fraction of molecules reflected *specularly* is $1 - \sigma$. Their recoil causes both a normal and a tangential force.

A second, nondimensional coefficient, the Smoluchowski–Knudsen accommodation coefficient, α, is introduced, which is defined by the equation

$$\alpha = \frac{E_i - E_{rd}}{E_i - E_w} \qquad (2.41)$$

where E_i is the energy flux (energy per unit time and unit area of the surface) of the incident molecules, E_{rd} the energy flux of the diffusely reflected molecules, and E_w the energy flux of these molecules if on the average they all had assumed the surface (wall) temperature. This coefficient is therefore a measure of the degree to which the energy of the diffusely reflected molecules has been "accommodated" to the surface temperature. For perfect accommodation, $E_{rd} = E_w$, therefore $\alpha = 1$. Depending on the types of surface and incident molecules, α generally ranges from 0.8 to 1.0. Evidently,

$$E_i = I \frac{mV^2}{2}, \qquad E_{rd} = \sigma I \frac{m\bar{\bar{v}}_{rd}^2}{2}, \qquad E_w = \sigma I \frac{3}{2} kT_w \qquad (2.42)$$

Since the Smoluchowski–Knudsen coefficient refers to the energy rather than to the momentum, a slight conversion is necessary for calculating the force exerted by the molecules on the surface. Solving (2.41) for E_{rd} and

2.4 Free Molecule Flow

making the substitutions indicated by (2.42),

$$\sigma \frac{m\bar{\bar{v}}_{rd}^2}{2} = (1-\alpha)\frac{mV^2}{2} + \alpha\sigma\frac{3}{2}kT_w \qquad (2.43)$$

From (2.39), $\bar{\bar{v}}_{rd}^2 = \frac{3}{8}\pi\bar{v}_{rd}^2$, therefore

$$\bar{v}_{rd} = 4\sqrt{\frac{1}{3\pi}\left[\left(\frac{1-\alpha}{\sigma}\right)\frac{V^2}{2} + \alpha\frac{3}{2}\frac{kT_w}{m}\right]} \qquad (2.44)$$

The free molecule **pressure**, p_{fm}, exerted on the surface is the sum of the pressures from the incident flow, from the recoil of the specularly reflected flow, and from the recoil of the diffusely reflected flow. Hence

$$p_{fm} = p_i + p_{rs} + p_{rd}$$

where $p_i = ImV\cos\phi_i$ and $p_{rs} = (1-\sigma)ImV\cos\phi_i$. The pressure resulting from the diffusely reflected flow is obtained by integrating the surface normal component of the velocity v_{rd} over the unit hemisphere above the surface element (Fig. 2.10). Hence

$$p_{rd} = \sigma Im\bar{v}_{rd}\iint \cos\phi\, d\Omega = \pi\sigma Im\bar{v}_{rd}$$

The final result for the pressure exerted by the molecules on the surface, obtained by summing the three terms, is

$$p_{fm} = Im[(2-\sigma)V\cos\phi_i + \pi\sigma\bar{v}_{rd}] \qquad \textbf{(2.45a)}$$

Similarly, the **shear stress**, s_{fm}, is

$$s_{fm} = s_i + s_{rs} + s_{rd}$$

where $s_i = ImV\sin\phi_i$, $s_{rs} = -(1-\sigma)ImV\sin\phi_i$, and, by symmetry, $s_{rd} = 0$. Therefore

$$s_{fm} = \sigma ImV\sin\phi_i \qquad \textbf{(2.45b)}$$

In general, there will be several species of molecules in the flow, with different masses, and different σ's and α's. Because in free molecule flow the molecules do not interact, the normal and tangential forces on the surface can be obtained simply by summing the contributions from the various species.

Depending on the geometry of the external surfaces, these pressures and shear stresses acting on the surface of a spacecraft will produce drag, lift, and a moment. Drag and lift are important in aerobraking into a planetary atmosphere and in estimating the remaining life of a spacecraft in the upper atmosphere before it reenters. Because of its effect on spacecraft attitude control, the moment also can be important. In the earth's atmosphere, at 400 km altitude, the free-molecular torque on a medium-size spacecraft can be of the order of 10^{-4} N m, which is larger than the solar radiation torque. At altitudes above 1000 km, it will tend to be smaller than either the solar radiation or gravity torques.

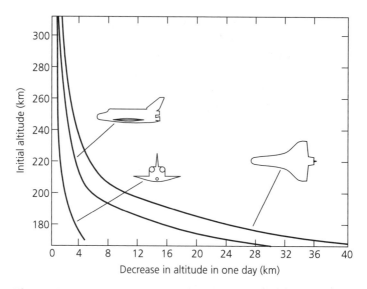

Figure 2.11 U.S. Space Shuttle Orbiter: decrease in altitude per day at no thrust. For the three basic attitudes. (Adapted from Ref. 14.)

It has been found that the upper atmosphere largely **corotates** with the earth. The motion in a nonrotating reference frame is therefore from west to east along circles of constant latitude, approximately with a velocity

$$V_a = 2\pi r_g \sin\theta / P_g$$

(r_g = earth radius, θ = colatitude, and the period P_g = 24 h). At the equator, $V_a = 0.465$ km/s, which is small compared with the orbital velocity V_{crc} of about 7.75 km/s of low-earth-orbiting satellites (Sect. 3.7).

The velocity, V, of the spacecraft relative to the atmosphere, that is, the velocity of the incident molecules in hypersonic flight, is therefore the vector difference of $\mathbf{V}_{crc}$ and $\mathbf{V}_a$. As is readily shown, the angle ϑ between the two vectors is given by $\cos\vartheta = \cos i_{eq}/\sin\theta$, where i_{eq} is the inclination of the orbit relative to the equator.

As an example of the aerodynamic forces in the earth's upper atmosphere, Fig. 2.11 shows the rate of orbital decay (at zero thrust) of the U.S. Space Shuttle Orbiter. At the range of altitudes shown, the atmospheric density decreases by about a factor of 10 for each 100-km increase in altitude. Free molecule flow in the case of the Orbiter applies down to about 280 km. Below this altitude, the regions of "transition flow" and "slip flow" apply.

Drag-free satellites in low earth orbits are of interest in geodetic research. Their proximity to the earth's surface makes their trajectories sensitive to the earth's mass distribution. Gravitational disturbances of the order of 10^{-12} times normal gravity have been detected. Spacecraft of this type use an internal, free-floating, proof mass, shielded from the atmosphere by an external shell. Small thrusters on the shell are controlled such that the relative position of the proof mass to the shell remains approximately constant. The shell therefore is constrained to follow the proof mass.

The method used to calculate the pressure and shear stress also lends itself to an estimate of the **heat transfer** in free molecule flows. The energy flux incident on the surface is $E_i = (1/2)ImV^2$; the energy flux leaving the surface is E_r. The amount of heat transferred into the surface per unit time and unit area is $q = E_i - E_r$. Therefore, from the definition (2.41) of the accommodation coefficient, assuming $\sigma = 1$,

$$q = \alpha I \left(\tfrac{1}{2} mV^2 - \tfrac{3}{2} kT_w \right)$$

In the general case in which several molecular species S_j, $j = 1, 2, \ldots$, are present,

$$q = \sum_j \left[\alpha_j I_j \left(\tfrac{1}{2} m_j V^2 - \tfrac{3}{2} kT_w \right) \right] \tag{2.46}$$

2.5 Solar Radiation Pressure

Solar radiation can exert an appreciable force and torque on a spacecraft. For instance, at an altitude of 500 km in the earth's atmosphere, the effects of solar radiation on an orbiting spacecraft are of the same order of magnitude as the atmospheric drag. At higher altitudes, the solar radiation pressure predominates.

The radiation pressure decreases with the inverse square of the distance from the sun. (Solar radiation needs to be distinguished from the solar wind, which is the name given to the particle stream, mostly protons and electrons, that emanates from the sun. The pressure produced by the solar wind is negligible by comparison with the radiation pressure.)

The principal interest in solar radiation pressure derives from the disturbing effects it produces over long periods of time on the trajectory and attitude of spacecraft, on earth satellites, as well as on spacecraft on missions in the larger solar system.

Although its practicality has not been proved as yet, many authors have proposed to utilize the radiation pressure for **solar sailing**, particularly for small, scientific payloads to the inner planets or near-earth asteroids. The advantage, of course, is that no propellant needs to be carried. As has been shown theoretically, trajectory changes can be accomplished with good control. For missions for which low thrust is acceptable, solar sails must compete with the more versatile means that are provided by electric propulsion, such as ion propulsion.

The momentum of a photon is E/c (E = energy, c = velocity of light in vacuum). The energy flux (energy per unit area and unit time) of the solar photons in the earth's vicinity (i.e., at 1 AU) has been measured repeatedly. Its value, 1353 W/m^2, is referred to as the **solar constant**, designated here by the symbol j_h. It follows that the corresponding momentum flux that is incident on a surface perpendicular to the rays is $j_h/c = 4.51 \, 10^{-6}$ N/m^2.

If the surface is nonabsorbing and specularly reflecting, as would be the case, for instance, for solar sails, the pressure on the surface is double this value because of the recoil of the photons. For instance, at the average solar distance of Mercury, the pressure on a unit area perpendicular to the rays would be 5.9 10^{-5} N/m^2.

The **energy**, J_h, of the photons, incident at an angle ϕ_i (measured from the surface normal) on a surface of unit area, per unit time is

$$J_h = j_h(r_g^2/r^2)\cos\phi_i \qquad (2.47a)$$

where r is the distance from the sun and $r_g = 1$ AU. It follows that the corresponding flux of **momentum** is

$$J_h/c = (j_h/c)(r_g^2/r^2)\cos\phi_i \qquad (2.47b)$$

In what follows, we consider two separate cases. In the first, the surface is assumed to be nonabsorbing and reflecting specularly. In the second, the surface is partially absorbing and reflects the remainder diffusely, following Lambert's empirical law for such surfaces.

Nonabsorbing and reflecting specularly: The total pressure, p_h, is

$$p_h = 2(j_h/c)(r_g^2/r^2)\cos^2\phi_i \qquad (2.48a)$$

Because of the cancellation of the effect of the incident by the reflected photons, the shear stress

$$s_h = 0 \qquad (2.48b)$$

Partially absorbing with diffuse reflection: The absorption of light by a surface is characterized by the absorption coefficient α_h for solar photons. It should be noted that, as usually defined, this coefficient refers to the absorption of energy, not of momentum. The conversion, however is simple:

The reflection of light by partially absorbing surfaces usually follows closely Lambert's law (Lambert, 1728–1777). This law states that the intensity of the reflection falls off with the cosine of the angle, say ϕ, from the surface normal. (It may be noted here that the diffuse reflection of photons differs from that of molecules, discussed in the preceding section. In the latter case, thermalization is assumed, with the result that the intensity of the flux is independent of the angle.) The flux of reflected energy is therefore

$$(1-\alpha_h)J_h = i_n \iint \cos\phi\, d\Omega$$

where i_n is the intensity (energy per unit area and unit time and per unit solid angle) that is reflected in the direction of the surface normal. The integration is over all solid angles Ω, that is, over the surface of the unit hemisphere above the reflecting surface element. Carrying out the integration and solving for i_n gives

$$i_n = \pi^{-1}(1-\alpha_h)J_h$$

The pressure, p_{hr}, that results from the reflected photons is obtained from

$$p_{hr} = (i_n/c)\iint \cos^2\phi\, d\Omega$$

Integrating over the solid angles as before and substituting for i_n and for J_h

from (2.47a) gives

$$p_{hr} = 2(3c)^{-1}(1 - \alpha_h) j_h (r_g^2/r^2) \cos \phi_i$$

Finally, adding to this the pressure from the incident photons then gives the total pressure from solar radiation,

$$p_h = (j_h/c)(r_g^2/r^2) \cos \phi_i [\tfrac{2}{3}(1 - \alpha_h) + \cos \phi_i] \quad \text{(2.49a)}$$

By symmetry, the shear stress that results from the reflected photons vanishes. Therefore the total shear stress, s_h, equals that produced by the incident photons. Hence

$$s_h = (j_h/c)(r_g^2/r^2) \cos \phi_i \sin \phi_i \quad \text{(2.49b)}$$

2.6 Atmospheric Entry

Reentry into the earth's atmosphere or entry into a planetary atmosphere calls for an understanding of the aerodynamic forces and moments on the vehicle and of the heat transferred to it.

The flight paths are determined by gravity, drag, and lift. The lift, and to a lesser extent the drag, can be modulated by changes of the angle of attack, either by small thrusters or by movable aerodynamic surfaces. Also, the trajectories do not need to be restricted to a plane: by banking the vehicle, without thrust, a limited amount of cross-range can be obtained. The purpose, most often, is to steer toward an intended touchdown point. Banking can also be used to slow a vehicle that might otherwise overshoot the touchdown point.

If the flight path is meant to be in a plane and no active control is intended, care needs to be taken so that small asymmetries of the vehicle—even if it is supposedly axisymmetric—do not produce forces perpendicular to the plane. By imparting a slow rotation to the vehicle, the effect produced by such side forces can be eliminated.

Flight path control of manned vehicles is particularly demanding. Not only is high accuracy required in steering toward the landing point, but also the maximum deceleration may not exceed the physiologically set limit.

A major uncertainty affecting flight paths is caused by the large daily, seasonal, and solar cycle–induced variability of the atmospheric density at high altitudes. Although the density there is low, the high entry speed produces appreciable drag and lift, which can be of a magnitude comparable to those at lower altitudes, where the density is much higher but also the velocity much lower.

Entry at orbital velocities into a planetary atmosphere causes intense heating. The difficulties in predicting and guarding against the heat transferred to the vehicle are very great compared with the more conventional heat transfer problems encountered in other engineering applications. At speeds comparable to orbital velocities, the air (or carbon dioxide in the case of Mars) is partially dissociated behind the gasdynamic shocks and in the boundary layers. The various chemical species that are generated will diffuse in the boundary layers and can recombine at the surface, giving up their heat of reassociation to the surface. Because the ionization potentials are typically higher than those for dissociation, the degree of ionization is weak by comparison, but often sufficient to cause a microwave communications blackout. Depending on the severity of the temperature rise, there can also be heat transfer by radiation from the gas, in addition to the convective heating.

Much research has been conducted to produce heat protection materials. These materials, for instance lightweight, foamed ceramics, must have low heat conductivity, yet sufficient mechanical strength to withstand the aerodynamic pressure and shear. Ablation of the heat protection materials prevents their use on more than a limited number of missions but has the advantage of providing a heat barrier against the hot gas by outgassing. If the surface is black, a major part of the heat incident on these materials is radiated into space.

In addition to depending on aerodynamic studies of heat and mass transfer, predictions of the heating and its effect on the materials depend much on studies in chemical kinetics. This field, which is aptly called **aerothermochemistry** is beyond the scope of this book. The interested reader is referred to Refs. 15 and 16 at the end of this chapter.

For illustrative purposes, calculated surface temperatures at reentry of the Space Shuttle Orbiter are indicated in Fig. 2.12 from Ref. 17. Also indicated are the critical temperatures that are reached at maximum yaw of the Orbiter during ascent.

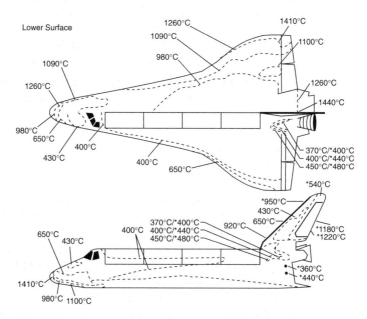

Figure 2.12 Calculated isotherms, Space Shuttle Orbiter, at reentry. (* denotes ascent temperatures at a yaw of 8.0°.) From Ref. 17, Peake, D. J. and Tobak, M. The original version of this material was first published by the Advisory Group for Aerospace Research and Development, North Atlantic Treaty Organization (AGARD/NATO) in "AGARDOgraph AG-252—Three-Dimensional Interactions and Vortical Flows with Emphasis on High Speeds," July 1980, by permission.

2.6 Atmospheric Entry

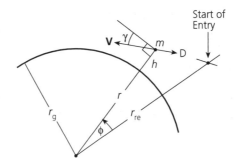

Figure 2.13 Entry into a planetary atmosphere, schematic.

Similarly, the materials research, much of it being conducted in arc jets, cannot be treated here. A good summary of the calculations of flight paths and estimates of the heat transfer is contained in Chap. 6 of Ref. 14.

In what follows, we illustrate the calculation of a reentry flight into the earth's atmosphere. This calculation, to stay simple, makes a number of assumptions that are only qualitative. The calculation, however, is sufficient to provide insight into the general problem.

As indicated schematically in Fig. 2.13, a flight path in the plane of the figure is assumed. The altitude above sea level is designated by h. The vehicle is supposed to have drag, designated by $D(h)$, but no lift or thrust. A constant drag coefficient, C_D, will be assumed. The velocity vector, $\mathbf{V}(h)$, can be taken as the velocity in inertial space, but the difference between it and the vehicle's velocity relative to the atmosphere will be neglected.

This assumption is quite crude: if the entry velocity, taken conventionally at about 120 km altitude, is the low-earth-orbit velocity, the vehicle velocity would be about 7.75 km/s, corresponding to a Mach number of about 20. But below this altitude, the drag rapidly increases, greatly reducing the vehicle's velocity. By contrast, the earth's, hence the atmosphere's, rotational velocity at the equator is 0.465 km/s, corresponding to a Mach number of about 1.35 at sea level.

The flight angle, $\gamma(h)$, that is, the angle between the velocity and the local horizontal, is taken positive for a downward path. Also let m the mass of the vehicle, $r(h)$ its distance from the earth's center, and r_g the earth's radius. The change with altitude of the gravitational acceleration, g, will be neglected.

The density $\varrho(h)$ is assumed to be an exponential function of the altitude. This can be roughly justified by assuming that the atmosphere satisfies the ideal gas law, at a temperature assumed to be independent of the altitude. From the hydrostatic equation, $dp = -g\varrho\, dh$ (p = atmospheric pressure), it then follows that

$$RT\, d\varrho = -g\varrho\, dh \tag{2.50}$$

Integration between the altitudes h_1 and h_2 and the corresponding densities ϱ_1 and ϱ_2 results in

$$h_2 - h_1 = (RT/g)\ln(\varrho_1/\varrho_2)$$

In particular, if the density ratio ϱ_1/ϱ_2 is chosen to be equal to the basis, e, of the natural logarithms, one obtains what is called the **scale height**, h_0, of the atmosphere, where

$$h_0 = RT/g \tag{2.51}$$

The change in density with altitude can therefore be expressed roughly by

$$d\varrho/\varrho = -dh/h_0 \tag{2.52}$$

The assumption that the scale height of the atmosphere is a constant in the altitude range of reentry trajectories is only qualitative. Thus, at 120 km altitude (where the temperature is about 360 K, substantially higher than the atmosphere's minimum temperature of 190 K), the scale height is 9.5 km, compared with only 5.4 km at the temperature minimum at 90 km altitude, but again 9.2 km at sea level. Nevertheless, because the density changes so drastically from its value of 1.22 kg/m^3 at sea level to 2.22 10^{-8} kg/m^3 at 120 km altitude, the concept of a scale height, more or less constant, is a useful one for approximate calculations of reentry trajectories.

The equation of motion tangential to the flight path is

$$m\frac{dV}{dt} = mg\sin\gamma - D \tag{2.53a}$$

The corresponding equation for the direction perpendicular to the path is most easily calculated from the path's radius of curvature, r_c, and from this the magnitude of the centrifugal force. One shows easily that

$$r_c = \left(\frac{d\gamma}{ds} + \frac{1}{r}\cos\gamma\right)^{-1}$$

where ds is the element of path length. (The first term comes from the change of the flight angle, the second term from the change of the orientation of the local horizontal.) Balancing the centrifugal force against the gravity component perpendicular to the path, and setting $ds = V\,dt$, results in the second equation of motion

$$V\frac{d\gamma}{dt} + \left(\frac{V^2}{r_g} - g\right)\cos\gamma = 0 \tag{2.53b}$$

where the approximation $r = r_g$ has been used.

It turns out to be advantageous to use the density rather than the time as the independent variable [18]. Making use of (2.52),

$$dt = \frac{h_0}{V\sin\gamma}\frac{d\varrho}{\varrho}$$

so that (2.53a) becomes

$$\varrho\frac{d(V^2/gr_g)}{d\varrho} = \frac{2h_0}{r_g} - \frac{\varrho h_0 AC_D}{m\sin\gamma}(V^2/gr_g)$$

with $D = C_D A(\varrho/2)V^2$ (A = cross-sectional area of the vehicle).

2.6 Atmospheric Entry

It is convenient here to introduce the constant

$$k = h_0 A C_D / m \tag{2.54}$$

and to define in place of the density the nondimensional quantity

$$\sigma = \ln(k\varrho) \tag{2.55}$$

With this, the final forms of the equations of motion become

$$\frac{d(V^2/gh_0)}{d\sigma} = 2 - \frac{\exp(\sigma)}{\sin \gamma}(V^2/gh_0) \tag{2.56a}$$

$$(V^2/gh_0)\frac{d\gamma}{d\sigma} = \left(1 - \frac{h_0}{r_g}(V^2/gh_0)\right)\cotan \gamma \tag{2.56b}$$

These two equations, with specified initial conditions, suffice to determine the dependent variables V^2/gh_0 and γ as functions of the independent variable σ. To solve this system of equations, a numerical method of integration is required.

The initial conditions are formulated at a reentry height, h_{re}, where the effect of the drag on the trajectory starts to become appreciable. For most applications, a convenient choice for entry into the earth's atmosphere is 120 km. Accordingly, the initial conditions are given by

$$\sigma = \sigma_{re} = \ln(k\varrho_{sl}), \quad V = V_{re}, \quad \gamma = \gamma_{re} \tag{2.57}$$

Once solved for V and γ, the altitude to which they apply can be found by integrating (2.52) between $h = 0$ and h, with the result that

$$h = h_0(\sigma_{sl} - \sigma) \tag{2.58}$$

where $\sigma_{sl} = \ln(k\varrho_{sl})$ and ϱ_{sl} the sea level density.

Figure 2.14 presents an example, calculated from (2.56), of an earth reentry trajectory. The start of the reentry is taken (arbitrarily) at 120 km altitude, with a velocity of 7.75 km/s and an angle of 10.0° from the local

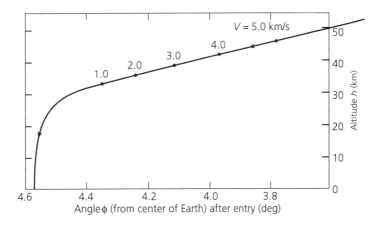

Figure 2.14 Reentry (shown below 50 km altitude) into the earth's atmosphere. Calculated from Eq. (2.56). Entry altitude, 120 km; entry angle γ, 10.0°; entry velocity, 7.75 km/s; k, 100 m³/kg.

horizontal. The trajectory, which is for $k = 100 \text{ m}^3/\text{kg}$, is shown as a function of the geocentric angle ϕ (shown in Fig. 2.13), which is a measure of the horizontal distance covered by the vehicle.

For a more shallow entry angle, higher entry velocity, and/or with lift, the vehicle can again skip out of the atmosphere. The continuation of the trajectory is then a Kepler-type orbit, which may terminate in another entry, with the vehicle losing velocity each time.

The earliest example of a repeated entry–skip–entry maneuver is provided by the NASA Magellan mission to Venus in 1993. The solar panels were turned into the wind to provide aerodynamic braking. After having completed its scientific tasks and entering and skipping about 700 times, with the attitude of the vehicle controlled intermittently by small thruster impulses, the vehicle finally reached the surface and was destroyed (as had been intended).

Nomenclature

c	velocity of light in vacuum
$f(v)$	Maxwell–Boltzmann velocity distribution
g	gravitational acceleration
g_0	standard gravitational acceleration
h	altitude
h_0	scale height
j_h	solar constant = 1353 W/m^2
k	Boltzmann constant = 1.3806 10^{-23} kg m^2/(s^2 K); also constant defined in (2.54)
m	mass
$\tilde{m}$	mass flow rate
n	unit vector, normal to surface
p	pressure
r	position vector; r, radius
s	shear stress
A	area
C_D	drag coefficient
F	force
G	universal gravitational constant = 6.673 10^{-11} m^3/(kg s^2)
I	incident flux
I_1, I_2, I_3	moments of inertia; **J**: inertia tensor
I_{sp}	specific impulse
J_n	zonal harmonics; C_{nm}, S_{nm}: tesseral harmonics
K	Knudsen number
M	moment
N	number of molecules per unit volume
P_n	Legendre polynomials; P_n^m: Legendre functions
V	velocity
$(\ )_{ex}$	nozzle exit
$(\ )_{fm}$	free molecule flow
$(\ )_g$	gravity; also Earth

( )$_h$	solar
( )$_i$	incident
( )$_r$	reflected
( )$_{re}$	reentry
( )$_{sl}$	sea level
( )$_t$	thrust
α	Smoluchowski–Knudsen coefficient;
α_h	solar radiation absorption coefficient
λ	mean free path
μ	gravitational parameter
ϱ	density
ϕ_i	angle of incidence
Φ_g	gravitational potential

Problems

(1) Consider the vertical takeoff of the first stage of a launch vehicle, neglecting the aerodynamic drag, the variation of the gravitational acceleration with height, and the earth's rotation. The liftoff mass is 75,000 kg. The thrust is constant up to an altitude of $h_1 = 25.0$ km and is 2.00 times the liftoff weight. The specific impulse is 370 s.

Find the time t_1 after liftoff when the vehicle has reached the altitude h_1. Also find the vehicle's mass m_1 and velocity V_1 at that time.

(2) A single-stage sounding rocket takes off vertically from sea level. The earth's rotation and the variation of the gravitational acceleration with altitude can be neglected. Aerodynamic drag is included with the simplification that the drag coefficient is assumed to be constant and that the atmospheric density varies exponentially with height. The following data are given:

Lift-off mass	200 kg
Propellant mass	100 kg
Thrust	6.00 kN
Specific impulse	250 s
Drag coefficient	0.80
Vehicle cross-sectional area	0.0250 m^2
Air density at sea level	1.225 kg/m^3
Atmospheric scale height	9.20 km

(a) By numerical integration of the equation of motion, find the altitude at which the propellant is exhausted.
(b) After propellant depletion, the vehicle will coast to a still higher altitude. Find the peak altitude.

(3) An asteroid has the approximate shape of an ellipsoid of revolution. Its semimajor axis is 6 km, its semiminor axis is 4 km. The average density is $2.70 \cdot 10^3$ kg/m^3.

Find the surface gravitational accelerations on the semimajor and the semiminor axes. (Universal gravitational constant $G = 6.673 \cdot 10^{-11}$ m^3 kg^{-1} s^{-2}.)

(4) An inflated Mylar sphere of 5.00 m radius is on a low earth orbit at 500 km altitude. The surface density of the skin is 0.70 kg/m². (The gravitational parameter of the earth is 3.986 10⁵ km³/s²; the earth's mean radius is 6378 km).

Find the gravity gradient torque about the center of the sphere.

(5)* A spacecraft in the form of a sphere with 2.00 m diameter is in a low-earth-equatorial orbit and is moving from west to east at 300 km altitude. At this altitude, the principal species present are atomic oxygen and molecular nitrogen. Free molecule flow and fully diffuse reflection from the spacecraft surface are assumed.

The spacecraft surface temperature is 300 K. The atmosphere corotates with the earth, which has the consequence that the velocity of the spacecraft relative to the atmosphere is 7.30 km/s (the orbital velocity less the earth's equatorial rotational velocity).

Using the following data, compute the drag.

	O	N_2
Number density (m^{-3}) at 300 km	7 10^{14}	9 10^{13}
Mass (atomic units)	16.0	28.0
Accommodation coefficient	0.80	0.80

Atomic mass unit = 1.6605 10^{-27} kg.
Boltzmann constant = 1.3806 10^{-23} kg m²/(s² K).

(6)* The problem is related to the production in space of semiconductor and other materials that require for their manufacture an extremely hard vacuum. Advantage is taken of the vacuum in the wake of a spacecraft operating in the free molecular flow regime of the earth's atmosphere.

Assumed is a spacecraft in the shape of a circular cylinder with radius $r = 1.00$ m. The axis is parallel to the direction of flight in the atmosphere. The velocity of the spacecraft relative to the atmosphere (i.e., relative to the center of mass velocity of the molecules) is 7.3 km/s. On the axis, at a distance of $0.50r$ behind the vehicle's base and parallel to it is a small disk with dimensions that are negligible compared with r.

The altitude is 400 km. The temperature is 1000 K and the number density of the atomic oxygen is 10^{14} m^{-3}.

From the assumed Maxwellian distribution, calculate the number of oxygen atoms per second and cm² that strike the disk's front surface.

References

1. Bursa, M. and Pec, K., "Gravity Field and Dynamics of the Earth," Springer Verlag, Berlin, 1993.
2. Kaula, W., "Theory of Satellite Geodesy," Blaisdell Publishing, Waltham, MA, 1966.
3. Kellog, O. D., "Foundations of Potential Theory," Springer Verlag, Berlin, 1929.

4. Morse, P. M. and Feshbach, H., "Methods of Theoretical Physics," Vol. 2, McGraw-Hill, New York, 1953.
5. Huzel, D. K. and Huang, D. H., eds. "Design of Liquid-Propellant Rocket Engines," *Progress in Astronautics and Aeronautics*, Seebass, A. R., ed., American Institute of Aeronautics and Astronautics, Washington, DC, 1992.
6. Meyer, R. X., "Inflight Formation of Slag in Spinning Solid Propellant Rocket Motors," *Journal of Propulsion and Power*, Vol. 8, No. 1, pp. 45–50, 1992.
7. Ashley, H. and Landahl, M. T., "Aerodynamics of Wings and Bodies," reprint, Dover Publications, New York, 1985.
8. Ashley, H., "Engineering Analysis of Flight Vehicles," Addison-Wesley, Reading, MA, 1974.
9. Martin, F. W. and Slotnick, J. P., "Flow Computation for the Space Shuttle in Ascent Mode Using Thin-Layer Navier-Stokes Equations," *Progress in Astronautics and Aeronautics*, Seebass, A. R., ed., American Institute of Aeronautics and Astronautics, Washington, DC, 1990.
10. Emmons, H. W., ed., "Fundamentals of Gas Dynamics," Section H, Vol. 1, Schaaf, S. A., "*Flow of Rarefied Gases*," Princeton University Press, Princeton, NJ, 1958.
11. Saksaganskii, G. L., "Molecular Flow in Complex Vacuum Systems," translated from the Russian, Gordon & Breach Science Publishers, New York, 1988.
12. Tribble, A. C., "The Space Environment," Princeton University Press, Princeton, NJ, 1995.
13. Hedin, A. E., "MSIS-86 Thermosphere Model," *Journal of Geophysics Research*, Vol. 92, No. A5, pp. 4649–4662, 1987.
14. Griffin, M. D. and French, J. R., "Space Vehicle Design," AIAA Education Series, American Institute of Aeronautics and Astronautics, Washington, DC, 2nd printing, 1991.
15. Incropera, F. P. and Dewitt, D. P., "Fundamentals of Heat and Mass Transfer," John Wiley & Sons, New York, 3rd ed., 1990.
16. Horton, T. E., ed., "Thermophysics of Atmospheric Entry," *Progress in Astronautics and Aeronautics*, Vol. 82, American Institute of Aeronautics and Astronautics, Washington, DC, 1982.
17. Peake, D. J. and Tobak, M., "Three-Dimensional Interactions and Vortical Flows with Emphasis on High Speeds," North Atlantic Treaty Organization Advisory Group for Aerospace Research and Development, AGARD-AG-252, Paris, 1980.
18. Loh, W. H. T., "Entry Mechanics," *Re-Entry and Planetary Entry Physics and Technology*, Loh, W. H. T. ed., Springer Verlag, Berlin, 1968.

3

Orbits and Trajectories in an Inverse Square Field

The principal forces that determine the path of a spacecraft are normally gravitation and thrust. In comparison, forces such as atmospheric drag or solar radiation pressure are small and can often be neglected. In this chapter the motion of space vehicles will be considered when gravity is the only force present, that is, at times when the vehicle is coasting without thrust or other forces acting on it.

In the vicinity of a planet or other astronomical body, the gravitational field is composed of an inverse square field directed toward the planet's center of mass and a much smaller gravitational perturbation field (Sect. 2.1.3). The latter is caused by the planet's mass distribution, which will generally deviate somewhat from exact spherical symmetry. The resulting perturbation terms, although small, are important in the application of satellites to geodesy and to high-precision navigation. They fall off, however, more rapidly with distance from the center of mass than is the case for the inverse square law force and become negligible at a large distance from the astronomical body.

If the astronomical body is a planet, it will orbit the sun and therefore be accelerated relative to a heliocentric ecliptic reference frame (Sect. 1.1.2). A similar case, which is particularly important, occurs for motions of spacecraft in the gravitational and acceleration field jointly produced by the earth and the moon.

The present chapter, however, will be limited almost entirely to applications in which the motion of a spacecraft can be described solely in terms of an inverse square law gravitational force produced by a *single* astronomical body isolated from the influence of other bodies. This approximation is often valid, when the high precision requirements of geodesic or navigational satellites may not be needed.

In a final section, some relativistic corrections to Kepler orbits will be discussed.

Although the mathematical relations presented in this chapter relate to spacecraft, it should be realized that they are merely a special case of the motion of planets about the sun and of moons about the planets. Nearly all these relations were known to astronomers in the 18th and 19th centuries and even earlier [1, 2]. Modern texts [3–8] on the motion of spacecraft make extensive use of the methods developed originally in astrodynamics.

Of course, the mass of a spacecraft can be neglected in comparison with the mass of a planet.

CHAPTER 3 Orbits and Trajectories in an Inverse Square Field

3.1 Kepler Orbits and Trajectories

Newton's second law therefore applies in the form

$$\frac{d^2\mathbf{r}}{dt^2} = -\frac{\mu}{r^3}\mathbf{r} \tag{3.1}$$

where $\mathbf{r}$ is the position vector from the center of attraction to the spacecraft and μ the gravitational parameter of the astronomical body. Since the position vector is in the plane defined by the center of attraction and the acceleration vector, it is evident that the motion is in a plane through the center of attraction.

Two constants of motion are the **orbital angular momentum**, designated by $\mathbf{h}$, of the spacecraft about the center of attraction, and the sum, designated by w, of the **potential and orbital kinetic energies.** (A third constant of motion is the eccentricity vector, perpendicular to the orbit plane and of magnitude e, as defined below later.) Both $\mathbf{h}$ and w are understood to be taken per unit mass of the spacecraft. Therefore, if $\mathbf{v}$ denotes the velocity of the center of mass of the spacecraft in an (approximately inertial) reference frame in which the planet is at rest,

$$\mathbf{h} = \mathbf{r} \times \mathbf{v} \tag{3.2}$$

$$w = -\mu/r + \tfrac{1}{2}v^2 \tag{3.3}$$

If, as indicated in Fig. 3.1, θ is the angle between $\mathbf{r}$ and a direction fixed in the reference frame, the radial and tangential components of the velocity are

$$v_r = \dot{r}, \qquad v_\theta = r\dot{\theta}$$

so that

$$h = r^2\dot{\theta} = \text{const.} \tag{3.4}$$

and

$$w = -\mu/r + \tfrac{1}{2}(\dot{r}^2 + r^2\dot{\theta}^2) = \text{const.} \tag{3.5}$$

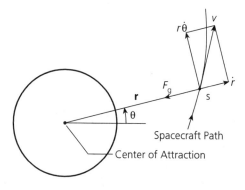

Figure 3.1 Spacecraft path in an inverse square central force field.

Substitution of $\dot{\theta}$ from (3.4) into (3.5) and differentiation with respect to time result in the relation

$$\ddot{r} + \mu/r^2 - h^2/r^3 = 0$$

It can be cast into a more standard form by introducing the reciprocal radius $u = r^{-1}$. Since

$$\dot{r} = \frac{dr}{d\theta}\frac{d\theta}{dt} = \frac{dr}{d\theta}\frac{h}{r^2} = -h\frac{du}{d\theta}$$

$$\ddot{r} = \frac{d\dot{r}}{d\theta}\frac{d\theta}{dt} = \frac{d}{d\theta}\left(h\frac{du}{d\theta}\right)\frac{h}{r^2} = -h^2 u^2 \frac{d^2 u}{d\theta^2}$$

u is seen to satisfy the harmonic oscillator equation

$$\frac{d^2 u}{d\theta^2} + u = \frac{\mu}{h^2} \qquad (3.6)$$

with the solution

$$u = \mu/h^2 + C\cos(\theta - \theta_0) \qquad (3.7)$$

where C and θ_0 are constants of integration.

The constant C can be expressed in terms of the constants of motion h and w by observing that from (3.4), (3.5), and the expression just found for u

$$w = -\mu u + \frac{1}{2}h^2\left[\left(\frac{du}{d\theta}\right)^2 + u^2\right] = \frac{1}{2}(C^2 h^2 - \mu^2/h^2)$$

hence

$$C = \pm\frac{\mu}{h^2}\sqrt{1 + \frac{2wh^2}{\mu^2}}, \qquad h \neq 0 \qquad (3.8)$$

For physically realizable systems, u, and therefore C, must be real. Hence

$$2wh^2/\mu^2 \geq -1 \qquad (3.9)$$

For positive C, u has a maximum, hence r a minimum at $\theta = \theta_0$. For negative C, the minimum r occurs when $\theta = \theta_0 + \pi$. This point on the trajectory where the spacecraft is closest to the astronomical body is, in conformance with astronomical custom, referred to as the **periapsis** (in the case of the earth usually referred to as **perigee**, in the case of the sun as **perihelion**). For closed orbits, the point of largest distance is called the **apoapsis** (or **apogee** in the case of the earth and **aphelion** in the case of the sun) (Greek: *peri...* = around, about; *apo...* = out of, from). It is customary to count the angle θ in the direction of motion starting from the periapsis. Therefore $\theta_0 = 0$ for positive C, $\theta_0 = -\pi$ for negative C. For a circular orbit, the origin of θ is usually taken at its intersection with either the equatorial or the ecliptic plane. In astronomy, θ is known as the **true anomaly** and is often given the symbol f in place of θ.

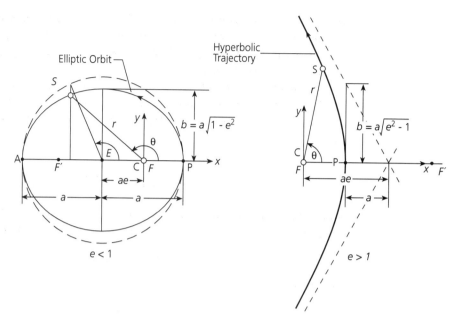

Figure 3.2 Geometry of the one-body problem. C, center of attraction; S, spacecraft; P, periapsis; A, apoapsis; F, F', focal points.

From (3.7) follows the final result for describing the path,

$$r = \frac{h^2}{\mu(1 + e \cos \theta)}, \quad h \neq 0 \tag{3.10}$$

(irrespective of the sign of C), where

$$e = \sqrt{1 + 2wh^2/\mu^2}, \quad e \geq 0 \tag{3.11}$$

Equation (3.10) is the expression in polar coordinates of a *conic section*, with *e* the eccentricity. For $0 < e < 1$, the path is an ellipse, for $e > 1$ a hyperbola, and in the two limiting cases a circular orbit and a parabolic path. It also follows that the center of the astronomical body is a focal point of the conic section. Figure 3.2 illustrates these geometrical properties.

It is of historical interest to note that expressions equivalent to (3.10) were derived by Johannes Kepler (1571–1630) on the basis of Tycho Brahe's (1546–1601) astronomical observations of the motion of the planets. Kepler formulated these laws: (1) The orbit of each planet is an ellipse in a plane containing the sun, with the sun at one focus. (2) The line connecting the sun and the planet sweeps out equal areas in equal time intervals. (3) The square of the orbital period of a planet is proportional to the cube of the semimajor axis of the ellipse.

Newton was the first to derive Kepler's three laws mathematically, starting from his law of universal gravitation. Although he had at that time already invented the calculus, he used instead geometrical arguments.

Kepler's laws are actually more accurate when applied to spacecraft than to planets. This is because planetary masses are not entirely negligible when compared with the sun's mass. (For Jupiter, the most massive of the planets, the mass ratio is approximately 0.955 10^{-3}.) As a consequence, the foci of the planetary orbits do not exactly coincide with the center of the sun, as had been assumed by Kepler, but more nearly with the center of mass of the solar system, the so-called barycenter.

Returning to the earlier development, one finds a remarkably simple relation between the semimajor axis, denoted by a, and the energy w. At the periapsis the velocity is purely transverse and $\theta = 0$, so that from (3.3) and (3.10)

$$w = -\frac{\mu}{r_p} + \frac{1}{2}\frac{h^2}{r_p^2}, \qquad r_p = \frac{h^2}{\mu(1+e)}$$

where r_p is the periapsis radius. Hence $w = -\mu(1-e)/(2r_p)$. From the geometrical relation indicated in Fig. 3.2, $r_p = \pm a(1-e)$, it follows that

$$w = \mp \mu/(2a) \tag{3.12}$$

and also

$$h^2 = \pm \mu a(1-e^2) \tag{3.13}$$

In these equations and in all others in this chapter where two signs are indicated, the upper sign stands for elliptic orbits, the lower one for hyperbolic trajectories. (The reader needs to be aware that with some authors the quantity a is negative for hyperbolas. In the present text, the convention is adopted that a is always positive.)

From (3.4) and (3.10) follows

$$\left. \begin{array}{c} v_r = (\mu/h)e\sin\theta, \qquad v_\theta = (\mu/h)(1 + e\cos\theta) \\ \\ v = \sqrt{\mu\left(\frac{2}{r} \mp \frac{1}{a}\right)} \end{array} \right\} \quad h \neq 0 \tag{3.14}$$

If dA designates the area enclosed between two radii separated by the angle $d\theta$, then from (3.4), $dA/dt = h/2$. Hence

$$\int_{t_1}^{t_2} dA = \frac{h}{2}(t_2 - t_1) \tag{3.15}$$

which states that for a given orbit the radius vector **r** sweeps out equal areas in equal time intervals. This equation is known as Kepler's "second law." (Actually, it was discovered by Kepler before the "first law.") An important conclusion drawn from it is that satellites on eccentric orbits move more slowly near their apoapsis than they do near their periapsis. Broadcasting systems, such as the Molniya (Russian: lightning) system, which are designed to cover primarily northern latitudes, take advantage of this by using a large eccentricity and by placing the apogee north of the equatorial plane. As a consequence, a typical ground receiver can have a direct line of sight to the spacecraft for a time that can exceed by more than one-half the orbital period.

Kepler's "third law" applies to closed orbits, that is, to *elliptic* and *circular* paths. Let P designate the orbital period. Because the area of an ellipse with

CHAPTER 3 *Orbits and Trajectories in an Inverse Square Field*

semimajor axis a and semiminor axis $b = a\sqrt{1-e^2}$ is πab, it follows from (3.15) that

$$P = 2\pi (a^2/h)\sqrt{1-e^2}$$

From (3.10) and the geometric definition of the eccentricity (Fig. 3.2),

$$r_p = h^2/[\mu(1+e^2)] = a(1-e)$$

so that

$$P = 2\pi\sqrt{a^3/\mu} \qquad (3.16)$$

proving Kepler's third law. Remarkably, the period depends only on the semimajor axis and not on the eccentricity.

Alternatively, the period can be expressed in terms of the total energy, w, per unit mass of the spacecraft, because from (3.12)

$$P = \pi\mu\sqrt{\tfrac{1}{2}(-w)^{-3}} \qquad (3.17)$$

again showing the result to be independent of the eccentricity. Figure 3.3 illustrates the relationship between the total energy and the orbital periods of several earth satellites. (Two of the examples shown have circular orbits; the Molniya satellites have strongly eccentric orbits; the moon's orbit also has an appreciable eccentricity: its distance from the earth's center varies from 356,400 to 406,700 km.)

Finally, we note that for *parabolic* paths, $e = 1$ and $w = 0$. Because at an infinite distance the potential energy tends to zero, it follows that the kinetic energy also tends to zero. The parabolic trajectory therefore represents the limiting case in which a spacecraft will escape the gravitational field, with a velocity that tends to zero at infinity.

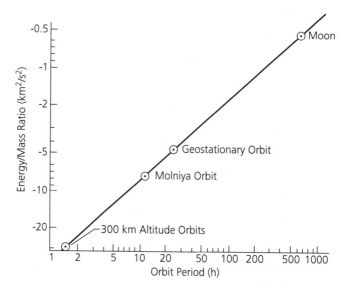

Figure 3.3 Total energy per unit mass of satellite versus orbital period.

The minimum velocity at periapsis that is needed for escape, if r_0 is the distance of the periapsis from the center of the planet, is, from (3.3),

$$v_{\text{esc}} = \sqrt{2\mu/r_0} \tag{3.18}$$

This can be compared with the velocity $v_{\text{crc}} = \sqrt{\mu/r_0}$ of a spacecraft on a circular orbit with the same radius r_0. Hence $v_{\text{esc}} = \sqrt{2}\, v_{\text{crc}}$. For the earth, assuming an initial circular orbit with $r_0 = 6678$ km (the mean equatorial radius plus 300 km altitude, a value that is fairly typical for parking orbits), v_{crc} is approximately 7.73 km/s, the escape velocity 10.9 km/s. For a west-to-east launch at the equator, the earth's circumferential velocity due to its diurnal rotation is 0.465 km/s, so that the minimum theoretical velocity increment that is needed for escape is 10.4 km/s.

3.2 Position as a Function of Time

So far, only the path, not the position of the spacecraft as a function of time, has been considered. Whereas the path can be expressed in terms of elementary functions, this is not the case for the temporal dependence of position. We limit the discussion at first to *elliptic orbits*.

It is customary to introduce the so-called **eccentric anomaly**, an angle that is traditionally designated by E. As shown in Fig. 3.2, the eccentric anomaly is obtained by projecting the satellite position on a circle of radius a that passes through the periapsis and apoapsis.

The needed relation is most easily demonstrated by introducing, as indicated in Fig. 3.2, the rectangular Cartesian coordinates of the satellite position

$$x = a(\cos E - e), \quad y = a\sqrt{1-e^2}\sin E \quad (e<1) \tag{3.19}$$

from which $r = a(1 - e\cos E)$ and

$$a(\cos E - e) = r\cos\theta, \quad a\sqrt{1-e^2}\sin E = r\sin\theta \quad (e<1) \tag{3.20}$$

(When using a single one of these equations to determine θ from E, or E from θ, an ambiguity arises as a consequence of the multivaluedness of the inverse trigonometric functions. In simple cases, the correct quadrant of the computed angle is often obvious. But when it is embedded in software, it is advisable to resolve the ambiguity by using both equations. This procedure also provides an easy check on the calculation.)

The angular momentum, h, per unit mass, therefore becomes

$$h = x\dot{y} - y\dot{x} = a^2\sqrt{1-e^2}\,\dot{E}(1 - e\cos E) = a\sqrt{1-e^2}\,r\dot{E}$$

Comparison with (3.13) results in $\dot{E}(1 - e\cos E) = \sqrt{\mu/a^3}$, or integrated

$$E - e\sin E = \sqrt{\mu/a^3}\,(t - t_{\text{p}}) \quad (e<1) \tag{3.21}$$

where t_{p} is the time of periapsis passage. This equation relates satellite position and time and is known as **Kepler's equation**.

In astronomy the (dimensionless) right-hand side of Kepler's equation is traditionally designated by M and is called the **mean anomaly**. The factor

$\sqrt{\mu/a^3}$ is designated by n and is called the **mean angular velocity** [since, as follows immediately from (3.16), n is the angular velocity averaged over the elliptic orbit]. Therefore (3.21) is often written in one of the forms

$$E - e\sin E = n(t - t_p) = M \quad (e < 1)$$

Before discussing methods for solving Kepler's equation, we complete this section by listing some other frequently used equations. Thus it follows from differentiating Kepler's equation, solving the result for $\dot{E}$, and substituting the result into (3.19) after differentiation that

$$\dot{x} = -\sqrt{\frac{\mu}{a}}\frac{\sin E}{1 - e\cos E}, \quad \dot{y} = \sqrt{\frac{\mu(1 - e^2)}{a}}\frac{\cos E}{1 - e\cos E} \quad (e < 1) \quad (3.22)$$

Analogous results can also be obtained for the *hyperbolic trajectories*. The eccentric anomaly, in this case called the **hyperbolic anomaly**, can be defined by means of the relation $r = -a(1 - e\cosh E)$ analogous to the equation preceding (3.20). We list these results:

$$x = a(e - \cosh E), \quad y = a\sqrt{e^2 - 1}\sinh E \quad (e > 1) \quad (3.23)$$

$$E - e\sinh E = -\sqrt{\mu/a^3}\,(t - t_p) \quad (e > 1) \quad \mathbf{(3.24)}$$

$$\dot{x} = -\sqrt{\frac{\mu}{a}}\frac{\sinh E}{e\cosh E - 1}, \quad \dot{y} = \sqrt{\frac{\mu(e^2 - 1)}{a}}\frac{\cosh E}{e\cosh E - 1} \quad (e > 1)$$

$$(3.25)$$

3.3 D'Alembert and Fourier–Bessel Series

The need frequently arises to compute a spacecraft's *position* in a known orbit as a function of time. This requires the solution of Kepler's equation (3.21) or (3.24), hence of a transcendental equation for the eccentric anomaly. (The inverse problem of finding the *time* at which the spacecraft is in a given position is straightforward because Kepler's equation is linear in time.) Numerically, the solution can be found by the Newton–Raphson or the more recent, robust method of Muller. Once the eccentric anomaly is found, the position follows directly, for instance, from (3.19) or (3.23).

Alternatively, the position of the spacecraft can be found by series expansions, originally developed for applications in celestial mechanics.

3.3.1 D'Alembert's Method

This series expansion (D'Alembert, 1717–1783) is useful for elliptic orbits with eccentricities not too close to unity. In astronomy, particularly because the orbits of the majority of planets and of their moons have only small eccentricities, D'Alembert's method has found very frequent applications. It is also useful in applications to space technology.

Let $\varepsilon = E - M$. Clearly, ε is a periodic, odd function of t (with $t = 0$ at a periapsis passage), hence of M. Making use of ε, Kepler's equation can be written

$$\varepsilon = E - M = e\sin E = e\sin(M + \varepsilon) = e(\sin M \cos\varepsilon + \cos M \sin\varepsilon)$$

Since $\varepsilon \to 0$ as $e \to 0$, this suggests expanding ε in a power series in e, $\varepsilon = \alpha_1(M)e + \alpha_2(M)e^2 + \alpha_3(M)e^3 + \cdots$ with coefficients α_i that are $\mathcal{O}(1)$ and depend periodically on M. Substituting the series into both sides of the preceding equation,

$$\alpha_1 e + \alpha_2 e^2 + \cdots = e[\sin M \cos(\alpha_1 e + \alpha_2 e^2 + \cdots) + \cos M \sin(\alpha_1 e + \alpha_2 e^2 + \cdots)]$$

The series arguments in the sines and cosines are close to 0 for sufficiently small e, which allows the sines and cosines to be expanded with rapid convergence by their own power series. Collecting the terms with the same power in e, one finds

$$\alpha_1 = \sin M, \qquad \alpha_2 = \alpha_1 \cos M = \tfrac{1}{2}\sin(2M)$$
$$\alpha_3 = -\tfrac{1}{8}\sin M + \tfrac{3}{8}\sin 3M, \qquad \cdots\cdots$$

The derivation is completed by substituting the coefficients α_i back into the power series for ε and collecting terms with the same frequency in M. The more complete result, carried out to higher orders in e than sketched here, is

$$E = M + e\left(1 - \frac{1}{8}e^2 + \frac{1}{192}e^4\right)\sin M + e^2\left(\frac{1}{2} - \frac{1}{6}e^2\right)\sin 2M$$
$$+ e^3\left(\frac{3}{8} - \frac{27}{128}e^2\right)\sin 3M + \frac{1}{3}e^4 \sin 4M + \frac{125}{384}e^5 \sin 5M + \cdots$$

(3.26)

The functions $\theta - M$ and r are also periodic functions of M, the first one odd, the second one even. We only state the results. Neglecting, as before, terms of order e^6 and higher,

$$\theta = M + e\left(2 - \frac{1}{4}e^2 + \frac{5}{96}e^4\right)\sin M + e^2\left(\frac{5}{4} - \frac{11}{24}e^2\right)\sin 2M$$
$$+ e^3\left(\frac{13}{12} - \frac{43}{64}e^2\right)\sin 3M + \frac{103}{96}e^4 \sin 4M + \frac{1097}{960}e^5 \sin 5M + \cdots$$

(3.27)

and

$$\frac{r}{a} = 1 + \frac{1}{2}e^2 - e\left(1 - \frac{3}{8}e^2 + \frac{5}{192}e^4\right)\cos M - e^2\left(\frac{1}{2} - \frac{1}{3}e^2\right)\cos 2M$$
$$- e^3\left(\frac{3}{8} - \frac{45}{128}e^2\right)\cos 3M - \frac{1}{3}e^4 \cos 4M - \frac{125}{384}e^5 \cos 5M - \cdots$$

(3.28)

It has been shown [9] that the three series (3.26), (3.27), and (3.28) converge for all values of M if $e < 0.6627$.

3.3.2 Fourier–Bessel Series

In addition to D'Alembert's series, a number of other methods, suitable for different ranges of the eccentricity, exist for solving Kepler's equation. Sketched here briefly is the Fourier–Bessel series. It converges rapidly for small values of e.

By differentiating Kepler's equation (3.21) for elliptic orbits, one obtains

$$dE = \frac{dM}{1 - e\cos E}$$

Since $(1 - e\cos E)^{-1}$ is a periodic, even function of M with period 2π, it can be expressed by the Fourier series

$$\frac{1}{1 - e\cos E} = A_0 + \sum_{k=1}^{\infty} A_k \cos(kM) \qquad (3.29)$$

where

$$A_0 = \frac{1}{2\pi}\int_{-\pi}^{+\pi} \frac{dM}{1 - e\cos E} = \frac{1}{2\pi}\int_{-\pi}^{+\pi} dE = 1$$

$$A_k = \frac{1}{\pi}\int_{-\pi}^{+\pi} \frac{\cos(kM)}{1 - e\cos E}\,dM = \frac{1}{\pi}\int_{-\pi}^{+\pi} \cos[k(E - e\sin E)]\,dE$$

$$= 2J_k(ke) \qquad k = 1, 2, \ldots$$

and where J_k is the Bessel function of the first kind of order k. The final result is obtained by integrating (3.29) over M term by term, resulting in

$$E = M + 2\sum_{k=1}^{\infty} \frac{1}{k} J_k(ke) \sin(kM) \qquad (3.30)$$

3.4 Orbital Elements

Up to this point it was sufficient to specify elliptic orbits or hyperbolic trajectories by their semimajor axis and eccentricity. However, for a complete description of the path of a spacecraft, relative to a fixed center of attraction, additional data are needed: Two angles are required to specify the orientation of the plane of the motion, and one angle is needed to specify the orientation in this plane of the major axis. Finally, the time at which the spacecraft passes a fixed point needs to be specified.

The six scalar quantities referred to are known as the **orbital elements** (also known as the "Keplerian elements"). Those most often chosen in space technology are:

a: the **semimajor** axis of elliptic orbits or hyperbolic trajectories.
e: the **eccentricity**.
i: the **inclination**, that is, the angle between the plane of the motion and the equatorial plane of the earth or planet. For motions about the

sun, in place of the equatorial plane the ecliptic plane is used. A more precise definition, which avoids a possible ambiguity in the sign of i, is to define the inclination as the angle i, $0 \le i < \pi$, between the angular momentum vector (therefore a vector perpendicular to the orbit plane) and the northward-pointing perpendicular to the equatorial or ecliptic plane.

Ω: the **right ascension of the ascending node**, which is the angle from the vernal equinox line to the ascending node, taken positive in the easterly direction.

ω: the angle from the ascending node to the periapsis, taken positive in the direction of motion.

t_p: the time at which the spacecraft passes the periapsis. (Sometimes t_p is replaced as an orbital element by the mean anomaly M at some reference time.)

Figure 3.4 illustrates the orbital elements. Point P in the figure is the **periapsis**, point AN the **ascending node**, that is, the point where the spacecraft crosses the equatorial or ecliptic plane from the southern to the northern hemisphere. The angles Ω, i, ω (in this order) correspond to the Euler angles ϕ, θ, ψ shown in Fig. 1.7. For historical reasons, and in conformity with the technical literature, they are denoted by these new symbols.

Some of the orbital elements, as listed, can be ill defined. Thus, for circular orbits, for which there is no distinct periapsis, and also in cases in which $e \approx 0$, it is customary to measure the position of the spacecraft instead from the ascending node. If $i = 0$, there is no distinct ascending node. In

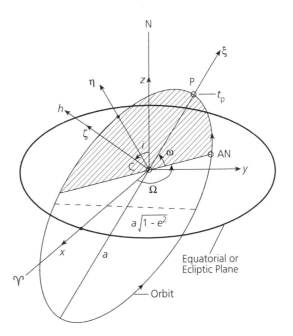

Figure 3.4 The six orbital elements a, e, i, Ω, ω, t_p. ♈, first point of Aries; N, north; AN, ascending node; P, periapsis; **h**, angular momentum vector.

this case, or when $i \approx 0$, one often uses the quantity $\bar{\omega} = \omega + \Omega$ as one of the orbital elements.

In the important case of circular orbits in the equatorial plane, $e = i = 0$. In this case just two orbital elements suffice: the orbit radius a and the time of passage of the vernal equinox line.

In the inverse square force field considered in the present chapter, the orbital elements are constants. Their importance, however, derives largely from their use in analyzing the effects of small perturbations from the inverse square field. Such perturbations may be caused, for instance, by the nonsphericity of the mass distribution of the earth, as discussed in Sect. 2.1.3. In these cases, the orbital elements are no longer constant but change slowly with time, producing secular terms in the equations.

Finally, the following expressions for the declination δ and the right ascension α (Figs. 1.3 and 1.4) in terms of the orbital elements i, Ω, ω and the true anomaly θ may be noted:

$$\sin \delta = \sin i \sin(\omega + \theta) \qquad (3.31a)$$

$$\cos(\alpha - \Omega) = \frac{\cos(\omega + \theta)}{\cos \delta} \qquad (3.31b)$$

They follow directly from considering the right angle spherical triangle bounded by the orbit, the equator, and the meridian through the spacecraft position. The same equations also apply when the symbols are interpreted to apply to the ecliptic as the reference plane.

Frequently used coordinate transformations between elliptic orbital elements and orthonormal coordinates are listed in Tables 3.1 and 3.2.

Table 3.1 Transformations from elliptic orbital elements to orthonormal coordinates and their derivatives

(For the definition of the coordinates, see Fig. 3.4)
Defining
$$l_1 = \cos \omega \cos \Omega - \sin \omega \sin \Omega \cos i$$
$$l_2 = -\sin \omega \cos \Omega - \cos \omega \sin \Omega \cos i$$
$$l_3 = \sin \Omega \sin i$$
$$m_1 = \cos \omega \sin \Omega + \sin \omega \cos \Omega \cos i$$
$$m_2 = -\sin \omega \sin \Omega + \cos \omega \cos \Omega \cos i$$
$$m_3 = -\cos \Omega \sin i$$
$$n_1 = \sin \omega \sin i$$
$$n_2 = \cos \omega \sin i$$
$$n_3 = \cos i$$
$$x = l_1 \xi + l_2 \eta, \qquad y = m_1 \xi + m_2 \eta, \qquad z = n_1 \xi + n_2 \eta$$
$$\dot{x} = (\mu/h)[-l_1 \sin \theta + l_2(e + \cos \theta)]$$
$$\dot{y} = (\mu/h)[-m_1 \sin \theta + m_2(e + \cos \theta)]$$
$$\dot{z} = (\mu/h)[-n_1 \sin \theta + n_2(e + \cos \theta)]$$

Table 3.2 Transformations from orthonormal coordinates and their derivatives to elliptic orbital elements

(For the definition of the coordinates, see Fig. 3.4)
Defining

$r = \sqrt{x^2 + y^2 + z^2}$

$a = \mu r/(2\mu - r(\dot{x}^2 + \dot{y}^2 + \dot{z}^2))$

$e^2 = (1 - r/a)^2 + (\mu a)^{-1}(x\dot{x} + y\dot{y} + z\dot{z})^2$

$h = \sqrt{\mu a(1-e^2)}$

$\cos i = (x\dot{y} - y\dot{x})/h \qquad (0 \le i < \pi)$

$\sin \Omega = (y\dot{z} - z\dot{y})/(h \sin i), \qquad \cos \Omega = (x\dot{z} - z\dot{x})/(h \sin i)$

$e \sin E = (\mu a)^{-1/2}(x\dot{x} + y\dot{y} + z\dot{z}), \qquad e \cos E = 1 - r/a$

$\theta = 2 \tan^{-1}\left(\sqrt{\frac{1+e}{1-e}} \tan \frac{E}{2}\right)$

$\sin(\omega + \theta) = \dfrac{z}{r \sin i}, \qquad \cos(\omega + \theta) = (y/r)\sin\Omega + (x/r)\cos\Omega$

3.5 Spacecraft Visibility above the Horizon

As an example of the use of orbital elements, we calculate the time at which a spacecraft becomes visible above the horizon of a ground-based observer or ground station. Problems of this type are commonly encountered, for instance, in designing a satellite navigational system in which several satellites (typically four) must be simultaneously visible to the observer. The same problem also arises in satellite communication systems, because at the microwave frequencies used in these systems, the communication paths are essentially straight line. In the following we calculate the limit of visibility as determined by the theoretical horizon, that is, by the plane tangent to the spherical figure of the earth at the location of the observer. In practice, atmospheric absorption and refraction at the carrier frequency, or simply obstacles such as mountain ranges or buildings, limit the useful paths for communications to some positive angle of elevation above the horizon. Unless special circumstances dictate otherwise, 10° to 15° above the horizon are considered to be a minimum for most microwave links.

Figure 3.5 indicates the geometrical relations. A spherical earth is assumed. The ground station, designated by O, is located at geographic latitude ϕ_0 and longitude λ_0 east of Greenwich. The orbital plane intersects the horizon plane of the ground station along the line H–H′. The spacecraft becomes visible at H and disappears at H′. Orthogonal coordinates are defined, with the origin at the earth's center, the x axis coincident with the vernal equinox line, and the z axis along the northward polar axis.

To find the direction of the line H–H′, we introduce the unit vectors $\mathbf{n}_0$ perpendicular to the ground station's horizon plane, pointing outward from the earth, and $\mathbf{n}_S$ perpendicular to the plane of the spacecraft orbit, pointing in the direction of the spacecraft's angular momentum vector. The vector

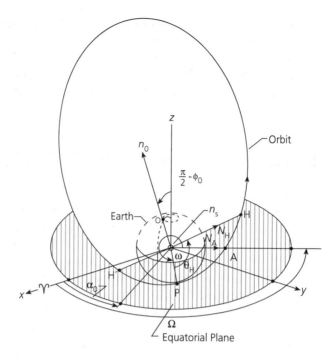

Figure 3.5 Visibility of a spacecraft for a ground observer at O. Spacecraft is visible above line H–H'.

$\mathbf{N}_H = \mathbf{n}_0 \times \mathbf{n}_S$ then determines the line of intersection of the two planes and points toward point H.

If α_0 designates the right ascension of the ground station's meridian and α_G the right ascension of Greenwich, then $\alpha_0 = \alpha_G + \lambda_0$. The components of $\mathbf{n}_0$ are

$$n_{0x} = \cos\phi_0 \cos\alpha_0, \qquad n_{0y} = \cos\phi_0 \sin\alpha_0, \qquad n_{0z} = \sin\phi_0$$

Making use of the orbital elements introduced in the preceding section, we obtain for the components of $\mathbf{n}_S$

$$n_{Sx} = \sin i \sin\Omega, \qquad n_{Sy} = -\sin i \cos\Omega, \qquad n_{Sz} = \cos i$$

Hence

$$\begin{aligned}
N_{Hx} &= \cos i \cos\phi_0 \sin\alpha_0 + \sin i \cos\Omega \sin\phi_0 \\
N_{Hy} &= \sin i \sin\Omega \sin\phi_0 - \cos i \cos\phi_0 \cos\alpha_0 \\
N_{Hz} &= -\sin i \cos\phi_0 \cos(\Omega - \alpha_0)
\end{aligned} \qquad (3.32)$$

The unit vector $\mathbf{N}_A$ pointing toward the ascending node A has the components

$$N_{Ax} = \cos\Omega, \qquad N_{Ay} = \sin\Omega, \qquad N_{Az} = 0 \qquad (3.33)$$

The angle $\omega + \theta_H$ in the plane of the orbit, where θ_H is the true anomaly of point H, is then obtained from the scalar product $(\mathbf{N}_H/N_H) \cdot \mathbf{N}_A$. Hence

$$(N_{Hx}/N_H)\cos\Omega + (N_{Hy}/N_H)\sin\Omega = \cos(\omega + \theta_H) \qquad (3.34)$$

If the orbital elements and the location of the observer are known, this then determines θ_H. From (3.10) and (3.20) follow the radius r_H and the eccentric anomaly E_H at point H, and from (3.21) the time t_H at which the spacecraft rises above the horizon. As is seen from this derivation, all six orbital elements need to be known when calculating t_H. The same method also provides the corresponding results for the descent of the spacecraft below the horizon.

3.6 Satellite Observations and the *f* and *g* Series

A frequently occurring task is to observe from a ground station the position and velocity of a spacecraft and from these to derive the orbital elements, for instance, after injecting a spacecraft into a low-altitude parking orbit. Injection errors, usually very small but inevitable, can then be corrected by programming the thrust direction and (for liquid propellant engines) the duration of the burn of the next rocket stage. Alternatively, the correction can be made by the spacecraft's own thrusters.

Precision observations of spacecraft orbits use a combination of optical and radar techniques. More accurate than any other **optical** technique is the technique of photographing the spacecraft against a star background. The photographs are repeated at precise instants of time, obtained from standard time signals. For high precision, a number of corrections may have to be made, which may include changes in the right ascension and declination of the stars caused by the earth's precession and nutation. These may have accumulated since the time of compilation of the star catalog that is being used. Other corrections may include the effect of differential atmospheric refraction between a star position and the spacecraft image, an effect that can become significant when the spacecraft is near the horizon.

The use of theodolites provides an alternative, although slightly less accurate, method. All optical techniques provide highly precise measurements of the true azimuth and elevation of the spacecraft at fixed times and of the velocity component perpendicular to the line of sight.

Because of the limited angular resolution of the beam, **radar** measurements cannot give information at right angles to the line of sight with accuracies comparable to optical methods. However, radar complements optical determinations by providing highly accurate range and range rate measurements.

The range, that is, the distance along the line of sight, is obtained from the time delay between pulses transmitted by the ground station and the return pulses received after reflection from the spacecraft ("skin tracking"). The radar pulses reaching the spacecraft can also be amplified by a repeater on the spacecraft and returned to the ground station.

Refraction in the troposphere and ionosphere causes the propagation velocity to differ slightly from the speed of light in vacuum. This effect depends on the angle of elevation and on tropospheric and ionospheric conditions but can be corrected by suitable calibration or by the use of two different, widely spaced, microwave frequencies.

The range rate, which gives the spacecraft's velocity component relative to the ground station along the line of sight, is obtained by a determination

CHAPTER 3 Orbits and Trajectories in an Inverse Square Field

of the Doppler shift between the transmitted and returned signals. The measurement consists of a count of the cycles over a fixed time interval and can be made extremely accurately. By using more than one ground station so as to obtain at the same instant several range and range rate measurements, radar can replace entirely, although with somewhat less accuracy, optical measurements.

The so-called *f* and *g* series that are discussed in the following are important in several methods of orbit determination. They are also useful in extrapolating over short time intervals a known spacecraft position and velocity, a task that occurs in planning a spacecraft rendezvous. The results apply to elliptic orbits as well as to hyperbolic trajectories.

In deriving the method, it is assumed that the position vector $\mathbf{r}_0$ and velocity vector $\mathbf{v}_0$ are known at some initial time t_0. The method then allows one to find $\mathbf{r}$ and $\mathbf{v}$ after a time interval Δt later. Expanding $\mathbf{r}$ in a Taylor series in Δt,

$$\mathbf{r} = \mathbf{r}_0 + \left(\frac{d\mathbf{r}}{dt}\right)_0 \Delta t + \frac{1}{2!}\left(\frac{d^2\mathbf{r}}{dt^2}\right)_0 (\Delta t)^2 + \frac{1}{3!}\left(\frac{d^3\mathbf{r}}{dt^3}\right)_0 (\Delta t)^3$$
$$+ \frac{1}{4!}\left(\frac{d^4\mathbf{r}}{dt^4}\right)_0 (\Delta t)^4 + \cdots \tag{3.35}$$

we calculate the coefficients up to order four. Expansions to higher order can be found in more specialized texts, for example, in Herrick [10].

Because a Keplerian orbit or trajectory is assumed, $d^2\mathbf{r}/dt^2$ can be replaced by the right-hand side of (3.1). Higher derivatives then follow by differentiation; thus

$$\frac{d^3\mathbf{r}}{dt^3} = \frac{3\mu\dot{r}}{r^4}\mathbf{r} - \frac{\mu}{r^3}\mathbf{v}$$

$$\frac{d^4\mathbf{r}}{dt^4} = \frac{\mu}{r^4}\left(\frac{\mu}{r^2} - \frac{12(\dot{r})^2}{r} + 3\ddot{r}\right)\mathbf{r} + \frac{6\mu\dot{r}}{r^4}\mathbf{v}$$

When substituted into the series, the as yet unknown term $(\ddot{r})_0$ will appear. By differentiating $\dot{r} = (\mathbf{r}/r) \cdot \mathbf{v}$ one finds

$$\ddot{r} = -r^{-1}(\mu/r + (\dot{r})^2 - v^2)$$

It will be found convenient to separate in the series terms that are proportional to $\mathbf{r}_0$ from those proportional to $\mathbf{v}_0$ and to write

$$\mathbf{r} = f\mathbf{r}_0 + g\mathbf{v}_0 \tag{3.36}$$

Upon substitution of these results into (3.35), one finds the final result

$$f = 1 - \frac{\mu}{2r_0^3}(\Delta t)^2 + \frac{\mu\dot{r}_0}{2r_0^4}(\Delta t)^3 - \frac{\mu}{24r_0^5}\left(\frac{2\mu}{r_0} + 15(\dot{r}_0)^2 - 3v_0^2\right)(\Delta t)^4 + \cdots$$

$$g = \Delta t - \frac{\mu}{6r_0^3}(\Delta t)^3 + \frac{\mu\dot{r}_0}{4r_0^4}(\Delta t)^4 - \cdots \tag{3.37}$$

Similar expressions can also be obtained for the velocity, because on differentiating (3.36) one finds

$$\mathbf{v} = (df/dt)\,\mathbf{r}_0 + (dg/dt)\,\mathbf{v}_0 \tag{3.38}$$

where

$$df/dt = -\frac{\mu}{r_0^3}\Delta t + \frac{3\mu \dot{r}_0}{2r_0^4}(\Delta t)^2 - \frac{\mu}{6r_0^5}\left(\frac{2\mu}{r_0} + 15(\dot{r}_0)^2 - 3v_0^2\right)(\Delta t)^3 + \cdots$$

$$dg/dt = 1 - \frac{\mu}{2r_0^3}(\Delta t)^2 + \frac{\mu \dot{r}_0}{r_0^4}(\Delta t)^3 - \cdots \tag{3.39}$$

The following numerical example illustrates the accuracy of these approximations. Assuming an elliptic orbit about the earth, of eccentricity $e = 0.5$ and altitude at perigee 300 km above the mean equatorial earth radius, the period $P = 15{,}337$ s. We furthermore assume that the initial position of the spacecraft was at the true anomaly $\theta_0 = 90°$ (at this point, as is easily shown, the angle between the tangent to the orbit and the radial direction is just $\cotan^{-1} e$).

If we assume $\Delta t = P/100$, the ratio of the last term listed in (3.37) to the first term is $-7.30\,10^{-6}$ for the f series and $1.13\,10^{-4}$ for the g series. If $\Delta t = P/10$, the corresponding values are -0.073 and 0.113, respectively.

It also follows from (3.36) that for $\Delta t = P/100$ the last term listed in the f series contributes to the final position a distance in the initial radial direction of -73 m, but -730 km if $\Delta t = P/10$. The last term listed for the g series contributes corresponding values of 122 m if $\Delta t = P/100$ but 1225 km if $\Delta t = P/10$.

For ease of reference, the more important equations derived up to this point in this chapter are summarized in Table 3.3.

3.7 Special Orbits

In this section some special Kepler orbits that have found frequent applications are considered.

3.7.1 Circular Orbits

From (3.14), with $a = r$, follows the velocity, v_{crc}, of a spacecraft on a circular orbit of radius r

$$v_{\text{crc}} = \sqrt{\frac{\mu}{r}} \tag{3.40}$$

For instance, for a **low earth orbit**, assuming an altitude of 300 km above the earth's mean equatorial radius (6378.1 km), $v_{\text{crc}} = 7.726$ km/s. The siderial period, from (3.16), is 5431 s or 90.5 min. To find, however, the time interval between two successive crossings of the same meridian, the rotation of the earth must be considered (see Fig. 1.9). Assuming that the orbit is in the

Table 3.3 Kepler trajectories, summary of formulas

(For the definition of the symbols, see Chap. 3 Nomenclature)

Ellipse	Hyperbola
$r = \dfrac{h^2}{\mu(1 + e\cos\theta)}$	Same
$e = \sqrt{1 + 2wh^2/\mu^2}$	Same
$w = -\mu/(2a)$	$w = +\mu/(2a)$
$h^2 = +\mu a(1 - e^2)$	$h^2 = -\mu a(1 - e^2)$
$v = \sqrt{\mu(2/r - 1/a)}$	$v = \sqrt{\mu(2/r + 1/a)}$
$P = 2\pi\sqrt{a^3/\mu}$	
$E - e\sin E = \sqrt{\mu/a^3}\,(t - t_p)$	$E - e\sinh E = -\sqrt{\mu/a^3}\,(t - t_p)$
$x = a(\cos E - e)$	$x = a(e - \cosh E)$
$y = a\sqrt{1 - e^2}\sin E$	$y = a\sqrt{e^2 - 1}\sinh E$
$r = a(1 - e\cos E)$	$r = -a(1 - e\cosh E)$
$\dot{x} = -\sqrt{\mu/a}\sin E/(1 - e\cos E)$	$\dot{x} = -\sqrt{\mu/a}\sinh E/(e\cosh E - 1)$
$\dot{y} = \sqrt{\mu/a}\sqrt{1 - e^2}\cos E/(1 - e\cos E)$	$\dot{y} = \sqrt{\mu/a}\sqrt{e^2 - 1}\cosh E/(e\cosh E - 1)$
$f = 1 - \dfrac{\mu}{2r_0^3}(\Delta t)^2 + \dfrac{\mu \dot{r}_0}{2r_0^4}(\Delta t)^3 - \dfrac{\mu}{24r_0^5}$	
$\quad\times \left(\dfrac{2\mu}{r_0} + 15(\dot{r}_0)^2 - 3v_0^2\right)(\Delta t)^4 + \cdots$	Same
$g = \Delta t - \dfrac{\mu}{6r_0^3}(\Delta t)^3 + \dfrac{\mu \dot{r}_0}{4r_0^4}(\Delta t)^4 - \cdots$	Same

equatorial plane and from west to east, this time interval in this example becomes $90.5 + 6.1 = 96.6$ min.

The velocity is seen to decrease with increasing radius. Thus the velocity of the moon on its orbit (which is at least very roughly circular) about the earth is on average only about 1.0 km/s.

It also follows from (3.40) that not only the orbiting velocity but also the corresponding angular velocity decreases with increasing radius. Hence, spacecraft orbiting at a larger distance from the center of attraction will appear to a ground observer to fall behind a spacecraft that orbits at less distance.

Circular earth orbits at altitudes between about 200 km and 1000 km are designated as **low earth orbits**. Below 200 km, the orbits of spacecraft without thrust decay rapidly by atmospheric drag, typically in less than 1 week. Above 1000 km, radiation damage of unshielded, unhardened electronic components can become significant.

There is the curious fact that the periods of spacecraft that orbit close to either the surface of Earth, the moon, Mercury, Venus, or Mars are all roughly the same, namely between 1.4 and 1.8 h. This is explained by the fact that the average densities, ϱ_{av}, of these celestial bodies do not differ greatly from one another. It follows from (2.4) for a spherical body of radius

R that $\mu = (4/3)\pi R^3 G\varrho_{av}$, hence from (3.16), with $a = R$,

$$P = \sqrt{\frac{3\pi}{G\varrho_{av}}}$$

independent of the radius.

3.7.2 Geostationary Spacecraft

A special case of circular orbits occurs for **geostationary satellites**. In this case the orbit elements are such that the satellite, for an earth-based observer, remains at a fixed point in the sky. To satisfy this requirement, the orbit must be in the equatorial plane, with the satellite moving from west to east with a siderial period equal to one siderial day (d_{si} = 23 h, 56 min, 24.09 s = 86,164.09 s). The radius of the orbit, r_{gs}, therefore follows from

$$\sqrt{\frac{\mu_g}{r_{gs}^3}} = \frac{2\pi}{d_{si}}$$

with the result that r_{gs} = 42,164.2 km, which is about 6.62 times the mean radius of the earth.

The discussion here has been limited to the case of a spherically symmetric earth. When taking into account the earth's actual figure, the orbit deviates slightly from the ideal circular orbit in the equatorial plane. The most important effect is an apparent slow change of the geographic latitude of the satellite. This effect, together with other perturbations from ideal Kepler orbits, will be considered in Sect. 5.9.1.

At a geostationary satellite's distance from the earth, a substantial portion of the total earth's surface is visible. Three such satellites, spaced 120° apart, cover the entire earth other than regions at high latitudes.

For **microwave communications**, either uplink or downlink, an important consideration concerns the atmospheric and ionospheric effects on signal propagation if the path is too close to the horizon. Particularly in the high range of microwave frequency, such as at K-band (10.9–36.0 GHz), communications at small link elevation angles can be affected by heavy rain, hail, or snow. The principal ionospheric effect at these angles is Faraday rotation. It can be detrimental to linearly polarized microwave links.

In practice, it is found that elevation angles above 10° to 15° ensure trouble-free operations. Figure 3.6 illustrates the earth coverage obtained by three geostationary spacecraft, located 120° apart, assuming a minimum elevation angle of 15°. Each spacecraft covers a spherical earth surface of 154 10^6 km^2. The gap at the intersection of adjacent regions is at a latitude of 37.4°.

Earth coverage by geosynchronous satellites, however, involves additional constraints beyond the requirement of minimum elevation angles. An important consideration is the traffic density, which may differ greatly for different pairs of countries or continents. Important improvements in the utilization of the available microwave spectrum can then be obtained

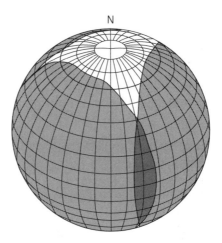

Figure 3.6 Earth surface covered by three equally spaced geostationary spacecraft, for a microwave link minimum elevation angle of 15°.

by shaping the antenna beams, as illustrated in Fig. 3.7 for the Intelsat VI satellite located over the mid-Atlantic. Shaping the antenna beam by the antenna feed makes it possible to concentrate the beam power on the regions where it is most needed and in such a way that the same frequency band can be used for communicating with nonoverlapping regions. Polarization discrimination provides the possibility of an additional doubling of the use of the same band.

The advantage gained by placing communications and other satellites into the geostationary orbit has led to a considerable crowding of this orbit. Over the Atlantic Ocean, where geostationary spacecraft handle much of the heavy communications traffic between North America and Europe, the density is particularly high. At some places on the orbit, the spacecraft are separated by less than 1°, forcing the uplinks to operate in different frequency bands and polarizations. The scarcity of the available bands is a major reason for the development of microwave components at ever higher frequency, limited as this approach is, however, by the problem of possible communications dropouts in severe weather.

At this time, geostationary spacecraft provide one of the most important modes of space communications. They are used in such applications as broadcasting voice and video, telephone service, computer links, and relay satellites for worldwide signal transmissions. An important advantage is that there is no need for slewing of the ground antennas. The apparent small satellite motions resulting from the lack of exact sphericity of the earth can be corrected by pulsing thrusters on the spacecraft so as to keep it within the main lobe of the transmitting and receiving ground antennas.

There is unavoidably some delay in the transmissions. This has some effect particularly on telephone connections. The altitude of the satellites above the earth at sea level is about 35,790 km, which corresponds to a

3.7 *Special Orbits*   79

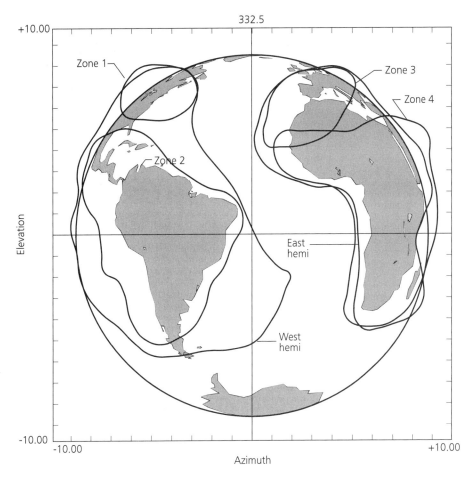

Figure 3.7 Atlantic ocean coverage by Intelsat VI spacecraft, with C-band, sixfold frequency reuse. (Courtesy of INTELSAT.)

time delay of about 120 ms. Taking into account the lengths of the up- and downlinks including the slant range, together with some (minor) electronic delays, typical one-way communications delays are of the order of a fourth of a second.

Other than microwave communications, there are several other applications of geostationary satellites. They include **meteorological satellites, earth and ocean observation, and surveillance satellites**. Their payloads may operate in many different spectral bands. Again, the large area that can be viewed is often an advantage. On the other hand, the resolution of surface features is necessarily more limited than is the case for low-altitude spacecraft. Also, the requirement for the propulsion to place the satellite into its orbit is more demanding.

An important consideration for geostationary spacecraft designed for long-term missions is that the geostationary orbit is beyond the outer Van Allen belt, hence in a region where the radiation environment is relatively benign.

3.7.3 Sun-Synchronous Orbits

Spacecraft on sun-synchronous orbits have a ground track that crosses a given meridian on the earth once a day ("24-hour satellites"). Also included among sun-synchronous orbits are those for which the spacecraft crosses a meridian every rational fraction or multiple (e.g., 12, 8, ... or 48, ... h) of a day. Because of the inclination of the ecliptic, different crossings of the meridian through a fixed point on the ground will, in general, have different declinations of the sun.

What makes sun-synchronous orbits interesting is that a given point on the earth can be viewed from the spacecraft repeatedly, although not necessarily on each orbit, at the same solar time. This feature is valuable for many tasks of earth observations for which solar illumination is essential. Because the repeated passes occur at similar conditions of solar illumination, such orbits greatly facilitate the correlations of observations of crop growth, drought conditions, changes in the polar ice caps, and other seasonal changes.

Sun-synchronous satellites also have great usefulness for communications receivers or transmitters that are located at large latitudes that cannot be reliably reached by links from geostationary satellites. Peak demands for communications and broadcasting tend to occur at the same hours of the day. The orbits are therefore chosen so that at times of peak demand the satellite is well above the horizon for ground transmitters and receivers.

An example is provided by the Russian domestic Molniya system. It is based on orbits with 12-h periods, large eccentricity, and large inclination (Molniya 1993 079A orbit period, 703.1 min; perigee altitude, 436 km; apogee altitude, 39,188 km; inclination to equatorial plane, 62.7°). The apogee is located far north and occurs alternatively over Russia and over North America. Figure 3.8 shows the ground track. Because of the large eccentricity, the time spent by the spacecraft at its apogee, hence at large northerly latitude, is the largest part of the orbital period. A minimum of two satellites, with ground tracks 12 h apart, are needed to provide continuous service.

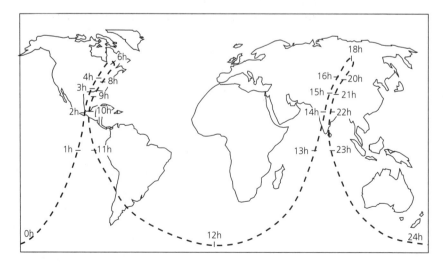

Figure 3.8 Molniya satellites (Russia) ground track.

In principle, a circular satellite orbit in the ecliptic plane could serve as a sun-synchronous orbit. Its period relative to the earth would be 24 h = 86,400 s (or a rational multiple thereof); its sidereal period, as obtained from (3.16), in a nonrotating reference frame would be one sidereal day, about 86,164 s. The elevation angle would change in accord with the annual seasons. Because points on the earth with higher latitudes would be poorly covered, such orbits are of only limited interest in comparison with high-inclination orbits, such as the Molniya orbits.

The orbital planes of Molniya-type sun-synchronous spacecraft must rotate relative to inertial space by one turn per year. This becomes possible by taking advantage of the lack of exact spherical symmetry of the earth's gravitational field. The large inclination of the orbit plane causes the spacecraft motion to be influenced by the zonal coefficients, most strongly by J_2. This is reinforced by the low altitude of the periapsis, where the effects produced by the higher harmonics, which fall off rapidly with distance, are still strong. Even though the perturbations are only modest for a single satellite orbit, they accumulate over many orbits. Of course, these orbits are no longer exact Kepler orbits, as had been assumed otherwise in this chapter.

3.8 Perturbations by Other Astronomical Bodies

In this section, motions of spacecraft are considered in which the inverse square force field of an astronomical body is perturbed by another or several other such bodies. Nevertheless, the spacecraft motions will be described, albeit in a very approximate way, by Kepler trajectories.

As indicated in Fig. 3.9, a spacecraft at point P_s is in the vicinity of an astronomical body P_0. Relative to P_0, the spacecraft motion is approximately a conic section; however, it is not exactly so, because not only is the spacecraft motion perturbed by other astronomical bodies P_j, $j = 1, 2, \ldots$, but also the motion of P_0 itself is perturbed by the bodies P_j. In some of the applications, P_0 may be a planet, with P_j the sun and possibly planets. In other applications P_0 may be the moon of a planet and P_j this planet.

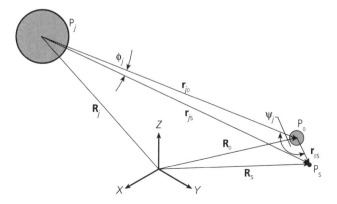

Figure 3.9 The derivation of Eq. (3.45).

Following (3.1) and referring to Fig. 3.9, the motion of the spacecraft in an inertial reference frame (X, Y, Z) is described by the equation

$$\frac{d^2 \mathbf{R}_s}{dt^2} = -\mu_0 \frac{\mathbf{r}_{0s}}{r_{0s}^3} - \sum_{j=1,2,\ldots} \mu_j \frac{\mathbf{r}_{js}}{r_{js}^3} \tag{3.41}$$

where μ_0 and μ_j are the gravitational parameters of the astronomical bodies. Similarly, the mass center of P_0 satisfies the equation of motion

$$\frac{d^2 \mathbf{R}_0}{dt^2} = -\sum_{j=1,2,\ldots} \mu_j \frac{\mathbf{r}_{j0}}{r_{j0}^3} \tag{3.42}$$

where $\mathbf{r}_{0s} = \mathbf{R}_s - \mathbf{R}_0$ and $\mathbf{r}_{js} = \mathbf{r}_{j0} + \mathbf{r}_{0s}$.

The motion of the spacecraft in a nonrotating reference frame, with its origin attached to the center of mass of P_0, is obtained by subtracting (3.42) from (3.41), resulting in

$$\frac{d^2 \mathbf{r}_{0s}}{dt^2} = -\frac{\mu_0}{r_{0s}^2}\left[\frac{\mathbf{r}_{0s}}{r_{0s}} + r_{0s}^2 \sum_{j=1,2,\ldots} \frac{\mu_j}{\mu_0}\left(\frac{\mathbf{r}_{js}}{r_{js}^3} - \frac{\mathbf{r}_{j0}}{r_{j0}^3}\right)\right] \tag{3.43}$$

The second term in the square bracket represents the perturbation caused by the bodies P_j. Even though the mass fraction μ_j/μ_0 may be very large (as is the case if one of the P_j is the sun, P_0 a planet), this second term is seen to be small compared with unity in two distinct cases, both being of practical interest in astrodynamics:

Case 1: If P_0 is a planet, the P_j stand for the sun (and possibly also planets), and $\mathbf{r}_{0s}$ is very small compared with the dimensions of the solar system (hence r_{0s}^2 is small compared with r_{js}^2 and r_{j0}^2), then $\mathbf{r}_{js}$ and $\mathbf{r}_{j0}$ are nearly the same and the two terms in the round bracket nearly cancel each other. The same case also arises when a spacecraft orbits the moon of a planet at a distance small compared with the planet–moon distance.

Case 2: If P_0 is the sun and P_j, $j = 1, 2, \ldots$, are planets, then it is the mass ratio μ_j/μ_0 that is small. Again, the second term in the square bracket represents a perturbation small compared with unity.

In **case 1**, denoting the magnitude of the spacecraft acceleration caused by P_0 by a_{0s}, and correspondingly the acceleration caused by P_j by a_{js},

$$a_{0s} = \frac{\mu_0}{r_{0s}^2} \tag{3.44a}$$

$$a_{js} = \mu_j \sqrt{\frac{1}{r_{js}^4} + \frac{1}{r_{j0}^4} - \frac{2\cos\phi_j}{r_{js}^2 r_{j0}^2}}, \qquad j = 1, 2, \ldots \tag{3.44b}$$

where from the triangle P_s–P_0–P_j and the angles ϕ_j and ψ_j indicated in Fig. 3.9,

$$r_{js}^2 = r_{j0}^2 + r_{0s}^2 - 2r_{j0}r_{0s}\cos\psi_j$$

$$\cos\phi_j = \frac{r_{j0} - r_{0s}\cos\psi_j}{r_{js}}$$

3.8 Perturbations by Other Astronomical Bodies

Defining the parameter of smallness $\varepsilon = r_{0s}/r_{j0}$, one obtains

$$\left(\frac{r_{js}}{r_{j0}}\right)^2 = 1 - 2\varepsilon \cos\psi_j + \varepsilon^2$$

$$\cos\phi_j = (1 - 2\varepsilon \cos\psi_j + \varepsilon^2)^{-1/2}(1 - \varepsilon \cos\psi_j)$$

Substitution into (3.44b) results in the equation

$$a_{js} = \frac{\mu_j}{r_{j0}^2}\sqrt{1 + \frac{1}{(1 - 2\varepsilon \cos\psi_j + \varepsilon^2)^2} - \frac{2(1 - \varepsilon \cos\psi_j)}{(1 - 2\varepsilon \cos\psi_j + \varepsilon^2)^{3/2}}}$$

Expanding by the binomial theorem the second and third terms in the square bracket and dropping all terms of order ε^3 and higher, we have

$$a_{js} = \frac{\mu_j r_{0s}}{r_{j0}^3}\sqrt{1 + 3\cos^2\psi_j}$$

Hence the relative disturbance acceleration a_{js}/a_{0s} of the spacecraft is given by

$$\frac{a_{js}}{a_{0s}} = \frac{\mu_j}{\mu_0}\left(\frac{r_{0s}}{r_{j0}}\right)^3\sqrt{1 + 3\cos^2\psi_j}, \qquad j = 1, 2, \ldots \qquad (3.45)$$

Depending on the position of the spacecraft, the minimum ratio of the disturbance acceleration to the zero-order acceleration is seen to occur for $\psi_j = \pm 90°$ and is

$$\frac{\mu_j}{\mu_0}\left(\frac{r_{0s}}{r_{j0}}\right)^3$$

The maximum value is twice this minimum value and occurs when $\psi_j = 0°$ or 180°. The relative importance of the effective disturbance force is seen to increase with the third power of the satellite distance.

Table 3.4 indicates the effects of disturbing astronomical bodies on a geostationary satellite. The ratios a_{js}/a_{0s} of the perturbing to the zero-order (in this case the earth's) gravitational accelerations are listed. These ratios are for the case in which the distance of the disturbing body from the earth is a minimum and $\psi_j = 0°$ or 180°. One concludes that the largest relative disturbances on a near-earth spacecraft are caused by the moon and the sun.

A different application of the disturbance calculation occurs when a spacecraft orbits the sun [denoted by the subscript ( )$_h$] which is now the primary body P_0, but the spacecraft orbit is disturbed by a planet. Here **case 2** applies.

A calculation analogous to that for case 1 gives the result for the relative disturbance acceleration in case 2 as

$$\frac{a_{\text{pl},s}}{a_{h,s}} = \frac{\mu_{\text{pl}} r_{h,s}^2}{\mu_h r_{\text{pl},s}^2} \qquad (3.46)$$

When, on the scale of the solar system, the spacecraft is very close to the planet, case 1 applies. Farther from the planet, case 2 applies. A useful

Table 3.4 Disturbance acceleration of geosynchronous satellites, caused by astronomical bodies, relative to earth gravity

Disturbing body	Mass ratio μ_j/μ_0	Relative disturbance acceleration a_{js}/a_{0s}
Moon	0.0123	$1.7\ 10^{-5}$
Sun	332,946	$0.8\ 10^{-5}$
Mercury	0.056	$0.9\ 10^{-11}$
Venus	0.815	$1.1\ 10^{-9}$
Mars	0.107	$0.5\ 10^{-10}$
Jupiter	317.9	$1.2\ 10^{-10}$
Saturn	95.2	$4.2\ 10^{-12}$
Uranus	14.6	$0.6\ 10^{-13}$
Neptune	17.2	$1.7\ 10^{-14}$
Pluto	0.11	$1.0\ 10^{-16}$
α Centauri A	$3.6\ 10^5$	$4.0\ 10^{-22}$

Source: Cornelisse, Schöyer, and Wakker. [11]

approximation, often sufficient for the preliminary planning of space missions, is then obtained by assuming that the two volumes corresponding to cases 1 and 2 are separated by a spherical boundary of radius $\bar{r}_{pl}$ centered on the planet. The resulting sphere is referred to as the **sphere of influence** of the planet.

A numerical estimate for this radius can be obtained by equating at the boundary, where $r_{pl,s} = \bar{r}_{pl}$, the relative magnitudes of the disturbances to each other, therefore

$$\frac{\mu_h}{\mu_{pl}}\left(\frac{\bar{r}_{pl}}{r_{h,pl}}\right)^3 = \frac{\mu_{pl}}{\mu_h}\left(\frac{r_{h,s}}{\bar{r}_{pl}}\right)^2$$

from which follows

$$\bar{r}_{pl} = r_{h,pl}\left(\frac{\mu_{pl}}{\mu_h}\right)^{2/5} \tag{3.47}$$

An approximation that can often be made is to assume that when the spacecraft is within the sphere of influence of the planet, the spacecraft is subject only to the attraction by the planet and outside it is subject only to the attraction by the sun. Use of this approximation is therefore referred to as the **patched conics** method.

Although case 2 has been discussed in terms of the combined action of the sun and a planet, it is often also applicable to the combination of a planet and one of its moons. In the case, for instance, of the earth and the moon, it is important, however, to note, as indicated in Table 3.4, that the disturbances of an earth-orbiting spacecraft by the sun and by the moon are of similar magnitude. In this case, it is no longer sufficient to consider only

Table 3.5 Radii of the spheres of influence of the planets

Planet	$\bar{r}_{pl}$ in 10^6 km	$\bar{r}_{pl}$ as a multiple of the planet's mean radius
Mercury	0.09–0.14	37–56
Venus	0.61–0.62	101–103
Earth	0.91–0.94	143–147
Mars	0.52–0.63	154–185
Jupiter	45.9–50.5	648–713
Saturn	51.6–57.7	859–961
Uranus	49.4–54.1	1940–2130
Neptune	85.7–87.6	3410–3490
Pluto	11.4–18.8	3560–5870

the magnitudes of the accelerations separately rather than to add them as vector quantities.

Because, with the exceptions of Mercury and Pluto, the orbit eccentricities are small, hence $r_{h,pl}$ approximately constant, the radii of the spheres of influence of most planets are also approximately constant, independent of their true anomaly. Table 3.5 lists these radii, corresponding to the minimum and the maximum heliocentric distance of the planet. It also lists them in terms of the planet's mean radius.

The concept of a sphere of influence goes back to Laplace (Laplace 1845), who used it in his studies of comets on trajectories passing close to Jupiter.

3.9 Planetary Flyby and Gravity Assist

One speaks of **flyby** or **swing-by** when a spacecraft on a locally hyperbolic trajectory passes close to an astronomical body. Among the purposes are scientific observations at close distance and the possibility of obtaining increased spacecraft speed at no expenditure of propellant. Still another purpose may be a desired change in the orientation of the flight path following the flyby.

The expression **gravity assist** is used to characterize flybys in which the spacecraft energy (in an inertial reference frame) is increased. Flybys passing one or several different planets, or even the same planet more than once, are often scheduled as a part of deep-space missions. In this way, energy increases can be achieved that would be prohibitive if the acceleration had to be obtained exclusively from rocket thrust.

Timing becomes critical: the planet, or in other cases several planets, must be in suitable positions in their orbits at the instant of the flyby. To be effective, the spacecraft must pass close to the planet. High accuracy in tracking the spacecraft's position on its course toward the planet is therefore

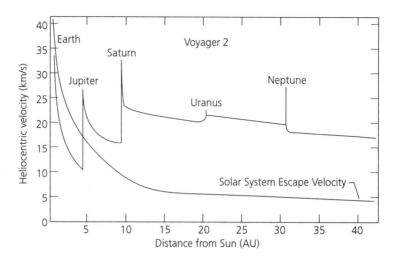

Figure 3.10 Planetary flybys by the Voyager 2 (United States) spacecraft. (Courtesy by NASA.)

essential. Typically, one or several midcourse corrections by short periods of thrust are made long before the vicinity of the planet is reached.

An example was provided by the U.S. NASA Galileo probe to Jupiter. Two earth gravity assist flybys were used. The second flyby passed the earth at 303.1 km, within 1 km of the intended altitude. It resulted in a saving of 5 kg of propellant for use at Jupiter.

As a further example, Fig. 3.10 illustrates the planetary flybys that were executed on the U.S. Voyager 2 mission. The graph shows the velocity in the heliocentric reference frame as it changed with each flyby. In each case, except in the encounter with Neptune, a velocity increase occurred.

Also shown in the figure is the solar system escape velocity as a function of the distance from the sun. As is seen by comparing the two curves, after the flyby of Jupiter the spacecraft already had enough speed to escape the solar system. (The other flybys shown were conducted for the exploration of these planets.)

For the purpose of preliminary mission planning, it is sufficient to consider the motion of planet and spacecraft within the planet's sphere of influence. The planet's local orbit can then be approximated by a rectilinear path. (This can be verified by a comparison of the planet's acceleration on its orbit with the gravitational acceleration near its surface. Thus, for the earth, the centripetal acceleration that results from the near-circular annual orbit about the sun is only $3.44 \; 10^{-3}$ times normal gravity.)

The path of the spacecraft in a *planetocentric* reference frame is a hyperbola (Fig. 3.11a). The semimajor axis is designated by a, the eccentricity by e, and the half-angle between the asymptotes by γ. μ_{pl} is the gravitational parameter of the planet and v_1 the velocity with which the spacecraft approaches the planet along one of the asymptotes before the path becomes appreciably deflected. Similarly, v_2 is the velocity at large distance after leaving the planet. The two velocities have equal magnitudes and we write

$$v = v_1 = v_2$$

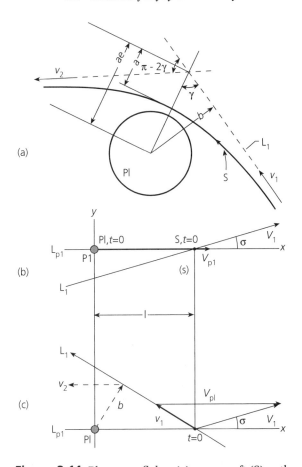

Figure 3.11 Planetary flyby: (a) spacecraft (S) path relative to planet; (b) in heliocentric reference frame; (c) in planetary reference frame. L_{pl}, planet's path; L_1, spacecraft asymptotic approach path; b, impact parameter. (Dashed lines are projections into the plane of the figure.)

The distance, b, from the approach asymptote to the planet's center is called the **impact parameter**, a term adopted from atomic and molecular physics.

From (3.14) and (3.10), letting the radius $r \to \infty$ (using the lower sign for hyperbolic trajectories), it follows that

$$a = \mu_{pl}/v^2$$

and $1 + e\cos(\pi - \gamma) = 0$, so that

$$e = 1/\cos\gamma$$

From the geometric relation shown in the figure, $b = ae\sin\gamma$, hence

$$\tan\gamma = bv^2/\mu_{pl} \tag{3.48}$$

The path of the spacecraft when viewed in the **heliocentric** reference frame is more complicated. The velocity in this frame is the vector sum of

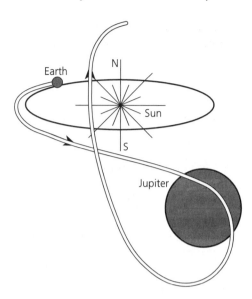

Figure 3.12 Ulysses (European Space Agency) solar mission. After a Jupiter gravity assist, spacecraft flies over the sun's south and north poles. The path is shown (schematically) as it would appear in the heliocentric reference frame.

the planet's velocity (in the general case out of the plane of the planetocentric spacecraft path) and the relative velocity of the spacecraft. In general, the path therefore cannot be embedded in a plane. To determine the path, in practice a numerical integration is required. Nevertheless, simple algebraic relations, to be derived in the following, suffice to calculate from the initial, heliocentric conditions the impact parameter and the change in energy.

As an example, the path taken by the Ulysses mission of the European Space Agency is shown in Fig. 3.12. After Earth escape, the path led to a flyby of Jupiter. The result was a large plane change and energy increment that allowed the spacecraft to leave the vicinity of the ecliptic plane and to cross the sun's north and south poles.

Figure 3.11b shows the paths of the planet and of the spacecraft's asymptotic approach in the *heliocentric* reference frame. The plane of the figure is a plane parallel to these paths. The separation distance of the paths at the crossover point in this projection is designated by s and the angle between the paths by σ. The origin of time is taken as the instant when the spacecraft is at the crossover point. At that time, the planet is assumed to lag behind the spacecraft by a distance l. The initial and final velocities are $\mathbf{V}_1 = \mathbf{V}_{pl} + \mathbf{v}_1$ and $\mathbf{V}_2 = \mathbf{V}_{pl} + \mathbf{v}_2$.

The same projection is also shown in Fig. 3.11c, which represents the *planetocentric* frame of reference where the planet appears at rest. The vector $\mathbf{b}$, the magnitude of which is the impact parameter, and the final relative velocity $\mathbf{v}_2$ are shown in projection.

In the Cartesian coordinate system (x, y, z) indicated in the figure, the components of the relative velocities are

$$v_{1x} = V_1 \cos \sigma - V_{pl}, \quad v_{1y} = V_1 \sin \sigma, \quad v_{1z} = 0 \\ v^2 = V_{pl}^2 + V_1^2 - 2V_{pl}V_1 \cos \sigma \quad\quad (3.49)$$

The components of the impact parameter vector can be found from

$$b_x = l + \lambda v_{1x}, \quad b_y = \lambda v_{1y}, \quad b_z = s$$

where λ is a scalar multiple of v_1. Since the scalar product $\mathbf{b} \cdot \mathbf{v}_1 = 0$, it follows that $\lambda = -lv_{1x}/v^2$, hence

$$b_x = lv_{1y}^2/v^2, \quad b_y = -lv_{1x}v_{1y}/v^2, \quad b_z = s$$

In terms of the initial conditions, formulated in the heliocentric reference system, the impact parameter is therefore

$$b = \sqrt{\frac{l^2 V_1^2 \sin^2 \sigma}{V_{pl}^2 + V_1^2 - 2V_{pl}V_1 \cos \sigma} + s^2} \quad\quad (3.50)$$

3.9.1 Capture by the Planet

The impact parameter is the principal quantity that determines whether or not the spacecraft will *collide* with the planet. It follows from (3.10) that the distance between spacecraft and planet has a minimum, r_{min}, at the periapsis and that $r_{min} = h^2/(\mu_{pl}(1 + e))$, where $h = bv_1 = bv$ is the angular momentum per unit mass. From (3.48) then follows

$$r_{min} = \frac{b^2 v^2}{\mu_{pl}(1 + e)}, \quad e = \sqrt{1 + \tan^2 \gamma} = \sqrt{1 + \frac{b^2 v^4}{\mu_{pl}^2}}$$

Designating the effective planetary radius that must be avoided by R'_{pl}, the condition for collision avoidance then becomes

$$r_{min} = \frac{b^2 v^2}{\mu_{pl}\left[1 + \sqrt{1 + (b/\mu_{pl})^2 v^4}\right]} > R'_{pl} \quad\quad (3.51)$$

where v^2 and b^2 in terms of the heliocentric initial conditions are obtained from (3.49) and (3.50).

When the inequality is opposite, the spacecraft will collide or, with application of rocket motor thrust and/or aerobraking, be *captured* in an orbit about the planet.

3.9.2 Change in Heliocentric Energy

The velocity of the spacecraft after the flyby is also easily calculated: $\mathbf{v}_2$ has a component in the direction of $\mathbf{v}_1$ of $-v\cos(2\gamma)$ and a component in the direction of $\mathbf{b}$ of $-v\sin(2\gamma)$. Therefore

$$\mathbf{v}_2 = -v\cos(2\gamma)\mathbf{v}_1/v - v\sin(2\gamma)\mathbf{b}/b$$

or

$$\left.\begin{aligned} v_{2x} &= -v_{1x}\cos(2\gamma) - \frac{vb_x}{b}\sin(2\gamma) \\ v_{2y} &= -v_{1y}\cos(2\gamma) - \frac{vb_y}{b}\sin(2\gamma) \\ v_{2z} &= -\frac{vs}{b}\sin(2\gamma) \end{aligned}\right\} \quad (3.52)$$

The potential energy of the spacecraft before and after the flyby is the same. Therefore the change in energy per unit mass in the heliocentric frame of reference is $(1/2)(V_2^2 - V_1^2)$. From

$$V_{1x} = V_{pl} + v_{1x}, \quad V_{1y} = v_{1y}, \quad V_{1z} = 0$$
$$V_{2x} = V_{pl} + v_{2x}, \quad V_{2y} = v_{2y}, \quad V_{2z} = v_{2z}$$

follows

$$\tfrac{1}{2}(V_2^2 - V_1^2) = V_{pl}(v_{2x} - v_{1x}) \quad (3.53)$$

which, after substitution from (3.52) and (3.48), results in

$$\tfrac{1}{2}(V_2^2 - V_1^2) = 2V_{pl}\cos^2\gamma\left[V_{pl} - V_1\cos\sigma \right.$$
$$\left. - \frac{lV_1^2}{\mu_{pl}}\sin^2\sigma\sqrt{V_{pl}^2 + V_1^2 - 2V_{pl}V_1\cos\sigma}\right] \quad (3.54)$$

A positive energy gain is therefore obtained if

$$V_{pl} > V_1\left[\cos\sigma + \frac{lV_1}{\mu_{pl}}\sin^2\sigma\sqrt{V_{pl}^2 + V_1^2 - 2V_{pl}V_1\cos\sigma}\right] \quad \mathbf{(3.55)}$$

3.9.3 Parallel Paths of Spacecraft and Planet

Particularly simple solutions are obtained when in the heliocentric reference frame the planet path and approach path of the spacecraft are parallel. In this case, $v = V_{pl} - V_1$. It then follows from (3.48) that, in terms of heliocentric initial conditions for the flyby, the minimum distance from the planet's center is

$$r_{\min} = \frac{b^2(V_{pl} - V_1)^2}{\mu_{pl} + \sqrt{\mu_{pl}^2 + b^2(V_{pl} - V_1)^4}} > R'_{pl} \quad (3.56)$$

The change in energy per unit mass, W, becomes

$$W_2 - W_1 = \tfrac{1}{2}(V_2^2 - V_1^2) = 2V_{pl}(V_{pl} - V_1)[1 + (b/\mu_{pl})^2(V_{pl} - V_1)^4]^{-1}$$

corresponding to a **relative** change

$$\frac{W_2 - W_1}{W_1} = 4(V_{pl}/V_1)((V_{pl}/V_1) - 1)[1 + (b/\mu_{pl})^2(V_{pl} - V_1)^4]^{-1} \quad (3.57)$$

Energy is therefore gained by the spacecraft if and only if $V_{pl} > V_1$.

The result also shows that for a given velocity ratio and impact parameter, the gain (or loss) is larger for massive planets. Comparing, for instance, Earth (mean equatorial radius = 6378 km, mean orbital velocity = 29.78 km/s, mass = 5.976 10^{24} kg) and Mars (3402 km, 24.13 km/s, 6.418 10^{23} kg) and assuming the spacecraft's heliocentric velocity to be one-half the planet's velocity and the impact parameter to be 1.5 times the planet's radius, one obtains

For the Earth flyby: $r_{min} = 7936$ km $(W_2 - W_1)/W_1 = 0.273$
For the Mars flyby: $r_{min} = 4817$ km $(W_2 - W_1)/W_1 = 0.026$

For flights to Mercury, Venus flybys are attractive. For flights to the outer planets, Jupiter flybys can provide a major boost to the spacecraft. Earth flybys can also be very useful. Thus, it is possible to design a spacecraft orbit that at its perigee is close to the Earth orbit and at its apogee close to the Mars orbit. Repeated Earth flybys can then be used to rotate the spacecraft's perigee–apogee axis to match partially the rotation of the Earth–Mars line and thereby provide for the spacecraft to pass repeatedly close to Earth and Mars (although, to provide for reasonably frequent such passes, some amount of thrust will still be needed to augment the effect of the flybys).

3.10 Relativistic Effects

In the applications of astrodynamics to spacecraft, relativistic effects can almost always be neglected.

To obtain a measure of the typical magnitudes of these effects, we consider here the minimum velocity v_{esc} at perihelion that is needed for the escape of a spacecraft from the solar system, assuming that the path starts at a distance of 1 AU from the sun. It then follows from (3.18), when applied to the sun, that $v_{esc} = 42.1$ km/s. From the relativistic equation for the change in mass

$$m = \frac{m_0}{\sqrt{1 - v^2/c^2}}$$

(m = mass, m_0 = restmass, v = velocity in an inertial frame, c = velocity of light in vacuum) follows

$$m/m_0 = 1 + 9.87\ 10^{-9}$$

Effects of this magnitude are detectable by high-precision Doppler navigational systems. Although of great interest for testing the theory of relativity by space experiments, relativistic effects are of only limited practical interest in today's space technology engineering. In deep-space missions, for instance, relativistic effects that may have accumulated over long periods of time are in practice eliminated by midcourse corrections, which can be based on direct measurements rather than on theory.

Because spacecraft are accelerated by gravitational forces, relativity experiments conducted with spacecraft must take into account the effects associated not only with special relativity (Einstein 1905) but also with general relativity, which is the generalization of the former and includes gravity (Einstein 1916).

As an example, it is of interest to consider a spacecraft on a circular orbit about the earth. By (3.14), the speed of the spacecraft is lower at higher altitudes. By special relativity, a clock on the spacecraft would then run faster than a clock on a lower orbit (but, depending on the orbit radius, generally slower than a comparable clock on the ground). This effect, however, is opposed by the effect of the gravity diminishing with altitude. The result is that at about 3200 km altitude the spacecraft clock runs at the same rate as a clock on the ground. Below this altitude, the spacecraft clock runs slow; above it the clock runs fast in comparison with the ground clock.

To improve earth coverage, most navigational satellite systems are deployed at relatively high altitudes where the satellite clocks run at a slightly higher rate than the ground clocks that are used to calibrate the system. In practice, this difference in rate is eliminated by the calibration procedures that are applied to the system.

In what follows, we will consider the **relativistic correction to Keplerian elliptic orbits**.

In the general theory of relativity, the dynamic equations are modified so that in place of (3.6) the equation for the orbit becomes [12]

$$\frac{d^2u}{d\theta^2} + u = \frac{\mu}{h^2} + \alpha u^2 \tag{3.58}$$

where, as before, h is a constant of motion, u the reciprocal of the radius, and the constant α is defined by

$$\alpha = 3\mu/c^2 \tag{3.59}$$

For the sun, with $\mu = 1.32712\ 10^{20}$ m^3/s^2, $\alpha = 4430$ m.

As α is very small compared with astronomical distances, the term αu^2 is much smaller than u and therefore represents a correction term that is added to the classical equation of motion.

Equation (3.58) is nonlinear. A standard way to solve it proceeds as follows. Write

$$u = u_0 + u_1$$

where

$$u_0 = \frac{\mu}{h^2}[1 + e\cos(\theta - \theta_0)]$$

is the corresponding expression (3.10) for the classical Kepler orbit and u_1 the relativistic correction. A close approximation to the solution of (3.58) is obtained by substituting u_0 for u on the right-hand side, resulting in the linear equation

$$\begin{aligned}\frac{d^2u_1}{d\theta^2} + u_1 &= \frac{\alpha\mu^2}{h^4}[1 + e\cos(\theta - \theta_0)]^2 \\ &= \frac{\alpha\mu^2}{h^4}\left[1 + \tfrac{1}{2}e^2 + 2e\cos(\theta - \theta_0) + \tfrac{1}{2}e^2\cos 2(\theta - \theta_2)\right]\end{aligned} \tag{3.60}$$

3.10 Relativistic Effects

The trial solution
$$u_1 = C_0 + C_1 \theta \sin(\theta - \theta_0) + C_2 \cos 2(\theta - \theta_0)$$
is seen to satisfy (3.60) with
$$C_0 = \frac{\alpha \mu^2}{h^4}\left(1 + \frac{1}{2}e^2\right), \quad C_1 = \frac{\alpha \mu^2}{h^4} e, \quad C_2 = -\frac{1}{6}\frac{\alpha \mu^2}{h^4} e^2$$
so that
$$u_1 = \frac{\alpha \mu^2}{h^4}\left[1 + \frac{1}{2}e^2 + e\theta \sin(\theta - \theta_0) - \frac{1}{6}e^2 \cos 2(\theta - \theta_0)\right] \tag{3.61}$$

As a consequence of the factor θ in the third term on the right, the amplitude represented by this term increases indefinitely with time. No such increase occurs in the other terms, and their effect on the orbit by comparison remains negligibly small. Dropping these terms and again adding the zero-order term, one obtains
$$u = u_0 + u_1 = \frac{\mu}{h^2}\left[1 + e\cos(\theta - \theta_0) + \frac{\alpha\mu}{h^2} e\theta \sin(\theta - \theta_0)\right]$$
or therefore
$$u = \frac{\mu}{h^2}\left[1 + e\cos\left(\theta - \theta_0 - \frac{\alpha\mu}{h^2}\theta\right)\right] \tag{3.62}$$
which can be verified by expressing the cosine of the difference by the cosine and sine of its parts, expanding these by their power series, and neglecting terms of order $(\alpha\mu/h^2)^2$.

This final result for the reciprocal of the instantaneous radius of the orbit indicates that the orbit is very nearly a classical ellipse. What is new is that the angular coordinate of the periapsis, hence the orientation in space of the ellipse, slowly rotates with time. In describing the motion of planets (particularly of Mercury, where the effect is most pronounced and more easily detectable), the effect is referred to as the *relativistic advance of the perihelion*.

For one full rotation of the orbit, $(\alpha\mu/h^2)\theta$ would have to increase by 2π. This would require $h^2/\alpha\mu$ revolutions of the planet or spacecraft along its orbit, with an elapsed time of $(h^2/\alpha\mu)P$, where P is the Keplerian orbital period as given by (3.16). The angular rate $\Delta\omega$ of the advance of the perihelion therefore becomes
$$\Delta\omega = 2\pi \frac{\alpha\mu}{h^2 P} \tag{3.63}$$

As found by Einstein, $\Delta\omega$ for the planet Mercury is $43''$ per century, in agreement with observational data.

The relativistic corrections to spacecraft orbits are at present too small to matter for engineering purposes. However, spacecraft could be used for deciding between Einstein's formulation of general relativity and competing theories. Spacecraft have the advantage over astronomical measurements that $\Delta\omega$ could be much larger than is the case for Mercury. This is seen from the formula

$$\Delta\omega = \frac{\alpha\mu^{1/2}}{(1-e^2)\,a^{5/2}} \qquad (3.64)$$

which follows easily from (3.63), combined with (3.16).

The relativistic effect is therefore seen to be more pronounced if the semimajor axis a is small and the eccentricity e close to one. Thus, for an earth satellite with a perigee altitude of 1000 km and apogee altitude above the earth's surface of 5000 km, $\Delta\omega = 1.03 \times 10^{-12}$ s^{-1}. Over 10 years, the apogee would be displaced by about 3700 m, a distance that could easily be measured. The practical difficulties, however, are great, because other, far larger effects, such as those caused by the irregular shape and mass distribution of the earth, would have to be taken into account with extreme accuracy.

Nomenclature

a	semimajor axis (Fig. 3.2); also acceleration [Eq. (3.44)]
b	semiminor axis (Fig. 3.2); also impact parameter (Fig. 3.11)
c	velocity of light in vacuum
e	eccentricity
E	eccentric anomaly (Fig. 3.2)
f, g	coefficients [Eq. (3.36)]
h	angular momentum per unit mass
i	inclination (Fig. 3.4)
M	mean anomaly
n	mean angular velocity
P	orbital period [Eq. (3.16)]
$\mathbf{r}, \mathbf{R}$	radius
$\bar{r}$	radius of sphere of influence
u	reciprocal of radius
$\mathbf{v}, \mathbf{V}$	velocity
w, W	energy per unit mass
α	right ascension (Fig. 1.3); also constant [Eq. (3.59)]
δ	declination (Fig. 1.3)
θ	true anomaly (Fig. 3.2)
μ	gravitational parameter
ϕ_j, ψ_j	angles referring to perturbation by third body (Fig. 3.9)
ω	angle from ascending node to periapsis (Fig. 3.4); $\Delta\omega$ angular rate of advance of periapsis
Ω	right ascension of the ascending node
$(\ )_{crc}$	circular
$(\ )_{gs}$	geostationary
$(\ )_G$	Greenwich
$(\ )_h$	sun
$(\ )_H$	horizon

( )$_p$	periapsis
( )$_{pl}$	planet
( )$_r$	radial
( )$_s$	spacecraft
( )$_{si}$	siderial
( )$_\theta$	tangential

Problems

(1) A geostationary satellite is located above the meridian 230° east of Greenwich. For a ground observer at longitude 202°03′ east of Greenwich and latitude 21°19′ north (Honolulu), compute the azimuth and elevation angles of the boresight of an antenna directed toward this satellite. (Note that the azimuth angle is conventionally counted starting from north in the easterly direction. The elevation angle is counted starting from the local horizontal.)

(2) Consider a Kepler elliptic orbit and on it the two points having true anomalies of 90° and 270°, respectively. Show that at these points the (acute) angle formed by the radius vector from the center of attraction and the tangent to the orbit is given simply by $\cotan^{-1} e$, where e is the eccentricity of the orbit.

(3) The ground trace (i.e., the projection downward on the earth's surface, assumed to be spherical) of an earth satellite is assumed to pass through a point with latitude 28°40′ (the U.S. Cape Canaveral launch site). The satellite's projection is assumed to move at an angle of 15°00′ counted from the east and increasing toward the north.

Find the maximum and the minimum latitudes reached by the ground trace.

(4) An earth-orbiting satellite is launched with a perigee altitude (the altitude above the earth's mean equatorial radius of 6378.1 km) of 1000 km and apogee altitude of 5000 km.

   (a) Compute the semimajor axis, the semiminor axis, the eccentricity, and the orbital period. Also compute the velocities at perigee and apogee relative to the geocentric, nonrotating reference frame.
   (b) Set up a computer program to print out the distance from the center of attraction and the true anomaly as functions of the eccentric anomaly at increments of 10°.

(5)* An earth satellite is on an orbit having a semimajor axis of 40,000 km and eccentricity of 0.150. The time after perigee passage of the satellite is 15,000 s.

   (a) Compute the eccentric anomaly of the satellite by D'Alembert's method, using the first four terms of the series. Also compute the radial distance from the center of attraction and the true anomaly.
   (b) Compute the same data from the Fourier–Bessel series, again using the first four terms. Compare the results.

(6) Jupiter has an equatorial radius of 71,490 km and a gravitational parameter of 126.71 10^6 km^3/s^2. A spacecraft executes a flyby with an impact parameter of 230,000 km. The asymptotic speed of approach of the spacecraft in the planetocentric, nonrotating reference frame is 20.0 km/s.

Utilizing the approximation discussed in Sect. 3.9, compute the distance of closest approach from Jupiter's surface. Also compute the turning angle of the spacecraft trajectory.

(7)* The following six orbital elements (see Fig. 3.4) of a spacecraft are given: a = 30,000 km, $e = 0.500$, $i = 30°$, $\omega = 70°$, $\Omega = 150°$, time of perigee passage $t_p = 0$. An earth ground observer is located at geographic latitude 45° north and at the time of perigee passage has a right ascension of 240°.

Compute the time relative to perigee passage when the spacecraft first appears above the observer's (theoretical) horizon. Also compute the azimuth of the spacecraft's first appearance as it appears to the observer.

References

1. Stumpff, K., "Himmelsmechanik," Vol. 1, VEB Deutscher Verlag der Wissenschaften, Berlin, 1959.
2. Poincaré, J.-H., "New Methods of Celestial Mechanics," *History of Modern Physics and Astronomy*, Vol. 13, translated from the French, American Institute of Physics, New York, 1993.
3. Battin, R. H., "An Introduction to the Mathematics and Methods of Astrodynamics," American Institute of Aeronautics and Astronautics, Washington, DC, 1987.
4. Baker, R. M. L. and Makemson, M. W., "Astrodynamics," Academic Press, New York, 1960.
5. Danby, J. M. A., "Fundamentals of Celestial Mechanics," 3rd printing, Willmann–Bell, Richmond, VA, 1992.
6. Prussing, J. E. and Conway, B. A., "Orbital Mechanics," Oxford University Press, New York, 1993.
7. Wiesel, W. E., "Spaceflight Dynamics," McGraw-Hill, New York, 1989.
8. Gurzadyan, G. A., "Theory of Interplanetary Flights," translated from the Russian, Gordon & Breach Publishers, New York, NJ, 1996.
9. Plummer, H. C., "An Introductory Treatise on Dynamical Astronomy", Dover Publications, New York, 1990.
10. Herrick, S., "Astrodynamics", Vol. 1, Van Nostrand Reinhold, London, 1971.
11. Cornelisse, J. W., Schöyer, H. F. R., Wakker, K. F., "Rocket Propulsion and Spaceflight Dynamics," Pitman Publishing, London, 1979.
12. Eddington, A., "The Mathematical Theory of Relativity," Cambridge University Press, London, 1923.

4

Chemical Rocket Propulsion

Rocket propulsion is based on the principle that a vehicle can gain momentum by expelling mass stored in the vehicle itself. Rocket propulsion therefore differs, for instance, from the air-breathing propulsion of aircraft jet engines, where most of the mass involved in the propulsion is atmospheric air.

In the vacuum of space, a space vehicle can also gain momentum by a planetary flyby with gravity assist (Sect. 3.9) or by solar radiation pressure ("solar sailing"). For flight within a planetary atmosphere, such as occurs during the early phase of a launch from the earth's surface, important advantages can also be gained by combining rocket and air-breathing propulsion, possibly in a single engine. Nevertheless, classical rocket propulsion is by far the most important means for propelling space vehicles of all types.

The source of energy in rocket propulsion is almost always the *chemical energy* released by the combustion of a fuel with an oxidizer. (For a class of small thrusters, chemical energy is released by the decomposition of a *single* propellant, referred to as a "monopropellant," most often hydrazine). Ion or colloid engines, thermal and magnetoplasmadynamic arc jets, resisto-jets, and related devices use *electrical energy* for their principal energy source. In space, this energy may be derived from solar cells or conversion from the thermal energy produced by radioisotope sources or by nuclear reactors. Electrical propulsion can have a far larger specific impulse, albeit at the cost of a much smaller thrust, than chemical propulsion.

To combine a high specific impulse with a high thrust, rockets with several other modes of propulsion have been proposed. They include nuclear fission thermal rockets in which the propellant, hydrogen, is heated by heat transfer from a reactor. Entirely speculative at this time are rockets that derive their energy from reactions of free radicals or from the combination of *para-* and *ortho-*hydrogen, from magnetically or inertially confined fusion, or from antimatter. For comparison, the maximum theoretical energy release per unit mass for some of these reactions is listed in Table 4.1. None of these reactions, with the possible exception of the fission thermal process, are considered practical today for rocket propulsion.

In addition to thrust and specific impulse, there are several other characteristics of rocket motors that are of great practical importance. They include, among others, the weight-to-thrust ratio of the rocket motor, the capability for throttling to adjust the thrust level, restart or multiple pulsing capability, storability of the propellant, nozzle designs that provide an optimal compromise between performance in the atmosphere and in the vacuum of space, and provision for expanding in space the nozzle exit area ("extendable nozzle").

Table 4.1 Comparison of the Theoretical Energy Release of Some Speculative Rocket Energy Sources

Source	Maximum theoretical energy release (J/kg)
Free radicals (H + H → H_2)	2.20×10^8
Nuclear fission	7.12×10^{13}
Nuclear fusion	7.62×10^{14}
Matter annihilation	9.0×10^{16}

For comparison, the energy release from the stoichiometric combustion of hydrogen with oxygen, at pressures and temperatures typical of rocket motors, is about 1.50×10^7 J/kg.

Approximate values of the specific impulse and thrust of rocket motors and thrusters, when operating in vacuum, are shown in Fig. 4.1. Areas in the diagram that are shaded indicate schematically the current state of the art. Separately shown are (1) chemical propulsion main engines (i.e., the rocket motors of launch vehicles and their boosters, lunar or planetary descent or ascent motors, etc.), (2) chemical thrusters (such as are used for attitude control and station keeping of satellites, some of them making use of monopropellants, e.g., hydrazine), (3) electric propulsion motors (including ion engines, arc jets, magnetohydrodynamic devices, pulsed Teflon thrusters), and (4) chemical–electric thrusters (e.g., hydrazine thrusters augmented by electric resistance or microwave heating). Points (1) to (12) designate specific motors and thrusters.

Significantly different in specific impulse, but not in the achieved thrust, are liquid and solid (including hybrid) chemical propellant motors. As shown in the figure, the regions overlap in terms of the specific impulse. The highest specific impulse is achieved by hydrogen-rich combustion with oxygen. Reactions with beryllium and fluorine can theoretically result in a still higher specific impulse. However, such propellants are too hazardous to be of practical value.

In principle, high thrust combined with high specific impulse could be obtained with nuclear–thermal rockets. In these motors, hydrogen, heated by heat transfer from a nuclear reactor, serves as the propellant. The predicted performance of the U.S. Nerva project of the 1960s, since abandoned, is indicated by point (8) in the diagram.

Electric propulsion comprises ion and colloid propulsion, arc jets, pulsed Teflon thrusters, and various magnetohydrodynamic devices. Ion thrusters have high electric efficiency and represent the most mature technology among these devices. They are being used on some spacecraft.

The area under the dashed curve shows the higher thrusts that could be obtained by clusters of electric thrusters, particularly ion propulsion motors. Still lacking, however, are space-borne nuclear reactors and thermal–electric conversion devices with high enough electric power to drive these thrusters. (The required reactor thermal power in watts is indicated in Fig. 4.1 and is

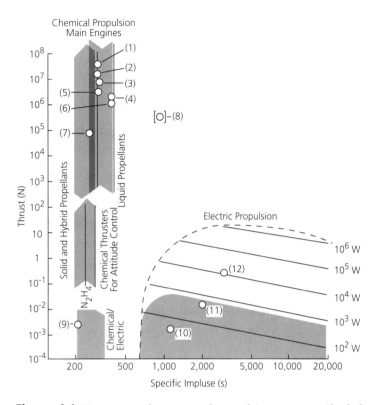

Figure 4.1 Vacuum performance of propulsion systems. Shaded areas are state of the art. (1) Saturn V—Apollo (USA), 5 boosters F-1, RP1/O_2 and solid boosters. (2) Ariane V (ESA) at liftoff, Vulcain HM-60 and solid boosters P-230. (3) Proton (Russia), UDMH/N_2O_4, 6 boosters. (4) Space Shuttle main engine SSME (USA), H_2/O_2. (5) DELTA 7925 (USA) at liftoff, RP1/O_2 and solid boosters. (6) Ariane V (ESA), Vulcain HM-60, H_2/O_2. (7) PAM STS STAR 48 (USA), solid motor. (8) Nerva (USA) nuclear–thermal, project abandoned. (9) Resisto-jet, NH_3. (10) Pulsed Teflon. (11) MPD arcjet, Ar. (12) Ion-bombardment engine, Hg, 30 cm (NASA).

based on an assumed thermal–electric conversion efficiency of 30% and an electric-to-beam power efficiency of 90%.)

The treatment in this chapter has profited from and follows to a considerable extent the book by Sutton and Ross [1] and the book by Huzel and Huang [2].

4.1 Configurations of Liquid-Propellant Chemical Rocket Motors

A typical design of a liquid-propellant chemical motor is illustrated in Fig. 4.2. The fuel, after passing a metering valve, enters a circumferential manifold and from there flows through radial passages in the **injector plate**. The oxidizer, after its valve, enters a dome where it is distributed over the top

100   CHAPTER 4 *Chemical Rocket Propulsion*

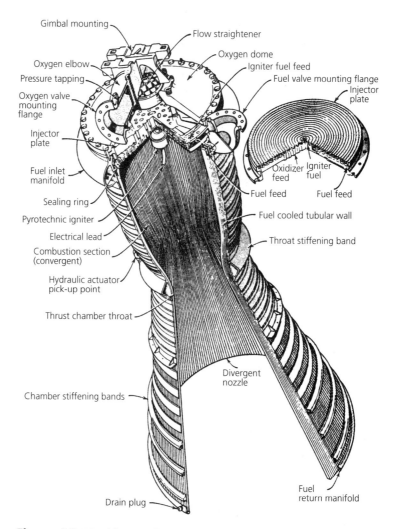

Figure 4.2 Liquid-propellant chemical rocket motor. From Ref. 1, Sutton, G. P. and Ross, D. M., "Rocket Propulsion Elements," 5th ed. Copyright ©1986. Reprinted by permission of John Wiley & Sons, Inc. Courtesy of Rocketdyne Division of Rockwell International, U.S.A.

surface of the injector plate. It continues its flow through axial passages in the plate, separate from the radial ones. Fuel and oxidizer are then injected at high pressure through separate, narrow injection holes into the **combustion chamber**, where they mix and ignite. The resulting combustion gas then accelerates in the **nozzle**, which consists of a converging (subsonic) and a diverging (supersonic) section. At the **throat** where the converging and diverging parts meet, the gas velocity equals its local sonic velocity. The aft end of the nozzle is referred to as the **exit plane**. The combined structure of combustion chamber and nozzle is frequently called the **thrust chamber**.

Typical of many large motors, such as the one shown in the figure, is the tubular construction of the thrust chamber. It can be manufactured by brazing together a large number of tubes, shaped to result in the configuration required by the nozzle. Before entering the injection plate, all or a portion of the fuel—or alternatively, but more rarely, the oxidizer—flows through the tubes to cool the thrust chamber walls. Thrust chambers of this type are referred to as being **regeneratively cooled**.

Circumferential stiffening bands around the nozzle are often used, particularly around the throat, where, in the illustrated motor, hydraulic actuators are attached. Their purpose is to provide for steering the vehicle by deflecting the motor structure through small angles in the two orthogonal directions, about 5° to 7° being common. To allow for this motion, the motor is supported on gimbals in mountings that transmit the thrust. The fuel and oxidizer lines must be flexible; stainless steel bellows are often used for this purpose.

Although the fuel and oxidizer will usually self-ignite when mixed, a separate pyrotechnic **igniter** is provided for starting the motor. Its use prevents a dangerous accumulation of unreacted propellant at the start of engine operation. In some cases the igniter is not designed to survive the motor firing and is allowed to be ejected through the nozzle.

4.2 Configurations of Solid-Propellant Motors

The principal feature of solid-propellant motors is that fuel and oxidizer, together with a binder and other additives, are already **premixed**. They are contained in this form in a **motor case**, rather than in separate tanks as would be the case for liquid propellants. At normal temperature, after curing, the propellant is a solid, rubber-like mass. At a moderately higher temperature it has the consistency of a highly viscous fluid and can be cast into the motor case.

Combustion occurs on the inner surface of the propellant and proceeds outward. The geometrical configuration of the propellant is such that the burn surface stays roughly constant; this ensures that the gas pressure and thrust stay approximately constant. At the end of the burn, the burn surface reaches the wall of the motor case, which is thermally protected by a noncombustible liner.

In casting the propellant into the motor case, a mandrel is used to provide an inner channel, not filled with propellant, to conduct the gas to the nozzle. Often, a second purpose of the mandrel is to form an initial burn surface that differs from a strictly circular cylindrical shape. The purpose is to increase the initial burn surface, making its area more nearly equal to the area of the final burn surface.

Solid-propellant motors are very frequently used as **upper stage motors**. Often, they also form an integral part of the spacecraft that contains the payload. The motor case, in this case, is usually near-spherical or ellipsoidal in shape, hence structurally nearly optimal to contain the gas pressure. An example of this type of motor is shown in Fig. 4.3.

Solid-propellant motors are often the main motors of smaller launch vehicles, such as **rocket probes**. Very large solid-propellant motors are frequently used as **boosters**, strapped to the sides of major launch vehicles. Boosters operate at least in part in the atmosphere and are therefore given some aerodynamic shaping. This requirement, and the need to conform externally to the main body of the launch vehicle, calls for a solid motor that

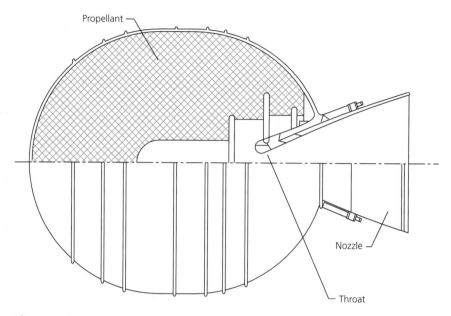

Figure 4.3 PAM-STS, Star 48 solid-propellant rocket motor. (Courtesy of Thiokol Corporation, U.S.A.) By permission.

has the form of a long cylinder. At the end of the burn, the empty motor cases are dropped off. Often, they are recovered to be used again in a later launch.

Boosters are often so large that it would no longer be practical to cast the propellant as a single mass. Reasons are the difficulty of casting such masses in the time before they cool and the difficulty of transporting such large objects to the launch site. The propellant and containing motor case are therefore often subdivided into segments that are stacked lengthwise and assembled only at the launch site. Maintaining the integrity of the joints between segments and guarding against gas leaks are critical. O-rings are used in pairs or in triplicates to obtain reliability by redundancy. In turn, they are protected from the heat and chemical attack of the combustion gas by zinc chromate putty or similar materials.

Figure 4.4 illustrates this type of booster. Here the major part of the propellant is cast in three cylindrical segments. At the end of the burn, the boosters are severed from the main vehicle by pyrotechnic mechanisms and separated from the vehicle by small rockets or springs.

Frequently, large boosters contain a parachute package for recovery of the empty case and nozzle from the ocean. The case, which is either steel or a composite, can then be used after refurbishment for another launch.

In comparison with liquid-propellant motors, solid-propellant motors are simpler in concept and construction and lower in cost. Generally, they have a lower specific impulse.

In a number of applications, a disadvantage of solid-propellant motors is that, once ignited, they produce thrust until all the propellant has been consumed. Because the exact performance of the propellant depends on several factors that are difficult to control to very high precision, the thrust cutoff is less precise than with liquid-propellant motors. To obtain

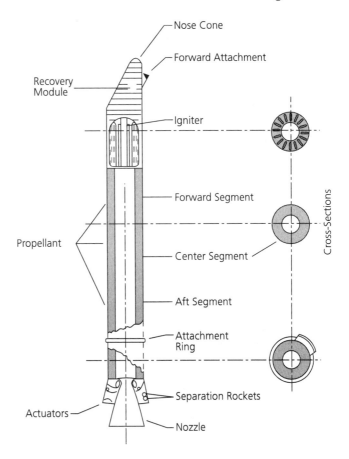

Figure 4.4 Ariane 5, solid-propellant booster. (Courtesy of CNEF, France.)

the accuracy required in maneuvers such as orbit insertion, small liquid-propellant thrusters on the spacecraft are needed for correction. (Proposals to terminate the burn at a precise time by explosively removing parts of the nozzle have not found general acceptance.)

This disadvantage is overcome in **hybrid rocket motors**. Here the fuel and oxidizer are stored separately, the former in a solid mass in a configuration similar to that in solid-propellant motors. Cryogenic oxygen, or a gaseous or liquid or gelled oxidizer, is injected into the motor case. Because the oxidizer flow can be precisely controlled, precise thrust termination, thrust throttling and multiple engine starts are possible.

4.3 Rocket Stages

The large rockets used for space missions are built and operated in stages, arranged one on top of the others. Strap-on solid-propellant boosters, firing simultaneously with the main liquid-propellant motors, in combination, can also be considered to constitute a single stage, providing thrust from booster ignition to booster termination.

By definition, each stage is a complete propulsion system. After completion of the burn, the stages are separated one by one from the remaining stages or payload.

The advantage that can be gained by staging follows from the fact that after the separation of a burned-out stage the remaining vehicle is lighter and therefore easier to accelerate to a still higher velocity. As is evident from the rocket equation (2.28), the rocket's initial mass, needed to accelerate a given payload mass to a given velocity, has an exponential dependence on the required velocity increment. The resulting theoretical gain obtained by staging is somewhat reduced, however, by the added mass of the stage separation mechanism and by the unfavorable mass scaling factor of smaller, multiple tanks, feed systems, and motors as contrasted with a larger, single system.

For launch vehicles, another advantage of staging is the possibility of designing the motors to perform optimally for the particular atmospheric pressure in which they are principally operating.

Stages are usually connected to each other by bolted flanges. Separation is then initiated by the use of pyrotechnic bolt cutters. Another method of connecting stages is by so-called Marman clamps, in which a circumferential steel band presses inverse wedges against pairs of wedges that are parts of the respective two stages. Separation is initiated by cutting the steel band pyrotechnically. The principle of the functioning of these clamps is illustrated in Fig. 4.5.

Springs provide a simple means of pushing off the separated stage. To prevent a difficult-to-predict interaction of the separated stage with the rocket motor plume of the next stage, some distance is needed between the separated stage and the remaining vehicle. A second reason for waiting

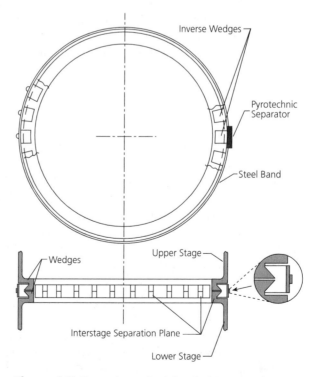

Figure 4.5 Operating principle of a Marman clamp.

for a minimum distance before firing the next stage is that the separated stage will usually experience a slight residual acceleration caused by gas that evolves from the still hot motor. This acceleration could possibly result in a collision of the separated stage with the remaining vehicle. In place of springs, small rocket motors are also being used.

In what follows, an optimal distribution of the masses of the stages of an n-stage rocket will be computed. The calculation, which follows in the main Ref. 3, provides insight into such questions as the optimal sequence of stages that may have different mass, specific impulse, and thrust. It should be noted, however, that most often the designer of a space mission has available only a limited choice of rocket motors, most of them having already been developed for earlier space missions. The calculated, theoretical optimal stage mass distribution therefore can provide only a general and qualitative guide.

Let $m_{0,i}$, $i = 1, 2, \ldots, n$, designate the masses, before ignition, of the stages of an n-stage vehicle and $m_{1,i}$ their masses at the end of the burn. Thus $m_{0,1}$ and $m_{1,1}$ refer to the first (lowest) and $m_{0,n}$ and $m_{1,n}$ to the last (uppermost) stage. The mass of the payload is designated by m_{pl}. (The expression "payload" is frequently used ambiguously; sometimes it refers to the scientific instruments and communication systems carried by a spacecraft; in other instances it may refer to the total spacecraft. In the present context, "payload" will be understood to be a complete spacecraft, exclusive, however, of main propulsion motors that are sometimes integrated with it; the latter are counted as an additional "stage.")

Each stage mass before the burn can be regarded as the sum of its initial, useful propellant mass $m_{\text{pr},i}$ and a residual mass $m_{\text{rs},i}$. The latter consists of the sum of the masses of the empty propellant tanks or motor case, excess propellant mass carried as a margin of safety, the propellant feed system, thrust chamber, nozzle, actuators, avionics, and stage structural elements. The **stage residual mass fraction**, ε_i, is then defined by

$$\varepsilon_i = m_{\text{rs},i}/m_{\text{pr},i}$$

so that

$$\left.\begin{array}{l} m_{0,i} = m_{\text{pr},i}(1 + \varepsilon_i) \\ m_{1,i} = m_{\text{rs},i} \end{array}\right\} \quad i = 1, 2, \ldots, n \quad (4.1)$$

A convenient concept is that of **subvehicles**, defined by the sum of all stages not yet fired plus the payload. Therefore, including the complete launch vehicle and the payload, there are $n + 1$ such subvehicles. The mass of the ith subvehicle before the burn of its lowest stage is designated by $M_{0,i}$, and after the burn—but before separation of the residual mass—by $M_{1,i}$. With these definitions

$$\left.\begin{array}{l} M_{0,i} = m_{\text{pl}} + \sum_{j=i}^{n} m_{\text{pr},j}(1 + \varepsilon_j) \\ M_{1,i} = m_{\text{pl}} - m_{\text{pr},i} + \sum_{j=i}^{n} m_{\text{pr},j}(1 + \varepsilon_j) \end{array}\right\} \quad i = 1, 2, \ldots, n \quad (4.2)$$

In particular, $M_{0,1}$ is the **launch mass** of the complete vehicle.

Useful parameters for characterizing the subvehicle masses are the **subvehicle mass fraction**

$$\nu_i = M_{1,i}/M_{0,i}$$

and the **sequential subvehicle mass fraction**

$$\psi_i = M_{0,i+1}/M_{0,i}$$

In particular, let $\Psi = m_{\text{pl}}/M_{0,1}$. Ψ, which is called the **payload mass fraction**, is the most important characteristic for judging the benefit that can be obtained from staging.

It follows directly from these definitions that Ψ is the product of the sequential mass fractions, that is,

$$\Psi = \prod_{i=1}^{n} \psi_i \tag{4.3}$$

The stage masses before the burn can be expressed by the same parameters. It is easily shown that

$$\left.\begin{aligned} m_{0,1} &= M_{0,1}(1 - \psi_1) \\ m_{0,i} &= M_{0,1}(1 - \psi_i) \prod_{j=1}^{i-1} \psi_j, \quad i = 2, 3, \ldots, n \end{aligned}\right\} \tag{4.4}$$

The three parameters ν_i, ψ_i, and ε_i are not independent of each other, because from

$$\nu_i = \frac{M_{0,i} - m_{\text{pr},i}}{M_{0,i}} = 1 - \frac{m_{\text{pr},i}}{M_{0,i}}$$

and

$$\psi_i = \frac{M_{0,i} - m_{\text{pr},i} - m_{\text{rs},i}}{M_{0,i}} = 1 - \frac{m_{\text{pr},i}(1 + \varepsilon_i)}{M_{0,i}}$$

follows

$$\psi_i = \nu_i(1 + \varepsilon_i) - \varepsilon_i, \quad i = 1, 2, \ldots, n \tag{4.5}$$

Next, we consider the theoretically optimal distribution of stages in two special cases. These suffice to answer, at least qualitatively, most questions related to the general case.

4.3.1 Optimal Distribution; Negligible Gravitational Acceleration

It will be assumed that during the period of thrust the gravitational acceleration is negligible compared with the acceleration caused by the thrust. This assumption is approximately valid in many space missions, exceptions being the ascent or descent from a planetary surface and low-thrust, electric propulsion.

The thrust is assumed to be parallel to the trajectory, which therefore is rectilinear. The different stages may have different, but constant, specific impulse.

4.3 Rocket Stages

Let $F_{t,i}$ be the magnitude of the thrust of the ith stage. For brevity, we define the characteristic velocity $c_i = g_0 I_{\text{sp},i}$. In accordance with the definition (2.24) of the specific impulse and (2.26), (2.27),

$$F_{t,i} = \tilde{m}_i g_0 I_{\text{sp},i} = -c_i\, dM_i/dt = M_i(t)\, dV_i/dt$$

where $M_i(t)$ and $V_i(t)$ are the mass and velocity, respectively, of the ith subvehicle. Integrating from the beginning to the end of the burn results in the velocity increment

$$(\Delta V)_i = V_{1,i} - V_{0,i} = -c_i \ln \nu_i$$

for the ith subvehicle.

The velocity at the start of the burn of subvehicle $i+1$ is the same as the velocity at the end of the burn of subvehicle i. Therefore the velocity gain ΔV of the payload is obtained by summing all the separate increments, with the result that

$$\Delta V = V_{\text{pl}} - V_{0,1} = \sum_{i=1}^{n} c_i (\ln(1+\varepsilon_i) - \ln(\psi_i + \varepsilon_i)) \qquad (4.6)$$

If the specific impulse, hence the characteristic velocities c_i, and the stage residual mass fractions are kept constant, the *velocity gain of the payload will be maximal* when the (positive) term

$$-\sum_{i=1}^{n} c_i \ln(\psi_i + \varepsilon_i)$$

has an absolute maximum, subject to the side condition (4.3).

Following Vertregt [3], the desired result is obtained by the method of Lagrange multipliers. Under mild conditions (not all Jacobians of the side conditions vanish, and certain inequalities among the second derivatives must be satisfied to outrule saddle points) the method will calculate the relative extrema.

We define the functions

$$f(\psi_1, \psi_2, \ldots, \psi_n) = -\sum_{i=1}^{n} c_i \ln(\psi_i + \varepsilon_i)$$

and

$$h(\psi_1, \psi_2, \ldots, \psi_n) = \ln \Psi - \sum_{i=1}^{n} \ln \psi_i$$

where h vanishes when the side condition (4.3) is satisfied. There is a constant multiplier, say λ, such that

$$\frac{\partial f}{\partial \psi_i} = \lambda \frac{\partial h}{\partial \psi_i} \qquad (4.7)$$

Evaluating the partial derivatives

$$\frac{\partial f}{\partial \psi_i} = -\frac{c_i}{\psi_i + \varepsilon_i}, \qquad \frac{\partial h}{\partial \psi_i} = -\frac{1}{\psi_i}$$

and substituting them into (4.7), together with the side condition, results in the $n+1$ algebraic equations for $\psi_1, \psi_2, \ldots, \psi_n$ and λ

$$\left.\begin{array}{l} c_i \psi_i = \lambda(\psi_i + \varepsilon_i), \qquad i = 1, 2, \ldots, n \\ \displaystyle\prod_{i=1}^{n} \psi_i = \Psi \end{array}\right\} \qquad (4.8)$$

Solving the first of these equations for ψ_i in terms of the Lagrange multiplier gives

$$\psi_i = \frac{\lambda \varepsilon_i}{c_i - \lambda} \qquad (4.9)$$

Finally, substitution of ψ_i from (4.9) into the side condition (4.3) results in

$$\lambda^n \prod_{i=1}^{n} \frac{\varepsilon_i}{c_i - \lambda} = \Psi \qquad \textbf{(4.10)}$$

which is a polynomial equation of degree n for λ.

The method of Lagrange multipliers, as discussed, gives the extremal values. To find the *absolute maximum* of the function (4.6) it is best, in applications, to examine numerically and separately the real solutions of (4.10).

To summarize the entire calculation: After solving (4.10) for the Lagrange multiplier, the values of the sequential subvehicle mass fractions are found from (4.9). The masses of the rocket stages follow from (4.4). Finally, the velocity increment of the payload is found from (4.6). The stage mass distribution that results in the *maximum* velocity gain is most directly obtained by comparison of the values of the Lagrange multipliers.

A qualitative understanding of the optimal distribution of the stages in a multistage rocket is best obtained by numerical examples. Table 4.2 lists the pertinent data for a two-stage rocket with a payload mass fraction $m_{pl}/M_{0,1}$ of 0.100. One of the stages is assumed to use a solid-propellant motor with a specific impulse of 280 s. The other stage uses a hydrogen–oxygen motor with a specific impulse in vacuum of 455 s. The stage residual mass fraction in both cases is assumed to be 15%.

In case 1a, the high specific impulse stage is the *second* stage to fire; in case 1b it is the *first* stage. As noted from the table, the sequential subvehicle mass fractions ψ_i are reversed in the two cases and the payload velocity gains ΔV are the same.

The stage mass fractions $m_{0,i}/M_{0,1}$, however, differ. Case 1a is seen to be more advantageous because the hydrogen–oxygen engine, which is more costly, is smaller than in case 1b, and this for the same final velocity gain of the payload. This conclusion is not limited to two-stage rockets but is also valid in the more general case of n stages.

4.3.2 Optimal Distribution; Vertical Ascent

In the ascent or descent from and to a planetary or lunar surface, the thrust and gravity forces are comparable in magnitude. In the case of large, liquid-propellant launch vehicles, launched vertically from the earth's surface,

Table 4.2 Optimal Stage Distributions

Case 1

Two-stage rocket. Gravitational force neglected compared with thrust. Payload mass fraction $\Psi = 0.100$. Specific impulse $I_{sp,i} = 280$ (solid-propellant motor) and 455 s (hydrogen–oxygen motor, vacuum performance), respectively. Stage residual mass fractions $\varepsilon_i = 0.15$.

		Case 1a	Case 1b
Propellant	First stage	Solid	H_2–O_2
	Second stage	H_2–O_2	Solid
λ (m/s)[a]		2240	2240
ψ_1	First stage	0.662	0.151
ψ_2	Second stage	0.151	0.662
$m_{0,1}/M_{0,1}$	First stage	0.338	0.849
$m_{0,2}/M_{0,1}$	Second stage	0.562	0.051
ΔV (m/s)		6393	6393

Case 2

Two-stage rocket. Vertical ascent in constant gravity field. Payload mass fraction $\Psi = 0.100$. Specific impulse $I_{sp,i} = 280$ and 455 s, respectively. Stage residual mass fractions $\varepsilon_i = 0$. Thrust-to-gravity fractions $\phi_i = 1.50$.

		Case 1a	Case 1b
Propellant	First stage	Solid	H_2–O_2
	Second stage	H_2–O_2	Solid
λ (m/s)[a]		2472	2472
ψ_1	First stage	0.150	0.669
ψ_2	Second stage	0.669	0.150
$m_{0,1}/M_{0,1}$	First stage	0.850	0.331
$m_{0,2}/M_{0,1}$	Second stage	0.050	0.569
ΔV (m/s)		4460	4460

[a] Corresponding to the absolute maximum of ΔV.

commonly used thrust-to-gravity ratios at takeoff are about 1.2 to 2.0 [Space Shuttle, U.S.A., about 1.42; Proton, Russia, 1.29; Ariane 5, ESA, 2.27]. The principal reason for choosing these relatively small ratios is that for larger ratios the mass and power requirements of the liquid-propellant feed systems become excessive as a consequence of the very large propellant flow rates that heavy launch vehicles must handle.

For smaller vehicles, such as rocket probes with solid-propellant motors, this restriction does not apply. Relatively high thrust-to-gravity ratios at takeoff are not only possible but also advantageous because they minimize the **gravity loss** (Sect. 2.2.4).

Whereas in the problem treated in the preceding section, time delays between successive firings of the stages do not affect the optimization of the stage mass distribution, this is no longer the case here. The effect of delays is most directly seen in the extreme case of **hovering** as it occurs in the descent to a planetary or lunar surface when thrust and gravitational force just cancel each other. In this case there is no change in either the kinetic or potential energy of the payload, and the entire energy expended by the motor is invested in the gas of the rocket plume. It is evident, therefore, that in the case of motion in a gravitational field it is advantageous to fire the stages without delay. Some exceptions will occur, however, when dictated by atmospheric effects on the performance of different motors.

In what follows, we consider the vertical ascent of a multistage rocket in a constant gravitational field. For simplicity, the stage residual masses are neglected and the rocket motor mass flow rates and also the specific impulse of each stage are assumed to be independent of time. Time delays between the firing of successive stages are assumed to be zero.

Analogous to the equation of motion in the preceding section, but now adding the gravitational term,

$$M_i(t)\, dV_i/dt = -c_i\, dM_i/dt - gM_i(t), \qquad i = 1, 2, \ldots, n$$

By separation of variables and integration,

$$\int_{V_{0,i}}^{V_{1,i}} dV_i = -c_i \int_{M_{0,i}}^{M_{1,i}} \frac{dM_i}{M_i} - g \int_{t_{0,i}}^{t_{1,i}} dt$$

Since the mass flow rate $\tilde{m}_i$ of each stage is assumed constant,

$$\tilde{m}_i(t_{1,i} - t_{0,i}) = M_{0,i} - M_{1,i} = M_{0,i}(1 - \nu_i)$$

Also, since the stage residual masses are neglected,

$$M_{0,i+1} = M_{1,i}, \qquad \nu_i = \psi_i$$

Defining the **thrust-to-gravity fractions**

$$\beta_i = F_{t,i}/(gM_{0,i}), \qquad \beta_i > 1$$

and expressing the magnitude of the thrust of the ith stage by $F_{t,i} = c_i \tilde{m}_i$, where c_i is the stage characteristic velocity,

$$t_{1,i} - t_{0,i} = (1 - \psi_i)c_i/(\beta_i g) \tag{4.11}$$

The velocity increment $(\Delta V)_i$ produced by each firing is therefore given by

$$(\Delta V)_i = V_{1,i} - V_{0,i} = -c_i[\ln \psi_i + (1 - \psi_i)/\beta_i]$$

Because $V_{0,i+1} = V_{1,i}$, the velocity increment ΔV of the payload is obtained by summing all stage velocity increments. Hence

$$\Delta V = V_{pl} - V_{0,1} = \sum_{i=1}^{n} c_i \left(\ln \frac{1}{\psi_i} - \frac{1 - \psi_i}{\beta_i} \right) \tag{4.12}$$

where $V_{0,1}$ is the initial velocity of the rocket. In a planetocentric reference frame, $V_{0,1}$ is usually zero, but positive if the rocket is launched from an aircraft.

In what follows, we wish to maximize ΔV by an optimal distribution of the stages, assuming that the characteristic velocities c_i and thrust-to-gravity fractions β_i remain fixed. We define the functions

$$f(\psi_1, \psi_2, \ldots, \psi_n) = \sum_{i=1}^{n} c_i \left(\ln \frac{1}{\psi_i} - \frac{1-\psi_i}{\beta_i} \right)$$

$$h(\psi_1, \psi_2, \ldots, \psi_n) = \ln \Psi - \sum_{i=1}^{n} \ln \psi_i$$

The theoretically optimum stage distribution will therefore occur when f has an absolute maximum, subject to the side condition (4.3). Making use of the method of Lagrange multipliers, expressed by (4.7), it follows from the evaluation of the partial derivatives that

$$\psi_i = \beta_i(1 - \lambda/c_i), \qquad i = 1, 2, \ldots, n \tag{4.13}$$

Finally, substitution into the side condition results in the nth degree polynomial equation

$$\prod_{i=1}^{n} \beta_i(1 - \lambda/c_i) = \Psi \tag{4.14}$$

for the Lagrange multiplier. In turn, the stage masses are obtained from (4.13) and (4.4). The velocity gain of the payload will result from selecting among the real roots of (4.14) the one resulting in the absolute maximum of ΔV.

Table 4.2, case 2, illustrates the results of such a calculation. As in case 1, a two-stage rocket with a payload ratio of 0.100 is assumed. The advantage of the hydrogen–oxygen motor being the last rather than the first stage (stage mass to initial vehicle mass = 0.050 rather than 0.331) is even more pronounced than in the zero-gravity case.

An alternative way to optimize the stage distribution is to maximize the payload's total energy gain, that is, the sum of its kinetic and potential energy. Still another example of stage optimization is formulated in the Problems section.

4.4 Idealized Model of Chemical Rocket Motors

In this section, certain simplifying assumptions will be made to describe solid- and liquid-propellant rocket motors. With these assumptions, a general understanding of the main features of such motors can be obtained.

The theory based on these idealizing assumptions is not sufficient, however, to calculate the performance of such motors to the accuracy required for space mission planning. To obtain this accuracy, recourse must be made to additional theoretical developments — the principal ones will be considered in subsequent sections — and to testing.

The assumptions that will be made presently, roughly in order of their importance, are that (1) the combustion gas satisfies the ideal gas equations and has constant specific heats; (2) the flow through the motor is steady state; (3) the velocity immediately following the combustion zone is low compared with the local sonic velocity, so that the thermodynamic conditions there are the ones of a stagnation point; (4) the flow properties depend only on a single spatial variable ("one-dimensional flow"); (5) heat losses to the boundaries and viscous drag are neglected, hence the flow is isentropic, with the possible exception of gas dynamic shocks that may occur in the supersonic region.

4.4.1 Flow Variables

Downstream of the combustion zone, conservation of the energy of the gas requires that

$$h(x) + \tfrac{1}{2}u^2(x) = \text{const.} = h_0 \qquad (4.15)$$

where x designates the spatial variable in the direction of the mean flow, $h(x)$ the enthalpy, and $u(x)$ the velocity of the gas relative to the motor. The subscript $(\ )_0$ indicates the stagnation condition.

Using conventional thermodynamic notation, it then follows with

$$h = c_p T, \qquad c_p = \frac{\gamma}{\gamma - 1} R$$

where $\gamma = c_p/c_v$ is the ratio of the specific heats and R the gas constant for the particular ideal gas, that

$$u = \sqrt{\frac{2\gamma}{\gamma - 1} R T_0 \left(1 - \frac{T}{T_0}\right)} \qquad (4.16a)$$

A more commonly used form of this equation is obtained by introducing the pressure in place of the temperature. For the assumed isentropic change of state,

$$T p^{-(\gamma-1)/\gamma} = \text{const.}$$

so that

$$u = \sqrt{\frac{2\gamma}{\gamma - 1} R T_0 \left[1 - \left(\frac{p}{p_0}\right)^{(\gamma-1)/\gamma}\right]} \qquad (4.16b)$$

It is often convenient to use the Mach number, $M = u/a$, as the independent variable, where $a = a(x) = \sqrt{\gamma R T}$ is the local speed of sound. Solving (4.16a) for the temperature then results in

$$T/T_0 = \left[1 + \tfrac{1}{2}(\gamma - 1)M^2\right]^{-1} \qquad \textbf{(4.17a)}$$

Making use of the isentropic relation among the thermodynamic variables, one obtains similarly

$$p/p_0 = \left[1 + \tfrac{1}{2}(\gamma - 1)M^2\right]^{-\gamma/(\gamma-1)} \qquad \textbf{(4.17b)}$$

$$\varrho/\varrho_0 = \left[1 + \tfrac{1}{2}(\gamma - 1)M^2\right]^{-1/(\gamma-1)} \qquad \textbf{(4.17c)}$$

where ϱ is the density. Also, from (4.16b)

$$u = \sqrt{\frac{2\gamma}{\gamma - 1} R T_0 \left[1 - \left(1 + \tfrac{1}{2}(\gamma - 1)M^2\right)^{-1}\right]} \qquad \textbf{(4.17d)}$$

4.4 Idealized Model of Chemical Rocket Motors

Finally, the velocity and the thermodynamic variables can be related to the cross-sectional area, $A(x)$, of the flow by means of the conservation of mass equation

$$\varrho(x)u(x)A(x) = \text{const.} = \tilde{m} \tag{4.18}$$

Taking the logarithm, and then differentiating, results in

$$d\varrho/\varrho + du/u + dA/A = 0 \tag{4.19}$$

In turn, from the isentropic relationship between density and temperature, after taking the logarithmic derivative,

$$d\varrho/\varrho = \left(\frac{1}{\gamma - 1}\right)\frac{d(T/T_0)}{T/T_0}$$

whereas taking similarly the logarithmic derivative of (4.16a),

$$du/u = -\frac{1}{2}\frac{d(T/T_0)}{1 - T/T_0}$$

Solving for $d(T/T_0)$, substituting the result into the preceding equation, and using (4.17a) gives for the first term in (4.19)

$$d\varrho/\varrho = -M^2 \, du/u$$

Hence the conservation of mass equation can also be written in the form

$$(1 - M^2)\,du/u + dA/A = 0 \tag{4.20}$$

From this it is seen that to increase the velocity of the flow in the subsonic region ($M < 1$), the cross-sectional area must *decrease*, and it must *increase* in the supersonic region ($M > 1$). The cross section where $M = 1$, hence $dA/dx = 0$, is referred to as the **throat**.

Although it is not directly related to rocket motors, it may be observed here that the condition $dA/dx = 0$ at the throat is merely a necessary but not a sufficient condition for $M = 1$, because, as indicated by (4.20), it may happen instead that $du/dx = 0$.

Flows that lead from a supersonic to a subsonic velocity, hence the reverse of normal nozzle flows, are also possible. However, they are far less stable and typically include gas dynamic shocks. This type of flow is related to the intake flow of jet engines in supersonic flight.

The conditions at the throat are commonly designated by an asterisk. Therefore, with this notation, $u^* = a^*$. The pressure p^* is obtained from

$$p^*/p_0 = \left(1 + \tfrac{1}{2}(\gamma - 1)\right)^{-\gamma/\gamma - 1} \tag{4.21a}$$

as follows immediately from (4.17b).

In applications to chemical rockets, the effective value of γ usually ranges from about 1.20 to 1.40, corresponding to a value of p^*/p_0 of about 0.56 to 0.53. This is the basis for the convenient mnemonic that the pressure at the throat is roughly one-half of the stagnation pressure, independent of all other variables.

Other useful expressions in terms of the conditions at the throat are, from (4.17c), (4.17d), and (4.18)

$$\frac{\varrho}{\varrho^*} = \left(\frac{\frac{1}{2}(\gamma + 1)}{1 + \frac{1}{2}(\gamma - 1)M^2} \right)^{1/(\gamma - 1)} \tag{4.21b}$$

$$\frac{u}{u^*} = \left(\frac{\frac{1}{2}(\gamma + 1)M^2}{1 + \frac{1}{2}(\gamma - 1)M^2} \right)^{1/2} \tag{4.21c}$$

$$\frac{A}{A^*} = \sqrt{\frac{1}{M^2} \left(\frac{1 + \frac{1}{2}(\gamma - 1)M^2}{\frac{1}{2}(\gamma + 1)} \right)^{(\gamma + 1)/(\gamma - 1)}} \tag{4.21d}$$

Figure 4.6 shows the pertinent flow variables, calculated from the preceding equations, for the nozzle contour illustrated in the figure.

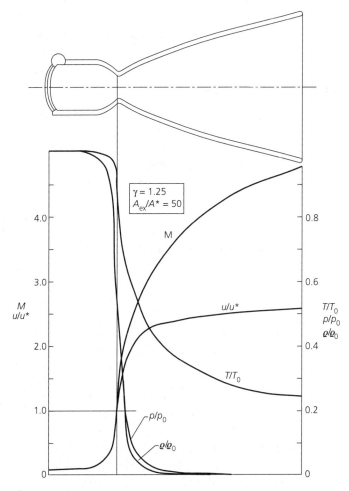

Figure 4.6 Flow variables in a typical rocket nozzle, assuming an ideal gas with ratio of specific heats $\gamma = 1.25$ and a nozzle area ratio $A_{ex}/A^* = 50$.

4.4 Idealized Model of Chemical Rocket Motors

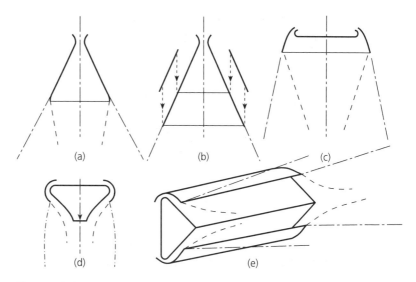

Figure 4.7 Schematics of thrust chambers: (a) bell nozzle; (b) extendable nozzle; (c) expansion–deflection nozzle; (d) axially symmetric aerospike nozzle; (e) planar aerospike nozzle. Dashed lines indicate jet boundaries at sea level, dashed–dotted lines those in vacuum.

4.4.2 Nozzle Contours

Configurations in which the cross-sectional area following the combustion zone first decreases, then increases, are typical of all chemical rocket motors. This concept for obtaining a supersonic velocity was first realized by De Laval (1845–1913), who applied it to the nozzles of steam turbines. In rocket motors, a common configuration is the **bell nozzle**, illustrated in Figs. 4.2 through 4.4 and 4.6.

The simple, one-dimensional analysis discussed in the preceding section applies equally to other configurations, including **extendable nozzles**, **expansion–deflection nozzles**, and **aerospike nozzles** (when without gas injection, also referred to as "plug nozzles"). They are shown schematically in Fig. 4.7. In place of the axially symmetric configurations, planar, two-dimensional geometries are also of interest. The aerospike and expansion–deflection nozzles are designed to obtain a more favorable compromise in nozzle efficiency when the motors must operate at changing atmospheric pressures as the launch vehicle rises through the atmosphere. The extendable nozzles are used on medium-sized motors that operate in the vacuum of space. They make possible large ratios of exit plane to throat areas, ratios that are higher than what is feasible for the large motors of launch vehicles that must operate in the atmosphere.

The combustion gas jet boundary (shown in the figure schematically by dashed lines for operation at sea level and by dash–dotted lines for vacuum condition) can vary greatly during ascent. Section 4.6 contains a fuller discussion of the change in nozzle performance as a function of altitude.

The aerospike and expansion–deflection nozzles, compared with the classical bell nozzle, tend to be somewhat heavier and present greater complexity in cooling. For upper stage motors operating in a near or complete vacuum, bell nozzles are therefore preferred. Especially in the case of small motors, the bell shape is often replaced by a straight cone, with little loss of efficiency.

As is evident from Fig. 4.7, quite abrupt changes in the nozzle contour near the throat are often employed. The purpose here is to keep the overall length of the nozzle short and to reduce the requirement for cooling. Such contour changes are possible without flow separation because the flow in a normally operating nozzle is strongly accelerating, resulting in a pressure drop that energizes the boundary layer.

4.5 Ideal Thrust

A generally applicable expression (2.23) for the thrust of a rocket was obtained in Chap. 2. For the idealized model discussed here, in place of integrating over the exit plane, it suffices to use the mean value, u_{ex}, of the gas velocity. Therefore the magnitude of the thrust of the ideal rocket motor becomes

$$F_t = \tilde{m} u_{ex} + (p_{ex} - p_a) A_{ex} \quad \textbf{(4.22)}$$

The same relation can also be stated in terms of the specific impulse,

$$I_{sp} = \frac{1}{g_0} \left[u_{ex} + \frac{(p_{ex} - p_a) A_{ex}}{\tilde{m}} \right] \quad (4.23)$$

It is noted here that $\tilde{m}$, the mass flow rate through the rocket nozzles, in a number of cases can be slightly less than the time derivative of the vehicle mass. This occurs with engines that use gas generators to drive the propellant feed turbo pumps, after which the gas is dumped overboard. The result is a reduction of specific impulse, typically by 1 or 2%.

In the normal operation of a rocket motor, the second term on the right of (4.22) or (4.23) is considerably smaller than the first term. This second term is often referred to as the **pressure thrust**, the first (and larger) one as the **velocity thrust**.

For a given mass flow rate, it follows from (4.20) with $M > 1$, that the exit plane velocity u_{ex} and hence the thrust become larger as the exit plane area A_{ex} is increased. There is an incentive, therefore, to make this area as large as possible, compatible with weight and size. For rocket motors that operate in the atmosphere, the ultimate size of the exit plane area is also limited by aerodynamic drag.

Already mentioned have been the extendable exit cones. As indicated schematically in Fig. 4.7b, they consist of one or several nested cones that can be mechanically extended in the axial direction and locked to each other. They have been used successfully in space to increase the exit plane area beyond the limitations imposed by the physical size of the rest of the vehicle.

Theoretically, the maximum exit plane velocity that could be obtained (with an infinite exit plane area) would be

$$u_{ex, max} = \sqrt{\frac{2\gamma}{\gamma - 1} R T_0}$$

as follows from (4.17d) with $M \to \infty$. The gain in thrust obtainable from an increased exit plane area can be considerable; for instance, in the case illustrated in Fig. 4.6, the theoretical gain from an infinite exit plane area would be 17%.

As seen from this relation, ideally, for a given mass flow rate, the thrust could be increased if the combustion temperature (i.e., the propellant chemical reaction energy) could be increased, combined with a large gas constant (i.e., a low molecular weight) and a ratio of the specific heats close to 1. In fact, all these factors are interrelated and cannot be optimized simultaneously.

4.6 Rocket Motor Operation in the Atmosphere

When operating in the vacuum of space or in the earth's upper atmosphere, the flow downstream of the nozzle throat is entirely supersonic. Perturbations, as may be induced by the atmospheric back pressure at the exit plane, therefore cannot influence the flow in the nozzle or combustion chamber.

In Fig. 4.8, the curve labeled by "a" represents the ratio of the gas pressure to the stagnation pressure under normal conditions (case of "ideal expansion"). The bell nozzle and the flow variables are the same as those in Fig. 4.6.

In most applications to launch vehicles, it is sufficient to assume that the back pressure at the exit plane equals the ambient atmospheric pressure. The details of the interaction of the flow of air at the base of the vehicle with the combustion gas at the nozzle exit plane are therefore neglected.

If the ambient atmospheric pressure is above a certain limit (which depends on the nozzle geometry and the exit-plane-to-throat area), a gas dynamic shock will form, at first at the exit plane and at still higher ambient pressure in the nozzle interior. Downstream of the shock, the flow will be subsonic. (At still higher ambient pressure, the entire flow would be subsonic, including that at the throat; for rocket motors, this particular condition is of no practical interest, however.)

Figure 4.8 illustrates the several cases that can occur, depending on the ambient pressure. Consistent with the previously made assumption of one-dimensional flow, the shock, if one occurs, would be normal to the mean flow direction.

The flow variables, indicated by the subscript ( )$_1$, immediately preceding the shock, and the flow variables, indicated by the subscript ( )$_2$, immediately following the shock are related to each other by the conservation of mass, momentum, and total enthalpy (enthalpy plus kinetic energy). For ideal gases, they give rise to a large number of alternative, and well-known, equations (e.g., [4]). Some of these are listed in the following for reference.

To compute the data shown in Fig. 4.8, the relation

$$\frac{p_2}{p_1} = \left(\frac{2\gamma}{\gamma+1}\right) M_1^2 - \frac{\gamma-1}{\gamma+1}$$

was used. Because for a prescribed nozzle geometry the flow variables ahead of the shock are known from (4.21), and because the ambient atmospheric

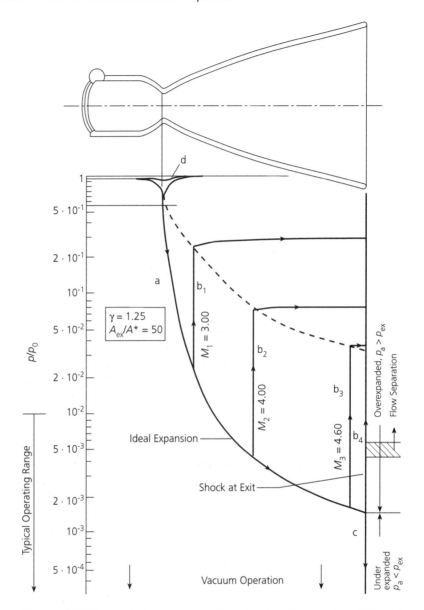

Figure 4.8 Theoretical pressure ratio p/p_0 for the nozzle of Fig. 4.6. (a) Ideal expansion; (b_1), (b_2), (b_3), (b_4) overexpansion with normal shock; (c) underexpansion; (d) subsonic flow throughout.

pressure is also assumed to be known, the location of the shock in the nozzle, as a function of the ambient pressure, can be readily calculated. Thus, the four vertical lines labeled b_1 to b_4 in the figure illustrate the shock pressure rise for four different assumed ambient pressures. The last, b_4, represents a shock that is located at the nozzle exit; it connects the pressure of the ideal expansion to the ambient pressure.

Other well-known relations that connect the flow variables ahead of and behind the shock are the Hugoniot and Fanno equations. The most generally useful equations in applications to rocket nozzles, however, are those that

4.6 Rocket Motor Operation in the Atmosphere

use as the independent variable the Mach number ahead of the shock. They are

$$\left.\begin{aligned}
M_2^2 &= \left(M_1^2 + \frac{2}{\gamma - 1}\right)\left(\frac{2\gamma}{\gamma - 1}M_1^2 - 1\right)^{-1} \\
\frac{p_2}{p_1} &= \frac{2\gamma}{\gamma + 1}M_1^2 - \frac{\gamma - 1}{\gamma + 1} \\
\frac{T_2}{T_1} &= \left(1 + \frac{\gamma - 1}{2}M_1^2\right)\left(\frac{2\gamma}{\gamma - 1}M_1^2 - 1\right)\left(\frac{(\gamma + 1)^2}{2(\gamma - 1)}M_1^2\right)^{-1} \\
\frac{s_2 - s_1}{R} &= \frac{\gamma}{\gamma - 1}\ln\left[\frac{2}{(\gamma + 1)M_1^2} + \frac{\gamma - 1}{\gamma + 1}\right] \\
&\quad + \frac{1}{\gamma - 1}\ln\left[\frac{2\gamma}{\gamma + 1}M_1^2 - \frac{\gamma - 1}{\gamma + 1}\right]
\end{aligned}\right\} \quad (4.24)$$

where s is the entropy.

When the ambient pressure is less than the exit plane pressure corresponding to the ideal expansion, additional expansion takes place downstream of the nozzle. This case is referred to as **underexpansion**. The reason is that if the exit plane area were larger, additional expansion with additional thrust could be had. In the vacuum of space, as a consequence of their finite exit plane area, all rocket motors necessarily operate in the regime of underexpansion.

The thrust, as conventionally defined, is given by (4.22). In the second term, the "pressure thrust," the term $p_a A_{ex}$, which is subtracted from the thrust, simply accounts for the difference of exit plane pressure and ambient pressure at the nose of the vehicle. (Omission of the term would violate the hydrostatic principle that in the absence of thrust the force from a uniform ambient pressure must be zero.)

The velocity thrust, $\tilde{m}u_{ex}$, for a given exit plane area is unaffected by underexpansion. The pressure thrust $(p_{ex} - p_a)A_{ex}$, in an underexpanded flow with a fixed exit plane area is seen to be increased in vacuum compared with operation in the atmosphere.

If the ambient pressure is larger than the ideal expansion exit plane pressure, there will be a gas dynamic shock, first at the exit plane and at still higher ambient pressures in the nozzle interior (Fig. 4.8). This condition is referred to as **overexpansion**. Since behind a normal shock the velocity is subsonic, there will be a further reduction of the velocity in the divergent section of the nozzle and at the exit plane, where the gas pressure matches approximately the ambient pressure. The result will be a diminished thrust.

In case there is a shock inside the nozzle, u_{ex} in the formula is no longer the value obtained from the ideal expansion. Also, the stagnation pressure downstream of the shock differs from that upstream. The exit plane velocity in this case is obtained by using the shock relations (4.24), followed by the ideal compression of the subsonic flow as discussed in Sect. 4.4.1.

In Fig. 4.9, the thrust, F_t, of a motor operating in the atmosphere is compared with the thrust, $F_{t,vac}$, of the same rocket motor in vacuum. The ratio $F_t/F_{t,vac}$ is shown as a function of the ratio of the combustion chamber

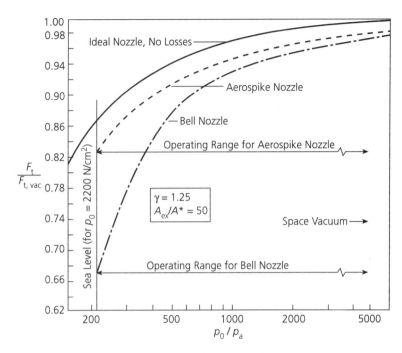

Figure 4.9 Performance comparison of comparable bell, aerospike, and ideal nozzles in the earth's atmosphere (the ideal nozzle as in Figs. 4.6 and 4.8).

stagnation pressure p_0 to the ambient pressure p_a. The data are for the same values of the effective ratio of the specific heats, $\gamma = 1.25$, and of the area ratio $A_{ex}/A^* = 50$ that were used to construct Figs. 4.6 and 4.8. The curve labeled "ideal nozzle" represents the case of one-dimensional flow of an ideal gas without losses.

4.6.1 Comparison of Different Types of Nozzles

Including all losses, typical thrust ratios of aerospike and of bell nozzles are shown in Fig. 4.9 for comparison with the ideal nozzle. The improved performance in the atmosphere of the aerospike (and, similarly, of the expansion–deflection nozzle) in comparison with the bell nozzle is evident.

Also indicated in Fig. 4.9 is the performance at sea level. In this case, an upstream stagnation pressure of 2200 N/cm² (the combustion chamber pressure of the U.S. Space Shuttle main engines) is assumed. It may be noted for comparison that if the same rocket motor and nozzle were operated in the atmosphere of Mars (ambient atmospheric pressure at Mars mean radius = 4.9 mbar = 490 N/m², hence about 0.5% of the pressure at earth sea level), the thrust of a bell nozzle would be close to 99% of the vacuum thrust.

When **bell nozzles** are used in launch vehicles, the throat-to-exit-plane area ratio needs to be chosen as a compromise between what would be optimal for performance in the lower atmosphere and optimal for the higher atmosphere or in vacuum.

At the earth's sea level, for the ideal bell nozzle of Figs. 4.6 and 4.8 and a combustion chamber stagnation pressure of 2200 N/cm², the thrust would

4.6 Rocket Motor Operation in the Atmosphere

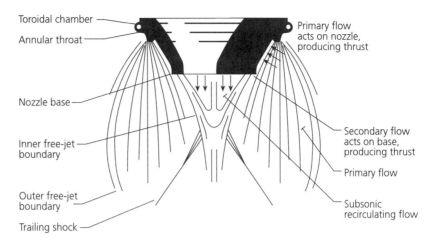

Figure 4.10 Flow field of an aerospike nozzle in the atmosphere. From Ref. 2, Huzel, D. K. et al., "Modern Engineering for the Design of Liquid Propellant Rocket Engines." Courtesy Rocketdyne Division of Rockwell International. Copyright © 1992, AIAA — reprinted with permission.

be 86% of the vacuum thrust. At the density scale height of the atmosphere (i.e., at 9290 m altitude), the thrust would already be 96% of the vacuum thrust.

Again taking as an example the Space Shuttle main engines, their nozzle exit pressure is about 0.08 atm. They do not approach the ideal expansion condition until an altitude of 18 km is reached. The loss of thrust at sea level compared with vacuum is about 20%. This loss of thrust of the main engines is tolerable because the solid-propellant boosters (which are designed for low-altitude performance) provide 80% of the liftoff thrust.

Aerospike nozzles have a more complex flow pattern, as indicated schematically in Fig. 4.10 for the case of an axially symmetric configuration. The combustion gas exits from the annular combustion chamber and throat to form a jet that is bounded partially by the surface of the spike and partially by the separation surface formed with the atmospheric air. The deflection of the combustion gas by the spike produces Mach compression cones that then coalesce into an oblique shock cone. The pressure of the gas on the spike contributes substantially to the total thrust.

Depending on altitude and vehicle velocity, the flow pattern changes and adapts to the variable air pressure. Because the combustion chamber and the throat have a large diameter, the aerospike nozzle tends to be somewhat heavier than a corresponding bell nozzle with its smaller throat diameter. Nevertheless, the aerospike nozzle is often preferred because of its superior adaptability to different atmospheric pressures. Truncation of the aerospike saves weight but produces a wake region with some attendant losses. With proper design, this loss can be kept small.

Expansion–deflection nozzles (Fig. 4.7) have configurations that are intermediate between those of bell and aerospike nozzles. The flow tends to close quickly behind the plug, creating a low-pressure wake. This reduces the thrust significantly, unless either gas from the turbo-pump system or air is bled into the wake.

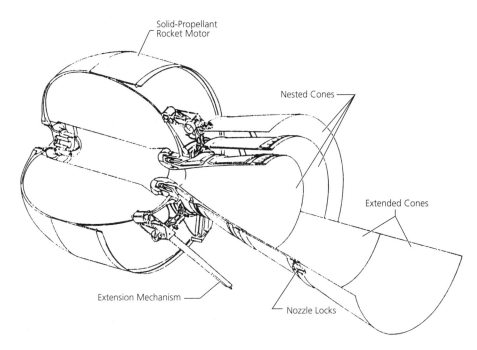

Figure 4.11 Solid-propellant rocket motor with extendable exit cone. (Courtesy of United Technologies Chemical Systems Division, U.S.A.)

Extendable exit cones (Fig. 4.11) have already been mentioned. They are suitable for ablatively or radiatively cooled nozzles; regenerative cooling as illustrated in Fig. 4.2 is evidently not practical.

Engines that are designed exclusively for operating in space in principle would require testing in a vacuum chamber. Because of the practical impossibility of providing ground test facilities with the required pumping rates, large space engines therefore must be tested at atmospheric conditions. If the nozzles designed for space operations were used in such tests, the nozzles might operate in a strongly overexpanded mode with internal shocks and flow separation. It may then be necessary to ground test the engines with nozzles that have an exit-plane-to-throat area ratio smaller than the design value and to rely on analytical methods to predict the thrust that will be obtained in space.

4.7 Two- and Three-Dimensional Effects

4.7.1 Nozzle Internal Shocks

The occurrence of internal shocks in the nozzle at high overexpansion was already mentioned in Sect. 4.4.4. The discussion there, however, was only qualitative. In fact, rather than a normal shock, there tends to be flow separation, coupled to one or more **oblique shocks**. Wakes at approximately ambient pressure will be formed downstream of the line of separation so that the nozzle will no longer flow full (Fig. 4.12). The flow will be unsteady, with a separation line that oscillates and with shocks and wakes that often rotate. The engines cannot be operated in this range of overexpansion,

4.7 Two- and Three-Dimensional Effects

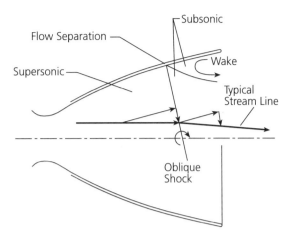

Figure 4.12 Schematic of strongly overexpanded bell nozzle with flow separation and oblique shock.

not only because of the drastic loss in thrust but also because of the often destructive engine vibrations that are induced by the nonsteady flow.

An empirical relation, applicable to bell nozzles, states that flow separation in the nozzle is likely to occur when

$$\alpha(p_a/p_0) > (p_{ex}/p_0) \qquad (4.25)$$

where α varies from about 0.25 to 0.35. This condition is known as the **Summerfield criterion**. The range for which it applies at typical operating conditions is indicated in Fig. 4.8.

Of lesser practical significance are the two- and three-dimensional flow patterns downstream of the nozzle exit plane. If the flow is *underexpanded*, the plume will contain a series of Mach cones (the characteristics of the hyperbolic equation that governs the flow) that interact with each other and reflect at the gas–air boundary. In the *overexpanded* case, oblique shocks, combined with expansion and compression Mach cones, occur that interact with each other and reflect at the jet boundary, forming an approximately diamond-shaped pattern. The gas in the plume may have cooled to the point where, because of partial condensation of the combustion gas, the shocks become visible.

4.7.2 Plume Deflection

At or near the vacuum condition, parts of the plume near the nozzle exit may be so strongly deflected that they impinge on nearby surfaces of the spacecraft. The resulting contamination by the combustion gas can increase the absorption of solar radiation, thereby raising the temperature of the spacecraft. Often more critical is the contamination by the plume of the surfaces of optical scientific instruments and of spacecraft components such as earth sensors.

An estimate of the effect can be obtained by using well-known results from the continuum, supersonic flow of the turning of a gas at the edge of a boundary. Figure 4.13 illustrates this case for an assumed nozzle exit plane Mach number of 2.00 and a ratio of the specific heats of 1.25. When

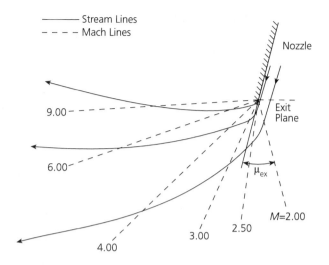

Figure 4.13 Backflow of rocket plume with potential contamination of spacecraft surfaces. Figure is drawn for $M_{ex} = 2.00$, $\gamma = 1.25$.

exhausted into vacuum, some small fraction of the nozzle flow can be deflected by more than 90°.

The fan of lines emanating from the edge of the nozzle are lines of constant Mach number (and therefore also of constant pressure, density, etc.). Also indicated are three typical streamlines. Only the streamlines that issue from the nozzle at points very close to the nozzle boundary can reach nearby spacecraft surfaces.

The angles formed between streamlines and lines of constant Mach number are the Mach angles

$$\mu = \sin^{-1}(1/M) \qquad (4.26)$$

The angle θ through which the flow turns, starting from θ_{ex} at the nozzle exit plane, is given by

$$\theta - \theta_{ex} = \nu(M) - \nu(M_{ex}) \qquad (4.27)$$

where $\nu(M)$, for two-dimensional flow, is the *Prandtl–Meyer* function

$$\nu(M) = \int \frac{\sqrt{M^2 - 1}}{1 + \frac{\gamma-1}{2}M^2} \frac{dM}{M}$$

$$= \sqrt{\frac{\gamma+1}{\gamma-1}} \tan^{-1} \sqrt{\frac{\gamma-1}{\gamma+1}(M^2 - 1)} - \tan^{-1}\sqrt{M^2 - 1} \qquad (4.28)$$

The constant of integration has been chosen such that at $M = 1$, $\nu = 0$. It follows that the maximum value of ν is

$$\nu_{max} = \frac{\pi}{2}\left(\sqrt{\frac{\gamma+1}{\gamma-1}} - 1\right) \qquad (4.29)$$

The corresponding maximum possible turning angle, given by (4.27), in addition depends on the exit plane Mach number.

Conversely, for a given turning angle, the Mach number follows from (4.27) and (4.28). Important for estimates of the level of contamination from the gas flow is the density, which can be calculated from the same relation (4.21b) that was derived for the isentropic flow through the nozzle.

4.7.3 Thrust Correction for the Nozzle Divergence Angle

In principle, nozzles could be designed that have a parallel, uniform gas flow at the nozzle exit. Such designs are used in supersonic wind tunnels. They apply the method of characteristics to the supersonic flow of the gas, usually assumed to be ideal, to construct the wall contour. For rocket propulsion, the theoretical advantage would be that the lateral velocity components at the nozzle exit would be zero. These components do not contribute to the thrust, yet they subtract from the available energy.

In practice, one deviates somewhat from this optimal design. In most bell nozzles, the nozzle contour at the exit plane makes a small angle, called the **divergence angle**, with respect to the axis of symmetry. Typical divergence angles range from 10 to 18°.

There are several reasons for this type of design. One is that nozzles with positive divergence will be shorter and less heavy. Without divergence, a short nozzle would require a pronounced concave curvature next to the exit plane, which could lead to nozzle internal shocks and resultant losses. Also, in solid-propellant motors, a strong concave curvature could lead to excessive erosion of the wall from the impingement of liquid droplets or solid particles, such as aluminum oxide, which are swept downstream by the gas stream.

Shorter nozzles with positive divergence also have the advantage of a lower total heat flux into the nozzle structure. This translates into a reduced requirement for cooling. For these reasons, a design compromise is usually made by allowing a small, nonzero divergence angle.

The resulting loss of thrust is small. A simple estimate can be obtained by assuming that the flow of the combustion gas can be approximated by a spherically symmetric, inviscid flow that is tangential to the wall at the nozzle exit. The magnitude of the exit velocity and the thermodynamic quantities such as pressure and density are therefore assumed to be constant on the spherical cap that spans the nozzle exit.

Let r_{ex} the radius of the cap, S_{ex} its area, ϕ the polar angle, and β_{ex} the divergence angle. Then

$$S_{ex} = 2\pi r_{ex}^2 \int_0^{\beta_{ex}} \sin\phi \, d\phi = 2\pi r_{ex}^2 (1 - \cos\beta_{ex})$$

The mass flow rate is now given by

$$\tilde{m} = 2\pi r_{ex}^2 \varrho_{ex} u_{ex} (1 - \cos\beta_{ex}) \qquad (4.30)$$

Designating the magnitude of the thrust by F'_t, it follows from conservation of momentum that in place of (4.22)

$$F'_t = \int_{S_{ex}} \int u_{ex} \cos\phi \, d\tilde{m} + \int_{S_{ex}} \int (p_{ex} - p_a) \cos\phi \, dS_{ex} \qquad (4.31)$$

where

$$\int_{S_{ex}} \int u_{ex} \cos\phi \, d\tilde{m} = 2\pi r_{ex}^2 u_{ex}^2 \varrho_{ex} \int_0^{\beta_{ex}} \sin\phi \cos\phi \, dt = \pi r_{ex}^2 u_{ex}^2 \varrho_{ex} \sin^2\beta_{ex}$$

$$= \frac{\tilde{m} u_{ex} \sin^2\beta_{ex}}{2(1-\cos\beta_{ex})} = \frac{\tilde{m} u_{ex}(1+\cos\beta_{ex})}{2}$$

The second integral in (4.31) represents the pressure thrust, which, because it is already small in comparison with the velocity thrust, does not need correction. Hence the loss in thrust is given by

$$F_t - F_t' = \tfrac{1}{2} \tilde{m} u_{ex}(1-\cos\beta_{ex}) \qquad (4.32)$$

For example, as this equation indicates, the relative loss of thrust is 0.8% for $\beta_{ex} = 10°$ and 2.4% for $\beta_{ex} = 18°$.

4.8 Critique of the Ideal Model

The most important deficiency of the ideal gas model considered in the preceding sections is the assumption of a thermally and calorically perfect gas. In fact, the composition of the gas, following the combustion, can change greatly as a consequence of the chemical reactions among the various species present. Partial dissociation in the combustion zone and partial recombination of the dissociated species in the downstream parts of the nozzle also play an important role. As the gas rapidly expands, not only will there be a shift in the chemical composition, but also, because the density in the high-Mach-number section of the nozzle is low, chemical equilibrium itself may be lacking.

For these reasons, thermochemical calculations of rocket motors are often performed for two extreme cases that are intended to bracket the true process: one is a calculation based on the assumption of a **shifting equilibrium**, that is, a local chemical and thermodynamic equilibrium that changes from station to station in the nozzle; the other is a calculation based on the assumption of **frozen species**, that is, the assumption that the composition attained after combustion remains the same throughout the expansion.

Another important phenomenon is the excitation, at high temperature, of the vibrational modes of the molecules. One consequence is that the ratio of the specific heats can be quite different from its value at normal temperature. A result of statistical mechanics shows that

$$\gamma = \frac{n+2}{n}$$

where n is the number of degrees of freedom of the molecule. Vibrational excitation provides an additional degree of freedom, hence γ decreases. As is evident from (4.16b) when applied to the velocity, u_{ex}, at the nozzle exit, even a small change of γ can result in a substantial change in u_{ex} and hence in the thrust.

Even if chemical equilibrium may be lacking, it is still possible to define at each location in the thrust chamber a single temperature for all species present. This is because, as a consequence of the high combustion chamber pressure at which modern rocket motors operate, the molecular collision rates are sufficiently high for quick thermalization to take place.

Less critical than the assumption of an ideal gas is the neglect, in the preceding sections, of the heat lost to the walls of the thrust chamber. An estimate of this quantity is essential for judging the structural integrity of the chamber. It directly influences the required rate of cooling or, in uncooled nozzles, the rate of ablation. However, the energy lost in this way by the gas is typically quite small in comparison with the total enthalpy (enthalpy plus kinetic energy). The assumption of a constant total enthalpy that was made in (4.15), therefore, is usually well satisfied.

An exception occurs only in the case of very small motors, particularly those that are pulsed intermittently. Such motors are used for attitude control and station keeping of spacecraft. Their large surface-to-volume ratio and the fact that the walls are cool at the start of each pulse substantially reduce the temperature of the gas. In some cases, the time-averaged specific impulse for a pulse may be lowered from the theoretical value by as much as a factor of 2.

All newly developed rocket motors go through a long series of test firings. In addition to ensuring the structural integrity of the motor, the tests serve to check and improve the theoretical predictions of the thrust and related quantities. Measurements taken for this purpose include among others the thrust, mass flow rate, combustion chamber pressure and temperature, nozzle throat temperature, and nozzle exit total and static pressures.

Small and medium-size motors can be fired on a test stand in an evacuated or — so as to simulate the atmospheric pressure at various altitudes — a partially evacuated chamber. But for large motors, the pumping requirements needed to maintain an approximate vacuum would be so high that testing in a chamber is no longer feasible. These motors therefore must be tested in the open atmosphere. It then becomes necessary to infer the thrust that would occur in vacuum or at altitude by the type of calculations that were discussed in Sect. 4.4.4.

Testing, as necessary as it is, of course does not obviate the need for a theoretical approach. Engineering calculations must go hand in hand with the design. They can eliminate early, less than optimal designs and can greatly reduce the number of tests that are ultimately needed.

4.9 Elements of Chemical Kinetics

To describe with some precision the processes of combustion of rocket propellants and of the subsequent expansion of the reaction products in the nozzle, it is necessary to apply the methods of chemical kinetics [5, 6].

It is beyond the scope of this book, however, to develop this topic to the point that would be needed to carry out calculations in detail. Only a general survey is provided, sufficient to give the reader some appreciation of the work done by specialists.

4.9.1 Chemical Thermodynamics of Ideal Gases

The basic principles that govern the properties of the gas after combustion as it expands in the nozzle are the same for all liquid-propellant and most solid-propellant motors.

The gas in the nozzle can be described as a mixture of chemical species that is homogeneous at any particular station in the nozzle. As the chemical species can react with each other, the composition of the mixture varies as it flows through the nozzle.

The species referred to can be molecules, atoms, or ions. In some cases, rotationally and/or vibrationally excited states are taken into account as separate species.

Sometimes more than a single phase of a species can occur. This happens in solid-propellant motors that contain aluminum as a part of the fuel. In these motors, liquid droplets and solid particles of aluminum oxide occur. This and other similar cases will not be covered in the following discussion.

Because of the relatively high pressures at which rocket motors operate, it can usually be assumed that at a specified station in the nozzle there exists a single temperature common to all species. Also, this temperature is sufficiently high so that the laws governing a thermally perfect (but generally not calorically perfect) gas apply. Therefore, if p is the pressure, ϱ the mass density, T the absolute temperature, R the gas "constant," (not really a "constant" here since it depends on the changing composition of the gas) and e the internal energy per unit mass, all referring to the **mixture** at some point in the nozzle,

$$\left. \begin{array}{l} p = \varrho RT \\ e = f(T) \end{array} \right\} \quad (4.33)$$

As indicated by the second equation, the internal energy of an ideal gas or a mixture of ideal gases is a function of the temperature. As a consequence, the enthalpy, h, and the specific heats c_v at constant volume and c_p at constant pressure of the mixture are also functions of the temperature.

Let $n_1, n_2, \ldots$ be the numbers in mole of the species $S_1, S_2, \ldots$ present in a mixture. It will generally be more convenient to express the quantity of each species by its **mole fraction**, defined by

$$x_1 = n_1 \bigg/ \sum_{i=1}^{k} n_i, \quad x_2 = n_2 \bigg/ \sum_{i=1}^{k} n_i, \quad \ldots, \quad x_k = n_k \bigg/ \sum_{i=1}^{k} n_i \quad (4.34)$$

Hence

$$\sum_{i=1}^{k} x_i = 1$$

The **partial pressure** is defined as the pressure that each gas in the mixture would have if it occupied the whole volume alone at the same temperature. The partial pressures are

$$p_1 = x_1 p, \quad p_2 = x_2 p, \quad \ldots, \quad p_k = x_k p \quad (4.35)$$

The sum of the partial pressures is therefore equal to the pressure p of the mixture (Dalton's law).

Similarly, the entropy of a mixture of ideal gases is the sum of the partial entropies (Gibbs' theorem). It can be shown that the entropy s of the mixture is given by

$$s = \int_{T_0}^{T} c_p \frac{dT}{T} - R \ln \frac{p}{p_0} \qquad (4.36)$$

The internal energy e and enthalpy h of the mixture are

$$e = \int_{T_0}^{T} c_v \, dT, \qquad h = \int_{T_0}^{T} c_p \, dT \qquad (4.37)$$

In most applications, including those in this text, only differences of s, e, and h matter. Therefore the constants T_0 and p_0 are arbitrary, although they are frequently fixed at 25°C and 1 atm.

The quantity by mass of species S_j in the mixture is expressed by the **mass fraction** g_j, $j = 1, 2, \ldots, k$. In accordance with (4.34),

$$g_j = \mu_j n_j \Big/ \sum_{i=1}^{k} \mu_i n_i \qquad (4.38)$$

where μ_j is the molecular weight of species S_j. It follows that

$$\sum_{j=1}^{k} g_j = 1$$

The specific heats and the gas constant of the mixture, when expressed in terms of the corresponding quantities for the separate species, become

$$c_v = \sum_{j=1}^{k} g_j c_{v,j}, \qquad c_p = \sum_{j=1}^{k} g_j c_{p,j}, \qquad R = \sum_{j=1}^{k} g_j R_j \qquad (4.39)$$

4.9.2 Degree of Reaction

Up to this point, the species in the mixtures were assumed to be nonreacting. In what follows, we introduce **reactions** among them, but as before, the discussion is limited to homogeneous mixtures that consist of single-phase species and that satisfy the ideal gas laws. When discussing reactions, the species in the mixture are often referred to as the **constituents** of the reaction.

Reactions are represented by the chemical equation

$$\cdots \nu_{-2} S_{-2} + \nu_{-1} S_{-1} \longleftrightarrow \nu_1 S_1 + \nu_2 S_2 + \cdots \qquad (4.40)$$

The reaction may proceed from left to right, in which case the left-side constituents are referred to as **reactants** and the right-side ones as **products**. Or, the reaction may proceed from right to left, with the designations for reactants and products reversed. The same species may appear on both sides (which then makes the distinction between a reactant and a product moot). Whether the reaction proceeds to the right or to the left, and how far, will depend on the temperature and pressure (or any other independent variables of state) of the mixture. In applications to rocket engines, the constituents are primarily molecules in their ground state or vibrationally and rotationally excited. However, in describing the **intermediate** steps in such reactions, free radicals and atoms often need to be taken into account.

The **stoichiometric coefficients**, $vj = \ldots -2, -1, 1, 2, \ldots$, indicate how the reactants can combine into products, restricted by the requirement that the number of each kind of atom needs to be conserved. Without any change in the meaning of (4.40), the stoichiometric coefficients can each be multiplied with the same arbitrary constant. A convenient convention that will also be used here is to choose the multiplicative constant such that the stoichiometric coefficients assume their smallest, yet integer value.

The numbers of moles of constituents S_j present inside a closed volume at some specified stage in the reaction will be designated by the **mole numbers** n_j, as defined before. In contrast to the stoichiometric coefficients, the mole numbers will be different at the initial, intermediate, and final states of a reaction. In general, they will depend on the variables of state, such as temperature and pressure. Additional complications (not considered here) arise when chemical and thermodynamic equilibrium is lacking.

Mole numbers can also apply to species that do not participate in the reaction, yet are present in the mixture. An example occurs in rocket motors that use liquid hydrogen and oxygen for their propellant. These motors are frequently operated hydrogen rich, that is, with a hydrogen/oxygen mixture ratio that exceeds the stoichiometric one. This lowers the molecular weight of the combustion gas, with the result that the specific impulse generally will be higher (Sect. 4.4.3). If N is the initial mole number of oxygen and N_e the excess mole number of hydrogen, and assuming that the reaction goes to completion, the combustion gas will contain $2N$ moles of water vapor, N_e moles of hydrogen, and no oxygen. Knowing the specific heats and the gas constant of the constituents, the corresponding quantities for the combustion gas then follow from (4.39).

In place of mole numbers, it is often more convenient to use the **mole fractions** defined in (4.34).

The **degree of reaction**, ε, is a measure of how far a reaction, either from left to right or right to left in (4.40), has gone in producing a **specified product species**. This particular species will be designated by S_1 if the reaction of interest goes to the right and by S_{-1} if to the left. The degree of reaction is defined such that $\varepsilon = 0$ before any of the specified product has been made and $\varepsilon = 1$ when at least one of the reactants is exhausted (so that no additional amount of this product can be made).

If ε is known, the mole fractions of all other constituents in the reaction can be found from their stoichiometric coefficients. Illustrated in Table 4.3 is the calculation for the dissociation of water vapor, at high temperature. Hydrogen is taken as the specified product S_1; N_1 is the final number of moles of hydrogen when the dissociation is complete.

4.9.3 Equilibrium and the Law of Mass Action

A mixture is in thermal and chemical equilibrium when at a fixed temperature and pressure there is no change in the amounts of the constituents. This is to be understood at the macroscopic level; on a microscopic scale, there will in general be chemical transformations in both directions, as

Table 4.3 Dissociation of Water Vapor: Mole Fractions as Functions of the Degree of Reaction

Constituent	Stoichiometric coefficients ν	Numbers of moles n	Mole fractions x
$S_{-1} = H_2O$	$\nu_{-1} = 2$	$n_{-1} = N_1(1-\varepsilon)$	$x_{-1} = \dfrac{n_{-1}}{\sum n} = \dfrac{1-\varepsilon}{1+\varepsilon/2}$
$S_1 = H_2$	$\nu_1 = 2$	$n_1 = N_1 \varepsilon$	$x_1 = \dfrac{n_1}{\sum n} = \dfrac{\varepsilon}{1+\varepsilon/2}$
$S_2 = O_2$	$\nu_2 = 1$	$n_2 = \tfrac{1}{2} N_1 \varepsilon$	$x_2 = \dfrac{n_2}{\sum n} = \dfrac{\varepsilon/2}{1+\varepsilon/2}$
		$\sum n = N_1\left(1+\tfrac{\varepsilon}{2}\right)$	

symbolically indicated in chemical equations such as (4.40). At equilibrium, at the macroscopic level, these transformations will cancel each other.

It can be shown that there will be equilibrium if the Gibbs function $G = H - TS$ (H = enthalpy, T = temperature, S = entropy of the system) has a minimum as a function of the degree of reaction.

If the rates of a reaction are very high in comparison with the changes of temperature and pressure, there will be an equilibrium at each moment. This equilibrium will generally differ from instant to instant.

Important examples are the gas flows in the nozzles of certain rocket motors. When the approximation of instantaneous, local equilibrium at each station in the nozzle is valid, the process is referred to as one of *shifting equilibrium*.

In other cases, a better approximation to the nozzle flow is obtained by assuming that downstream of the combustion zone recombination is sufficiently slow that there is no further change of the constituents. The process then is referred to as one of *frozen equilibrium*.

The **law of mass action**, stated next, assumes that the mixture is one of ideal gases in equilibrium. As can be shown, at a specified temperature there exists a so-called **equilibrium constant**, $K(T)$, which is characteristic of the particular reaction being considered. The "constant" in this expression refers to the fact that $K(T)$ is *independent of the pressure*. It depends, however, on the type of reaction and is a strong function of temperature.

The law reads as follows:

$$\left(\frac{x_1^{\nu_1} x_2^{\nu_2} \cdots}{x_{-1}^{\nu_{-1}} x_{-2}^{\nu_{-2}} \cdots}\right) \left(\frac{p}{p_0}\right)^{\nu_1+\nu_2+\cdots-(\nu_{-1}+\nu_{-2}+\cdots)} = K(T) \qquad (4.41)$$

and is valid for an arbitrary number of constituents, with mole fractions $\ldots x_{-2}, x_{-1}, x_1, x_2 \ldots$ and stoichiometric coefficients $\ldots \nu_{-2}, \nu_{-1}, \nu_1, \nu_2 \ldots$. The pressure of the gas mixture is designated here by p, whereas the symbol p_0 refers to a standard pressure, usually taken as 1 atm. [Clearly, if a different standard pressure, say p_0', were adopted, this would merely multiply K by (p_0/p_0') to the power $\nu_{-2}, \nu_{-1}, \nu_1, \nu_2 \ldots$.] The mass action law applies only when there is equilibrium.

Interchanging the algebraic signs of the subscripts results in the law of mass action for the reaction in the reverse direction. Therefore the two equilibrium constants are related by

$$K_1(T)K_{-1}(T) = 1 \qquad (4.42)$$

If the equilibrium constant is known for a specified reaction and temperature, the mass action law provides the additional information needed to determine the degree of reaction and from this the mole ratios of the constituents in the mixture. For instance, for the dissociation of water vapor, substituting into the law the mole fractions calculated in Table 4.3, one obtains

$$\left(\frac{\varepsilon}{1+\varepsilon/2}\right)^2 \left(\frac{\varepsilon/2}{1+\varepsilon/2}\right) \left(\frac{1-\varepsilon}{1+\varepsilon/2}\right)^{-2} \left(\frac{p}{p_0}\right)^{2+1-2} = K(T)$$

hence

$$\frac{\varepsilon^3}{(2+\varepsilon)(1-\varepsilon)^2} \frac{p}{p_0} = K(T) \qquad (4.43)$$

In principle, the equilibrium constants could be determined by quantum mechanics. Except in the simplest cases — which are of little interest for rocket propulsion — such calculations are far too difficult and time consuming, even with the most advanced computers available today. Semiclassical methods are being used, but for most rocket propellants even these methods are of only limited applicability because of their limited accuracy.

Equilibrium constants therefore are determined experimentally. The classical experimental method makes use of flow tubes filled with a gas mixture of known composition. In this tube the gas is heated and flows sufficiently slowly that equilibrium is established at the higher temperature. The gas then enters a capillary tube, where it is cooled very rapidly, in effect "freezing" the higher temperature equilibrium composition. The mole ratios corresponding to the frozen equilibrium are then determined by chemical analysis.

As an example, values of the equilibrium constants for the dissociation of hydrogen and oxygen are listed in Table 4.4 as functions of the temperature. Also listed is the degree of reaction at a pressure of the mixture of 1 atm.

In Table 4.5 are listed the mass action laws for several reactions among two and three constituents, designated by the letters A, B, C, D. Here the mass action laws are expressed in terms of the degree of reaction rather than in terms of the mole fractions as in (4.41). All of these equations, and similar ones for more complicated reactions, can be obtained by the method that was used to derive (4.43) for the dissociation of water vapor. Given the equilibrium constant and the pressure, the degree of reaction is the only unknown and can be obtained as a solution of these polynomial equations. Having found the degree of reaction, the mole fractions are found in the same manner as in Table 4.3 for the dissociation of water vapor.

Table 4.5 also demonstrates an important physical principle: If the sums of the numbers of moles on the two sides of the reaction are the same,

Table 4.4 Equilibrium Constants for the Dissociation of Hydrogen and of Oxygen and Degrees of Reaction at 1 atm

T(K)	H$_2 \longleftrightarrow$ 2H log K	ε	O$_2 \longleftrightarrow$ 2O log K	ε
5000	1.655	.958	1.712	.964
4000	0.449	.642	0.373	.610
3500	−0.410	.298	−0.580	.248
3000	−1.547	.839 10^{-1}	−1.854	.590 10^{-1}
2800	−2.115	.438 10^{-1}	−2.481	.286 10^{-1}
2600	−2.767	.207 10^{-1}	−3.215	.123 10^{-1}
2400	−3.528	.861 10^{-2}	−4.071	.461 10^{-2}
2200	−4.425	.307 10^{-2}	−5.081	.144 10^{-2}
2000	−5.496	.893 10^{-3}	−6.289	.360 10^{-3}

the degree of reaction at a given temperature is independent of the pressure. An increased (decreased) pressure drives the reaction toward the side with the fewer (larger) number of moles, the more so when this relative difference is large.

Table 4.5 The Mass Action Laws for Various Chemial Reactions

Reaction	$K(T) =$	Reaction	$K(T) =$
A $\longleftrightarrow$ C	$\dfrac{\varepsilon}{1-\varepsilon}$	2A $\longleftrightarrow$ C + 2D	$\dfrac{\varepsilon^3(p/p_0)}{(2+\varepsilon)(1-\varepsilon)^2}$
A $\longleftrightarrow$ 2C	$\dfrac{4\varepsilon^2(p/p_0)}{1-\varepsilon^2}$	3A $\longleftrightarrow$ C + D	$\dfrac{\varepsilon^2(3-\varepsilon)}{27(1-\varepsilon)^3(p/p_0)}$
A $\longleftrightarrow$ 3C	$\dfrac{27\varepsilon^3(p/p_0)^2}{(1-\varepsilon)(1+2\varepsilon)^2}$	3A $\longleftrightarrow$ C + 2D	$\dfrac{4\varepsilon^3}{27(1-\varepsilon)^3}$
2A $\longleftrightarrow$ C	$\dfrac{\varepsilon(2-\varepsilon)}{4(1-\varepsilon)^2(p/p_0)}$	3A $\longleftrightarrow$ 2C + 2D	$\dfrac{16\varepsilon^4(p/p_0)}{27(3+\varepsilon)(1-\varepsilon)^3}$
3A $\longleftrightarrow$ C	$\dfrac{\varepsilon(3-2\varepsilon)^2}{27(1-\varepsilon)^3(p/p_0)^2}$	A + B $\longleftrightarrow$ C	$\dfrac{\varepsilon(2-\varepsilon)}{(1-\varepsilon)^2(p/p_0)}$
A $\longleftrightarrow$ C + D	$\dfrac{\varepsilon^2(p/p_0)}{1-\varepsilon^2}$	A + B $\longleftrightarrow$ 2C	$\dfrac{4\varepsilon^2}{(1-\varepsilon)^2}$
A $\longleftrightarrow$ C + 2D	$\dfrac{4\varepsilon^3(p/p_0)^2}{(1-\varepsilon)(1+2\varepsilon)^2}$	A + 2B $\longleftrightarrow$ C	$\dfrac{\varepsilon(3-2\varepsilon)^2}{4(1-\varepsilon)^3(p/p_0)^2}$
A $\longleftrightarrow$ 2C + 2D	$\dfrac{16\varepsilon^4(p/p_0)^3}{(1-\varepsilon)(1+3\varepsilon)^3}$	A + 2B $\longleftrightarrow$ 2C	$\dfrac{\varepsilon^2(3-\varepsilon)}{(1-\varepsilon)^3(p/p_0)}$
2A $\longleftrightarrow$ C + D	$\dfrac{\varepsilon^2}{4(1-\varepsilon^2)}$	2A + 2B $\longleftrightarrow$ C	$\dfrac{\varepsilon(4-3\varepsilon)^3}{16(1-\varepsilon)^4(p/p_0)^3}$

Abbreviated from M. W. Zemansky, in American Institute of Physics Handbook, Chap. 4, Mc-Graw Hill, New York.

4.9.4 Reaction Rates

In the process of expansion in the nozzles of rocket motors, the chemical composition of the gas will differ depending on the station in the nozzle. The expansion can be so rapid that—depending on the type of propellant, pressure, and temperature—complete chemical equilibrium is not achieved, not even locally. The flow properties and the performance of the motor then depend on the **reaction rates** that govern the chemical reactions.

In these reactions there will almost always be intermediate species, for instance, free radicals, that participate. The speed with which chemical equilibrium is reached will then depend on many separate rates for the interaction of these species.

The seemingly simple recombination of hydrogen and oxygen to water in fact is the result of a chain of reactions. Omitting various excited states from the scheme, the most important steps are as follows:

$$\left. \begin{array}{r} H + O_2 \longrightarrow OH + O \\ O + H_2 \longrightarrow OH + H \\ OH + H_2 \longrightarrow H_2O + H \\ H + O_2 + X \longrightarrow HO_2 + X \\ H + H + X \longrightarrow H_2 + X \\ H + OH + X \longrightarrow H_2O + X \end{array} \right\} \quad (4.44)$$

The reactions are seen to be initiated in this case by atomic hydrogen and oxygen and by the formation of OH radicals. (X is an arbitrary species that can serve as a collision partner in a reaction.) There can take place simultaneously **consecutive reactions** and **concurrent reactions**.

Many reaction rates of the intermediate species in chemical rocket motors are only poorly known. Chemical kinetics, although a very old science with a very large literature, is still underdeveloped in terms of engineering needs. The theory is not yet developed well enough to calculate reaction rates *a priori* [6].

Laboratory methods are limited to relatively narrow ranges of temperature and pressure. Engineering applications must then rely on theoretical extrapolations. Especially with new propellants, it is not uncommon that some of the reaction rates of the intermediate species are not known within factors of 3 or more.

When all of the more important rates for the formation of the species, including the intermediate ones, are known at least approximately, a system of differential equations can be formulated for the numerical computation of the overall rate of the reaction. Because the individual rates in a chain can differ from each other by many orders of magnitude, there may be reactions that are particularly slow, which then result in "bottlenecks" in one or more paths of the reaction. If this is the case, simplifications in the system of differential equations are often possible.

4.10 Chemical Kinetics Applications to Rocket Motors

In the theoretical treatment of the reactions and expansion of the gas, starting from the mixture of propellants in gaseous form in the combustion zone and ending at the nozzle exit, one considers a (small) volume containing a fixed mass of gas moving downstream. The concentrations of the mixture's constituents as defined by (4.48) change because of the reactions among them.

In general, there could still be other causes for the concentrations to change. In the presence of gradients, there can be diffusion of the various constituents. In the very small rocket motors used for spacecraft attitude control or station keeping, diffusion to the walls with subsequent catalytic recombination on the walls can play an important role.

In the much larger motors of launch vehicles and upper stage vehicles, diffusion in the core of the flow can be neglected, although it plays an important role in the boundary layer, where it affects the heat transfer to the wall. For calculating performance parameters of large motors, such as the specific impulse, the heat lost to the walls by conduction and radiation and the losses due to shear in the boundary layer are negligible compared with the thermal energy of the gas. For these purposes, isentropic, steady-state flow can be assumed.

Most calculations [7, 8] are based on the assumption that the gas properties and velocity depend only on a single spatial variable ("one-dimensional flow"). Improvements obtained by including three-dimensional effects would in most cases be insignificant compared with the uncertainties produced by chemical kinetics effects.

If important reaction rates are not known or only poorly known, the performance can at least be bracketed by two cases: one is to assume that the chemical equilibrium among the constituents, once established in the combustion zone, remains unchanged in the subsequent expansion ("frozen equilibrium"). The other is to assume that chemical equilibrium is locally established at each station in the nozzle, although the equilibrium composition is allowed to vary from station to station ("shifting equilibrium").

Frozen Equilibrium Calculations

The basic notions are the same as those described in Sect. 4.4. The only difference is in dropping the assumption that the specific heats can be taken to be constant. In rocket motors, because of their high gas temperatures, various vibrational levels of the molecules can be excited. For this reason, the temperature dependence of the specific heats needs to be taken into account in all but the most coarse calculations.

As a consequence of the high temperature, or else low density, of the gas in rocket motor nozzles, the gas mixture can be assumed — to a good approximation — to be thermally (but not calorically) perfect. It therefore satisfies the ideal gas equation.

The specific heats $c_v(T)$ and $c_p(T)$ of the gas mixture are computed from the mole numbers and the constituents' specific heats from (4.38) and (4.39). In general, the latter must be found experimentally, with only partial support from thermodynamic theory. The internal energy and enthalpy of the mixture are calculated as functions of the temperature from (4.37).

The gas constant, R, of the mixture is also calculated from (4.39). Because the composition of the mixture, by assumption, remains the same, $R = c_p(T) - c_v(T)$ is constant throughout the flow in the nozzle.

Based on these approximations, the equilibrium flow in a rocket motor nozzle can be obtained from the set of equations (4.45a) to (4.45c) listed next. It is convenient to choose as dependent variables the temperature $T(x)$, pressure $p(x)$, and flow velocity $u(x)$ as functions of the spatial coordinate x.

Because viscous losses can be neglected and the flow is assumed to be one-dimensional and steady state, Euler's equation in fluid mechanics applies in the form

$$u\,du + \frac{1}{\varrho}dp = 0$$

Expressing the density $\varrho(x)$ by the perfect gas law, this becomes

$$u\,du + \frac{RT}{p}dp = 0 \qquad (4.45a)$$

Since the flow is assumed to be adiabatic, conservation of energy requires

$$c_p(T)\,dT + u\,du = 0 \qquad (4.45b)$$

With $A(x)$ designating the cross-sectional area of the nozzle, conservation of mass (the "continuity equation" in fluid mechanics) is expressed by $\varrho(x)u(x)A(x) = \text{const}$. Taking the logarithmic derivative,

$$\frac{d\varrho}{\varrho} + \frac{du}{u} + \frac{dA}{A} = 0$$

If the density is again expressed through the perfect gas law, this becomes

$$\frac{dT}{T} - \frac{dp}{p} - \frac{du}{u} = \frac{dA}{A} \qquad (4.45c)$$

If the geometry of the nozzle, hence $A(x)$ and dA/dx, is prescribed, all thermodynamic variables of state and the flow velocity can be computed from these equations as functions of the spatial variable x.

The initial conditions that are usually imposed are those in the combustion chamber. The gas velocity there is quite low, usually no more than 10 to 20 m/s, resulting in a very low Mach number. In effect, the conditions there can be equated approximately with the flow's stagnation condition. Following the combustion, equilibrium can usually be assumed in the combustion chamber.

Distinct from the idealized rocket model developed in Sect. 4.4, it is now no longer possible to obtain algebraic relations for the various quantities of interest. Instead, a numerical integration of the set of equations (4.45) is required.

The results are often expressed by stating them in terms of the temperature and pressure in the combustion chamber and of the expansion ratio of the nozzle or else the pressure at the nozzle exit plane (which usually differs from the ambient pressure, as discussed in Sect. 4.4.4). Alternatively, the results may be stated in terms of the ambient pressure. If the ratio of throat to exit plane areas is such that the thrust per mass flow rate is a maximum at the ambient pressure, it is then customary to refer to the specific impulse as the **specific impulse for the optimum expansion** at this pressure. If operating in vacuum, or near vacuum, the specific impulse is referred to as the **vacuum specific impulse**.

Shifting Equilibrium Calculations

Because the gas composition varies as a function of the spatial variable, there is now one additional dependent variable. Chosen for this variable will be the degree of reaction $\varepsilon(x)$.

The mole fractions of the chemical species that are present will vary from station to station in the nozzle. The calculation that was summarized in Table 4.3 for the dissociation of water vapor can easily be expanded to the general case. If, as before, the degree of reaction is defined as referring to the product species S_1, with stoichiometric coefficient ν_1 and mole number N_1 within a closed surface, it follows that the mole numbers of the various species are given by

$$n_i = N_1(\nu_i/\nu_1)\varepsilon, \qquad n_{-i} = N_1(\nu_{-i}/\nu_1)(1-\varepsilon), \qquad i = 1, 2, 3, \ldots$$

and therefore the mole fractions by

$$x_i = n_i / \sum n, \qquad x_{-i} = n_{-i} / \sum n$$

where

$$\sum n = \frac{N_1}{\nu_1} \left[\varepsilon \sum_j \nu_j + (1-\varepsilon) \sum_j \nu_{-j} \right], \qquad j = 1, 2, 3, \ldots$$

Hence

$$x_i = \frac{\nu_i \varepsilon}{\varepsilon \sum_j \nu_j + (1-\varepsilon) \sum_j \nu_{-j}}, \qquad x_{-i} = \frac{\nu_{-i}(1-\varepsilon)}{\varepsilon \sum_j \nu_j + (1-\varepsilon) \sum_j \nu_{-j}},$$
$$i = 1, 2, 3, \ldots \quad (4.46)$$

If the specific heats for the individual constituents are known as functions of the temperature, the specific heats for the gas mixture, $c_v(T, \varepsilon)$ and $c_p(T, \varepsilon)$ and also the gas constant $R(\varepsilon)$ can be obtained from (4.38) and (4.39) as functions of the degree of reaction.

Since, as before, viscous losses are neglected, and the flow is assumed to be one-dimensional, Euler's equation in the form

$$u\,du + \frac{RT}{P} dp = 0 \qquad \textbf{(4.47a)}$$

applies, with the only difference that the variables of state now depend explicitly on ε.

CHAPTER 4 Chemical Rocket Propulsion

Conservation of energy is expressed by

$$c_p(T,\varepsilon)dT + u\,du = 0 \qquad (4.47b)$$

and conservation of mass, as in (4.45c), by

$$\frac{dT}{T} - \frac{dp}{P} - \frac{du}{u} = \frac{dA}{A} \qquad (4.47c)$$

The law of mass action provides the final equation that is needed. As in (4.41), with the dependence of the mole fractions on the degree of reaction now made explicit,

$$\left(\frac{x_1(\varepsilon)^{\nu_1} c_2(\varepsilon)^{\nu_2} \ldots}{x_{-1}(\varepsilon)^{\nu_{-1}} x_{-2}(\varepsilon)^{\nu_{-2}} \ldots}\right)\left(\frac{p}{p_0}\right)^{\nu_1+\nu_2+\cdots-(\nu_{-1}+\nu_{-2}+\cdots)} = K(T) \qquad (4.47d)$$

Table 4.6 Optimum Mixing Ratio, Combustion Temperature, and Theoretical Specific Impulse of Liquid Propellants

		$p_c = 1000$ psi[a] to $p_{ex} = 1$ atm.			$p_c = 1000$ psi[a] to vacuum, with $A_{ex}/A^* = 40$		
Oxidizer	Fuel	$r_{m,opt}$	T_c (K)	I_{sp} (s)	$r_{m,opt}$	T_c (K)	I_{sp} (s)
O_2	H_2	4.13	2758	389	4.83	2996	455
	CH_4	3.21	3278	310	3.45	3308	369
	C_2H_6	2.89	3338	307	3.10	3369	366
	C_2H_4	2.38	3504	312	2.59	3539	371
	RP-1[b]	2.58	3421	300	2.77	3446	358
	N_2H_4	0.92	3149	313	0.98	3164	353
F_2	H_2	7.94	3707	412	9.74	4003	479
	N_2H_4	2.32	4479	365	2.37	4486	430
N_2O_4	MMH[c]	2.17	3140	289	2.37	3143	342
	N_2H_4	1.36	3010	292	1.42	3011	344
	B_5H_9	3.18	3696	299	3.26	3724	359
IRFNA[d]	MMH[c]	2.43	2971	280	2.58	2965	331
	UDMH[e]	2.95	3001	277	3.12	2995	329
H_2O_2	MMH[c]	3.46	2738	285	3.69	2725	337
	N_2H_4	2.05	2669	287	2.12	2663	338
N_2H_4	B_2H_6	1.16	2249	341	1.16	2249	403
	B_5H_9	1.27	2459	327	1.27	2459	390

Calculations based on shifting equilibrium of isentropic, one-dimensional expansion.
[a] 1000 psi = 689.5 N/cm^2.
[b] "Rocket Propellant-1."
[c] Monomethylhydrazine.
[d] Inhibited red fuming nitric acid.
[e] Unsymmetric dimethylhydrazine.
From Rocketdyne Division of Rockwell International, 1992 [2].

Table 4.7 Theoretical Specific Impulse of Solid and Hybrid Propellants. Expansion to Sea Level Pressure[a]

Propellant	I_{sp}
Solids	
PBAN, 18% Al, 70% NH_4ClO_4	265
DB, HMX, 20% Al, NH_4ClO_4	270
DB, 16% Be, NH_4ClO_4	280
PBA, 20% AlH_3, 65% NO_2ClO_4	290
Hybrids	
BeH_2, 81% F_2	382
H_2, O_2, 15% Al	390
H_2, O_2, 29% Li	405
H_2, O_2, 26% Be	456

[a] Expansion from $p_c = 1000$ psi ($= 689.5$ N/cm^2) to $p_{ex} = 1$ atm.
From Sutton and Ross [1].

Substituting here for the x_i the expressions found in (4.46), the law of mass action provides at each step of the integration a polynomial equation for ε as a function of p and T.

The complete set of equations (4.47), together with (4.46), provides the basis for a step by step numerical integration of the system of differential equations for finding T, p, u, and ε as functions of the spatial variable x.

4.10.1 Sample Results for Several Propellants

Calculations of the theoretical specific impulse of many propellant combinations have been carried out and published. In the United States they are largely based on JANAF thermophysical data. Most calculations are based on the assumption of either frozen or shifting equilibrium. Tables 4.6 and 4.7 contain data, based on shifting equilibrium calculations, for selected liquid, solid, and hybrid propellants.

The data in Table 4.6 for liquid propellants apply to mixing ratios (mass of oxidizer to mass of fuel) for which the theoretical specific impulse is a maximum. Separately listed are the performance at sea level and in vacuum (the latter for a nozzle exit plane to throat area of 40).

The data in Table 4.7 for solid and hybrid propellants apply to sea level conditions.

4.11 Liquid Propellants

All rocket propellants need to be handled with extreme care. Oxidizers in contact with combustible materials of all types can cause uncontrolled fires. In contact with the skin, they can cause severe burns. Fuels are similarly hazardous. The combustion products of most propellants are toxic.

Other than low-thrust electric thrusters (and, potentially, nuclear rocket motors), liquid-propellant rocket motors generally have the highest specific impulse. They can be designed for very high thrust, as needed for launch vehicles, but can also satisfy the low-thrust requirements of spacecraft attitude control and station keeping.

Excluding some propellants such as liquid fluorine (an oxidizer) and beryllium (a fuel, combined with liquid hydrogen), which are extremely hazardous, the highest specific impulse can be obtained by the combination of liquid oxygen and liquid hydrogen. An excess of hydrogen above the stoichiometric ratio lowers the molecular weight of the combustion gas, hence increases the specific impulse. As an example, the main engines of the U.S. Space Shuttle use a mixture ratio by mass of oxygen to hydrogen of 6.0 with a nominal specific impulse of 363 s at sea level.

The most common liquid propellant systems use a separate oxidizer and fuel. Such propellants are referred to as **bipropellants**. Distinct from them are **monopropellants**, such as hydrazine (N_2H_4), which release their chemical energy by decomposition, usually by means of a catalyzer. Monopropellants have the advantage of requiring only a single tank and feed but generally have a lower specific impulse. They are used principally on spacecraft. Propellant combinations that ignite at normal temperatures without the need for an igniter are called **hypergolic**.

A useful distinction is also made between **cryogenic propellants**, such as liquid oxygen and hydrogen, and noncryogenic ones. Propellants may be **earth storable**, that is, may be stored for prolonged periods at normal temperature, or **space storable**. The latter are usually defined as having a boiling point at 1 atm higher than 123 K.

In selecting suitable propellants, a number of other considerations are important. They include their fire and explosion hazards, toxicity of the boil-off, and corrosion of materials that are in contact with them. Several of the most common oxidizers, fuels, and monopropellants are described in the following. A more comprehensive list of their physical properties is found in Appendix D.

4.11.1 Oxidizers

Liquid oxygen, which boils at 90.0 K at 1 atm, is one of the most frequently used oxidizers. It is noncorrosive and, except for evaporation losses, can be stored indefinitely. When in contact with organic materials at normal pressure, it usually will not cause spontaneous combustion. However, at higher pressures, in the presence of lubricants and other inflammable materials, it can cause a violent explosion and fire. In contact with the skin, it can produce severe burns.

As with all cryogens, it is necessary to insulate thermally tanks, fluid lines, and valves to reduce the loss from evaporation. When used in large quantities, such as for launch vehicles, liquid oxygen is often economically produced directly at the launch site.

Nitrogen tetroxide (N_2O_4) at normal temperature and pressure is a yellow-brown liquid that produces reddish-brown, highly toxic fumes. In containers made from compatible materials, it is storable at normal temperature for prolonged periods. However, when not protected from contact with air, it will readily absorb moisture and form a strong, corrosive acid. It is hypergolic with many fuels. When in contact with organic materials, for instance wood, it will cause spontaneous ignition.

Nitrogen tetroxide is frequently used as the oxidizer on launch vehicles but also on spacecraft as one of the propellants in bipropellant systems. Its relatively high freezing point of $-12°C$ necessitates care in controlling the temperature, particularly on spacecraft.

Nitric acid (HNO_3) is commercially produced as a base material for fertilizers, dyes, and explosives. When it contains about 2% water, it is referred to as **white fuming nitric acid**. When containing about 5 to 20% nitrogen dioxide, it is a more powerful oxidizer and is then called **red fuming nitric acid** because of its orange to red color. So-called **inhibited nitric acid** (IFRNA) contains 15% NO_2, 2% H_2O, and 1% HF.

All variants are hazardous on skin contact and give off toxic vapors. Inhibited red fuming acid is less corrosive than the other variants and can be used in contact with stainless steel, some aluminum alloys, Teflon, and polyethylene.

4.11.2 Fuels

Hydrocarbons, which can be derived inexpensively from petroleum derivatives, also make useful rocket fuels. Examples are aircraft jet propulsion fuel, alcohol, and kerosene. In the United States, the rocket propellant RP-1 is frequently used. RP-1 in contact with air self-ignites at 240°C. It is compatible with aluminum, steel, copper, Teflon, and neoprene.

Unsymmetric dimethyl hydrazine [UDMH, $(CH_3)_2NNH_2$] is often used mixed with hydrazine (which is described later). Compared with pure hydrazine, it has the advantage of a lower freezing point ($-71°C$). It is toxic but can be stored for prolonged times. Materials that are compatible with it are titanium, stainless steel, several aluminum alloys, and Teflon.

Liquid hydrogen — when combined with fluorine (which, because of its inherent great hazard, has been used only in experimental motors) or liquid oxygen — gives the highest specific impulse that has been achieved with chemical rocket motors. Its main disadvantages are its low boiling point (20 K at 1 atm) and its low density (71 kg/m^3).

The low boiling temperature is the cause of inevitable boil-off during launch operations, even though the cryogenic tanks are thermally insulated as much as the mass of the insulation permits. In this connection, it should be noted that hydrogen–air mixtures explode over a wide range of mixture ratios. In the presence of an ignition source such as an electric spark, at 20°C and sea level atmospheric pressure, the lower and upper limits for deflagration are 4.0% and 76%, respectively, by volume fraction of hydrogen. The range for detonation is only slightly more narrow.

Formation of ice from air moisture on the surface of the launch vehicle's tanks can be a problem because of the increased mass of the vehicle.

Liquid hydrogen can be stored on spacecraft in space only for relatively short periods of time. In principle, liquid hydrogen could also be carried on space missions for much longer periods and be made useful for propulsion on planetary return missions. To limit the loss by boil-off, the vapor in that case would have to be recondensed. However, efficient cryogenic refrigerators of the size required for such missions still await their development.

As a consequence of the low density of liquid hydrogen, tanks are large, several times the size of the tanks that contain the oxidizer. The ratio in tank size becomes larger yet when, to augment the specific impulse, the rocket motors are run hydrogen rich. A small increase in density can be obtained by cooling the hydrogen below its normal boiling point (but so that it can still be pumped) to produce **hydrogen slush**.

4.11.3 Monopropellants

Hydrazine (N_2H_4) is the most commonly used monopropellant for spacecraft attitude control and station keeping. When in contact with a catalyst, it decomposes spontaneously, with a large release of thermal energy. The reaction is described by the two-stage process

$$3N_2H_4 \longrightarrow 4NH_3 + N_2$$
$$\swarrow$$
$$\text{approx. } 40\%$$
$$\swarrow$$
$$4NH_3 \longrightarrow 2N_2 + 6H_2$$

where ammonia is produced partly as an intermediary species in the reaction and partly as an end product.

A useful catalyst is iridium, finely dispersed on a substrate of alumina. The gradual deterioration of the catalyzer bed at the high temperature of the reaction is the main factor that limits the useful life of the thruster. The degradation is more rapid when the thrusters are repeatedly started cold. For this reason, electric heaters are used to maintain the catalyzer bed at temperatures from 100 to 300°C between pulses.

A common material for the construction of hydrazine tanks on spacecraft is titanium. Stainless steels are satisfactory for ground storage, but some steels can react over time with hydrazine and produce a solid residue that, if not filtered out, can plug the small propellant passages on spacecraft.

A disadvantage of hydrazine is its high freezing temperature (2°C), which usually makes it necessary to control the temperature of the tanks, valves, and fluid lines on the spacecraft by electric heaters. Inadvertent freezing can lead to rupture of the tubing, followed by a catastrophic failure of the entire spacecraft.

Hydrazine is a clear liquid, in appearance similar to water. It is toxic and flammable and, when in contact with skin, it will cause severe burns. It is stable up to 150°C and is compatible with titanium, stainless steel, aluminum 304 and 307, Teflon, and polyethylene.

Hydrogen peroxide (H_2O_2, normally at 95% concentration) starts to decompose at 140°C with release of thermal energy. It has found use as a monopropellant for attitude control and also for the propulsion of small, simple rockets.

4.12 Propellant Tanks

The propellant tanks of launch vehicles are often designed as integral parts of the vehicle structure. In addition to withstanding the internal fluid pressure, they must in this case also support the vehicle's static and dynamic launch loads. These include the thrust and the bending of the structure that results from transverse accelerations, steering control, and upper atmosphere winds. Still other such loads are those produced by propellant sloshing and the loads during ground transportation to the launch site.

Some large propellant tanks that have been designed have extremely thin, stainless steel skins that must be pressurized at all times to prevent collapse. More common, however, are tanks constructed from aluminum alloys with skins that are reinforced by internal stringers.

Oxidizer and fuel tanks on launch vehicles are arranged in tandem (Figs. 4.14 and 4.15) to reduce the maximum diameter of the vehicle. This arrangement lends itself to minimizing the aerodynamic drag and effectively transmitting the thrust through the structure.

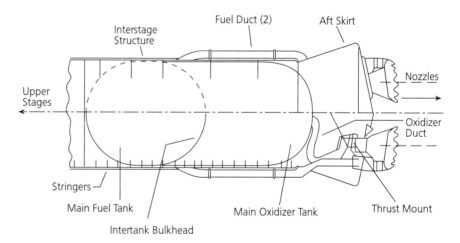

Figure 4.14 Typical tandem arrangement of propellant tanks on a launch vehicle. (Adapted from Ref. 2.)

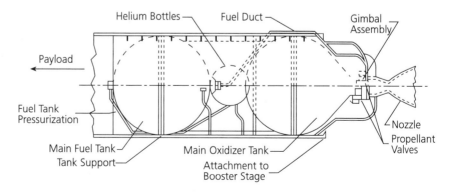

Figure 4.15 Typical arrangement of propellant tanks and bottles for pressurization gas on an upper stage vehicle. (Adapted from Ref. 2.)

The required internal volume of the propellant tanks is the nominal propellant volume at normal temperature, corrected

1. For the volume of the specified propellant **reserve**, in some cases as much as 5%. This reserve must be sufficient to compensate for errors produced by the propellant mixing ratio control, inaccuracies resulting from temperature differences in the propellant loading at the launch site, and fluctuations in the rocket motor performance.
2. For the volume of **trapped propellant**, that is, propellant that is not usable because the geometrical configuration is such that the tanks cannot be emptied completely.
3. In the case of cryogenics, for the volume of the propellant that is **boiled off** in the time between filling the tanks and launch.
4. For the tank's **ullage** (a term borrowed from wine making), that is, the volume that must compensate for the differential thermal expansion of propellant and tank walls and for deflection of the walls under pressure.

4.12.1 Noncryogenic Tanks

Figure 4.14 shows a typical arrangement of the fuel and oxidizer tanks on a lower stage of a launch vehicle. The large, cylindrical shape with ellipsoidal or hemispherical end caps lends itself to structural integration with the vehicle. To save weight, the fuel and oxidizer are often separated by what amounts to a *single* bulkhead, either forged or doubly welded to avoid all possibilities of leakage between the tanks. The tank walls are reinforced by internal stringers or else are fabricated from machined waffle grids. Internal structural reinforcements are also helpful in minimizing the sloshing of propellant induced by vehicle motion. For nonmetallic tanks, filament winding is often a preferred method to obtain the needed strength at minimum weight.

For a fixed volume and weight, spherical tanks have the least shell stress from internal pressure. Fill, drain, pressurization, and vent openings require local strengthening of the shell to compensate for stress concentrations. Spherical tanks, however, preclude the use of their walls as load-carrying members that can transmit the thrust.

Figure 4.15 illustrates spherical tanks arranged in tandem on an **upper stage vehicle**. Support rings around the midsection connect the tanks to the load-bearing skin and stringers of the vehicle. Large tanks are fabricated from welded gores, smaller ones often by forging the two half-shells.

Tank pressures typically range from 100 to 300 N/cm^2 for medium-size and smaller tanks. The pressure is maintained either by gas generators or by helium or dry nitrogen supplied from high-pressure bottles. These in turn are spherical, usually forged from titanium. The initial gas pressures in these bottles may be as high as 3500 N/cm^2 or even higher.

Launch vehicles usually use turbo pump–driven propellant feed systems. This allows the internal tank pressure to be relatively low, ranging typically from 20 to 70 N/cm^2 (absolute) pressure.

The materials used in the construction must be chemically compatible with the propellant. A brief list of such materials, together with the propellant that is in contact with them, is contained in Appendix E. Typical materials are aluminum alloys such as the 2000 and 6000 series used in the United States; steels such as the AISI series 300, A286; molybdenum and inconel; titanium alloys such as 6 Al-4V; and filament-wound composites.

In addition to the fill, vent, and pressurization lines and valves associated with the tanks are such auxiliary devices as pressure relief valves, sensors for the propellant quantity remaining in the tank, pressure and temperature sensors, and fluid flow baffles.

Fracture mechanics is an important theoretical tool in the analysis of high-pressure tanks. Different **safety factors** are applied depending on whether operating conditions are such as to present a hazard to personnel and vital components or do not do so. Tanks are proof tested with high-pressure water. This reduces the danger to personnel if the tank should burst. A commonly applied requirement is to proof test to 1.25 times the maximum expected operating pressure.

4.12.2 Cryogenic Propellant Tanks

The design principles described in the preceding section also apply to tanks intended to contain cryogens. Additional requirements are the choice of materials that are suitable at low temperatures and the need to thermally insulate.

An important required property of construction materials is that they remain ductile at the temperature of the stored propellant. Suitable are some specialty stainless steels and nickel alloys.

To prevent excessive boil-off, thermal insulation on the exterior of the tank is usually needed. For liquid hydrogen it is always needed. In this case, the temperature is so low that ambient air at the launch site will freeze on the exterior of the tank. Thermal insulation adds weight to the launch vehicle (but so does ice formed on the vehicle's surface). The need for and the degree of insulation will depend on such factors as the boiling point of the propellant at the tank's design pressure and the expected ambient conditions at the launch site.

Cork and various foamed materials have been found to be useful materials that not only thermally insulate but also withstand the aerodynamic forces encountered during launch. A typical material is a phenolic-impregnated fiberglass honeycomb filled with thermal insulation material.

The critical importance of insulating liquid hydrogen tanks, when compared with other cryogens such as liquid oxygen, is seen from the following example. The heat of vaporization of hydrogen at 1 atm pressure is $4.61 \; 10^5$ J/kg, that of oxygen $2.14 \; 10^5$ J/kg. The densities of the liquids are 70.8 kg/m^3 and 1131 kg/m^3, respectively. Assuming a stoichiometric mixture and geometrically similar tanks, the ratio of the volumes of the hydrogen to the oxygen tank is therefore 2.01, the surface ratio the 2/3 power of this, hence 1.59. If the thermal insulation is assumed such that the heat transfer rates per unit surface area are the same for both tanks, a simple calculation shows that the time to evaporate a specified fraction of hydrogen compared with the time to evaporate the same fraction of oxygen is only 0.170.

This rapid evaporation of hydrogen is an important consideration in launch preparations. Also, in the case of a delay, the need frequently arises to empty the vehicle tanks, only to refill them again from ground-based tanks for a later launch.

Thermal insulation is also needed for intertank common bulkheads, illustrated in Fig. 4.14. In the absence of insulation, the propellant with the lower boiling point could otherwise cause freezing of the other propellant.

During launch, spacecraft are protected by the launch vehicle's shroud and are therefore not exposed to aerodynamically induced forces. Multilayer thermal blankets (described in Chap. 7) can therefore be used to protect cryogen tanks from solar thermal radiation and from the radiation generated by hot engine parts.

4.12.3 Operation of Propellant Tanks in the Weightless Condition

The location of the drain port of tanks is normally such that acceleration of the vehicle will cause the fluid to move toward the port and cover it. However, during periods of zero thrust when in weightless condition, the propellants may be randomly distributed in the tanks, leaving the drain ports uncovered. If this were allowed to occur, the pressurization gas alone, or mixed with propellant, would enter the propellant feed system each time the rocket motor was started.

Several methods are available to prevent this. Conceptually, the simplest is to use the reaction control system that is ordinarily used for attitude control to provide a small, brief acceleration of the vehicle along the line of thrust, thereby settling the propellant at the drain port. (Clearly, there will be an analogous problem with the reaction control propellant in its own tank; this is discussed later.)

Another method makes use of the surface tension of the propellant relative to a **wire screen** or other mesh-type material located in the tank near the port. The objective is to confine at all times at least a portion of the propellant at the drain port. To mitigate against fluid sloshing, the screen is often placed in a small compartment, open to the propellant at one end.

Especially useful for the smaller tanks on spacecraft are **diaphragms**. They serve to separate the propellant from the pressurization gas. An example of such a tank, used for containing attitude control propellants, is shown in Fig. 4.16. The diaphragm is designed such that when the tank is full, it lies against the tank wall on the gas side. As propellant is being expelled, the diaphragm, in a rolling motion in contact with the wall, moves toward the drain port.

Diaphragms are frequently made from elastomers. Some propellants, particularly nitrogen tetroxide, however, tend to degrade them over long periods of exposure. For this reason metallic diaphragms are often preferred. To prevent corrosion and to facilitate welding of the edge of the diaphragm directly to the tank, an alloy similar to the one used for the tank shell is often chosen. To minimize the pressure difference between gas and propellant and the stress induced in the diaphragm by its bending, metallic diaphragms as thin as 0.25 mm have been used.

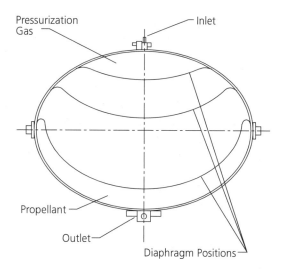

Figure 4.16 Tank with diaphragm for expulsion of propellant in a weightless condition.

In contrast to diaphragms, elastomeric **bladders** contain the entire propellant. They are compressed by the gas that is fed to the space between tank wall and bladder.

4.13 Propellant Feed Systems of Launch Vehicles

In the early days of experiments with liquid-propellant rockets, a very simple system that relied exclusively on gravity and acceleration was used to force the propellants into the rocket motor. The American rocket pioneer Robert H. Goddard, who in 1926 flew the first successful liquid-propellant (liquid oxygen and gasoline) vehicles, demonstrated rockets of this type. Although simple in concept, this system, however, cannot provide the large propellant flow rates that are required by today's heavy launch vehicles. To obtain these, **turbo pumps** are needed.

4.13.1 Feed System Cycles

Shown schematically in Fig. 4.17 are several variations of propellant feed systems for launch and upper stage vehicles. In all of these versions, the fuel and oxidizer are brought by pumps from the relatively low pressure in the tanks to the much higher pressure required for injection into the thrust chamber. This high pressure is also needed to produce a fine spray as the propellants enter the chamber, thereby promoting rapid mixing of the propellants and ensuring stable combustion.

The pumps are driven by high-speed, supersonic gas turbines. A single turbine and the two pumps are sometimes arranged on a single shaft. But more often a gear train between turbine and pumps is preferred. The aim is to reduce the high rotational speed of compact, efficient turbines to the lower speed that is tolerated by fluid pumps.

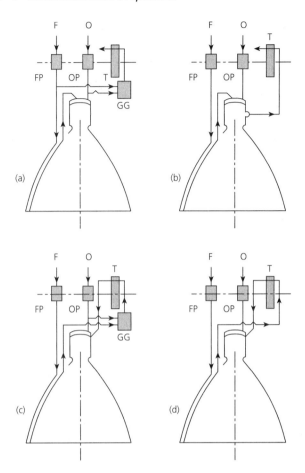

Figure 4.17 (a) Gas generator system; (b) thrust-chamber bleed-off; (c) dual combustion system; (d) cryogenic fuel expander cycle. F, fuel; O, oxidizer; FP, fuel pump; OP, oxygen pump; T, turbine; GG, gas generator.

In version (a) shown in the figure, relatively small flows of fuel and oxidizer are diverted to a **gas generator**, which produces the gas needed to drive the gas turbine. The exhaust from the turbine is dumped overboard in a direction to add to the thrust. It can also be used for injection into the base region, thereby improving the performance of aerospike nozzles (Fig. 4.10).

The main flow of the fuel, before being injected into the thrust chamber, flows through the nozzle tubular structure (Fig. 4.2), thereby cooling it.

Other arrangements are possible, such as separate turbines for each pump, with gas supplied from a gas generator either in series or in parallel. In place of the fuel, the oxidizer has sometimes been used as the coolant.

The schema shown in (b) is similar to (a), except that in place of the gas generator a **hot-gas bleed-off** from the thrust chamber drives the turbine. In (c) is shown an arrangement known as a **dual** or **staged combustion** system. Here, the fuel, after cooling the thrust chamber, is combined in the gas generator with a much smaller flow of oxidizer. The resulting fuel-rich combustion gas is designed to be at a temperature compatible with the turbine's temperature limitation. This gas drives the turbine and then is led to the thrust chamber, where it combines with the balance of the oxidizer. In (d) is shown the **cryogenic fuel expander cycle**. For hydrogen–oxygen propulsion, this cycle has become one of the preferred arrangements. Here the gaseous hydrogen that exits from the nozzle wall tubes expands first in the turbine and then enters the thrust chamber, where it combines with the oxygen.

Common to all systems that are illustrated in Fig. 4.17 is so-called **regenerative cooling** of the nozzle ("regenerative" because the heat added to the propellant is not lost but added to the subsequent combustion). Typically, the propellant is routed through the nozzle tubes from the nozzle exit plane to the thrust chamber injection plane ("single-pass cooling") or else starts at the injection plane and returns to it in adjacent parallel tubes ("double-pass cooling"). The tubes, typically of a nickel alloy, are deformed from their circular cross section to fit the nozzle contour but also to obtain maximum flow velocity, hence maximum heat transfer rate, near the throat, where the heat transfer from the combustion gas tends to be highest.

A consideration that sometimes decides the relative advantages in regenerative cooling of using either the fuel or the oxidizer is potential pinhole-size leaks from the coolant tubes into the thrust chamber. If the combustion gas is fuel rich, as is usually the case for hydrogen, a small leak of fuel into the nozzle interior may be harmless, since the leaking fuel will tend to cool the damaged spot rather than cause a burn-through, as would be likely to occur if the oxidizer had been employed as the coolant.

To supplement regenerative cooling, **film cooling** of the thrust chamber interior surface in the combustion zone is often used. For this purpose, a portion of the fuel is injected in a direction to wet the wall. The resulting fluid film, before its evaporation, can in some cases cover the thrust chamber surface nearly up to the nozzle throat. The fuel flow used in this manner, however, must be kept relatively small so as to avoid incomplete combustion. Film cooling is also often provided to protect the walls of gas generators.

The **turbines** that drive the propellant pumps are similar to aircraft jet engine turbines. The reader is assumed to be familiar with their basic design or may wish to consult a standard reference (e.g., [9]). Figure 4.18 illustrates a typical design of a two-stage turbine for launch vehicles.

To achieve a compact and efficient design of a turbo pump, the turbine tip speed must be considerably higher than the tip speed of the pumps. This condition is often met by introducing **gears** between turbine and pumps. A reduction of the shaft speed from turbine to pump by a factor of about 3 is quite common.

The useful life of turbo pumps on expendable launch vehicles is counted in minutes. The gear loading (torque and speed) can therefore be chosen to be extremely high to ensure a compact and lightweight design. To avoid almost immediate destruction by overheating, a large flow of lubricant is pumped through the gear case. An example of a turbo pump with a two-stage gas turbine and a two-stage reducing gear train driving the fuel and oxidizer pumps on a common shaft is shown in Fig. 4.19.

In designs that omit gears, the turbine diameter is chosen to be substantially larger than the pump diameter. This allows a relatively high tip speed of the turbine in combination with a lower tip speed of the pumps, as needed to avoid excessive cavitation.

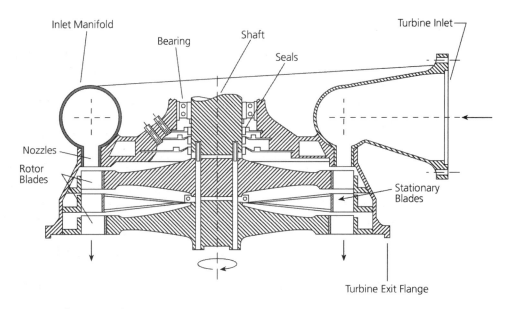

Figure 4.18 Two-stage gas turbine for driving propellant pumps. (Adapted from Ref. 2.)

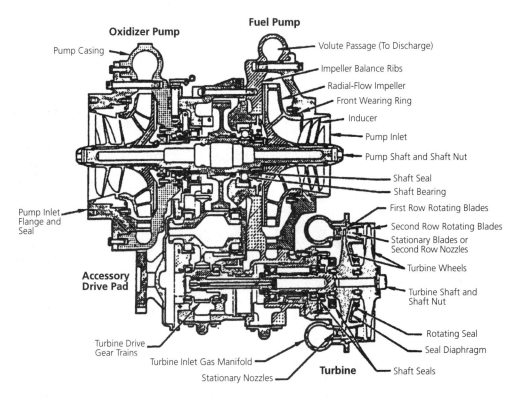

Figure 4.19 Geared turbo pump: oxidizer and fuel pumps on a common shaft driven by a two-stage turbine and two-state reducing gear. From Ref. 2, Huzel, D. K., et al., "Modern Engineering for the Design of Liquid Propellant Rocket Engines." Courtesy of Rocketdyne Division of Rockwell International. Copyright © 1992, AIAA — reprinted with permission.

4.13.2 Propellant Pumps

Because of the occurrence of cavitation, which depends on the vapor pressures of the propellants, pumps on launch vehicles deserve a somewhat more detailed discussion here.

Cavitation refers to the formation of vapor and gas bubbles in the low-pressure regions of the flow of a liquid. Bubble formation can also be excited acoustically, or simply by boiling; what is characteristic, however, of the phenomenon of cavitation is the rapid, and often destructive, collapse of the bubbles as they are transported by the fluid from a low- to a high-pressure region. In the collapse, the enclosed vapor and gas revert to the liquid phase or are absorbed by the fluid. The dynamic forces resulting from the collapse can be large enough to pit and destroy rapidly metallic surfaces in the vicinity.

In addition to its destructive effect, cavitation is deleterious to the mechanical efficiency of pumps and can induce a nonsteady flow resulting in large, possibly destructive, pressure fluctuations in the propellant ducts. An example of such a nonsteady flow is "rotating cavitation," which has some similarity to the rotating stall of compressors.

If a vapor bubble were in a state of equilibrium, the pressure in its interior would be the vapor pressure of the surrounding liquid, modified by the surface tension at the interface. For this reason, a rough estimate for the occurrence of cavitation can be obtained from the relation

$$p \lesssim p_v(T) \tag{4.48}$$

where p is the local pressure of the fluid and $p_v(T)$ its vapor pressure at the operating temperature.

Large departures from this simple criterion can occur in practice [10]. In very pure liquids, produced under laboratory conditions, the fluid can sustain several atmospheres of tension. Therefore the fluid pressure can be substantially below the vapor pressure without bubble formation.

On the other hand, in propellants and other engineering fluids, such large tensions cannot occur because sufficient numbers of nucleation sites will be present either in the fluid or on containing surfaces. Nuclei that are suspended in the fluid can be either small solid impurities or microscopic bubbles formed from gases in the fluid. Gas bubbles that are already present ahead of a pump can grow in size in the pump's low-pressure regions, even though the fluid pressure is well above the vapor pressure. This phenomenon has been called "pseudocavitation."

The vapor pressures of rocket propellants are well known, but the effect of particles and dissolved gases on the critical pressure must be determined case by case in the laboratory.

Cavitation, if present, will tend to occur near the pump inlet and on the suction side of the pump blades, where the pressure is lowest. As a consequence of the curvature in the meridional plane of the flow passage of centrifugal pumps (Fig. 4.20), the most prevalent location in these pumps for cavitation to occur will be near the tip of the blades at the inlet.

Propellant pumps on launch vehicles can be of the axial flow or centrifugal type. The latter are often preferred because they allow a highly compact arrangement and produce a large pressure ratio in a single stage.

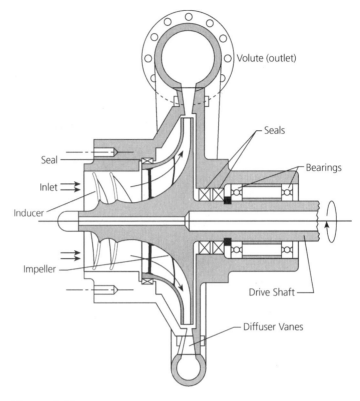

Figure 4.20 Typical propellant pump.

The centrifugal pump illustrated in Fig. 4.20 is quite typical. The fluid enters the pump in the direction parallel to the axis of rotation. It then turns toward an approximately radial direction, thereby increasing the pressure as a result of the centrifugal force. The rotating blades that are attached at their hub to the rotor have an approximately spiral shape. One sometimes distinguishes between the "inducer" and the "impeller" sections. In the former, the flow is nearly axial and the blades are designed to produce only a modest pressure rise. Most of this rise then occurs in the impeller section and in the diffuser, where stationary blades are arranged to turn the flow into the circumferential direction of the volute. The latter is given a cross-sectional area that increases approximately linearly with the azimuth. It discharges the fluid into the high-pressure duct. The pump shown in the illustration has an impeller with a "shroud." Its purpose is to minimize fluid leakage at the blade tips.

An approximate, theoretical method, sufficient for a first estimate of the pressure rise, is based on a model in which the blades and their action on the fluid are replaced by a circumferentially uniform body force (infinite number of blades model). One then considers a streamline, say $\bar{S}$, that is intermediate between the hub and the blade tips and spirals in conformity with the pitch angle of the blades.

If u and p are the fluid velocity and pressure, ϱ the density, and Φ the potential of a conservative force field that acts on the fluid, one form of Bernoulli's equation for the steady flow of an inviscous, incompressible

4.13 Propellant Feed Systems of Launch Vehicles

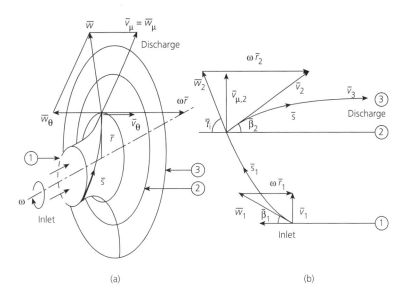

Figure 4.21 Schematic of centrifugal pump. (a) impeller; (b) velocity diagram. (1) inlet; (2) transition from impeller to stationary blades; (3) volute.

fluid is

$$\frac{1}{2}u^2 + \frac{p}{\varrho} + \Phi = \text{constant along streamlines} \quad (4.49)$$

The potential of the centrifugal force associated with the *rotating* parts of the pump is $-\omega^2 r^2/2$, where ω is the angular velocity and r the distance from the axis.

Applied to the axisymmetric surface defined by $\bar{S}$, this becomes

$$\frac{1}{2}\bar{w}_1^2 + \frac{\bar{p}_1}{\varrho} - \frac{\omega^2 \bar{r}_1^2}{2} = \frac{1}{2}\bar{w}_2^2 + \frac{\bar{p}_2}{\varrho} - \frac{\omega^2 \bar{r}_2^2}{2} \quad (4.50a)$$

where $\bar{w}$ is the magnitude of the velocity in the corotating frame of reference and the subscripts ( )$_1$ and ( )$_2$ designate the inlet to the inducer and the exit from the impeller, respectively. (It may be noted here that neither the Coriolis force nor the body force representing the action of the blades on the fluid contributes to the energy increment along the streamlines, hence to Bernoulli's equation, because both forces are perpendicular to the streamlines.)

The various velocity vectors at an arbitrary point on $\bar{S}$, on the rotating parts, are illustrated in Fig. 4.21a. Here ( )$_\mu$ and ( )$_\theta$ designate the meridional and circumferential components, respectively. The velocity components $\bar{v}$ and $\bar{w}$ relative to the stationary and rotating parts, respectively, of the pump are related by

$$\bar{w}_\theta + \omega \bar{r} = \bar{v}_\theta$$

For the *stationary* blades $\Phi = 0$. From (4.49),

$$\tfrac{1}{2}\bar{v}_2^2 + \frac{\bar{p}_2}{\varrho} = \tfrac{1}{2}\bar{v}_3^2 + \frac{\bar{p}_3}{\varrho} \quad (4.50b)$$

where ( )$_3$ designates the condition at the trailing edges of the stationary blades. As a further approximation, this latter condition is taken to be the same as the condition at the discharge end of the volute.

Figure 4.21b illustrates the velocity vectors at the three stations (1), (2), and (3) that are being distinguished in the present approximation. Conventionally, the angles β and γ, which are measured in the plane tangential to the axisymmetric stream surface, are called the "pitch angles" of the flow at the leading and trailing blade edges, respectively. In an actual pump, these angles differ somewhat from the true blade angles because of the effect of the finite number of blades on the cascade's inflow and discharge angles.

The pitch angles of the flow leaving the impeller must be compatible with the pitch angle of the flow entering the stationary blades. As is readily seen from the vector diagrams in the figure, this condition, for the streamline $\bar{S}$, is

$$\cot \bar{\gamma}_1 + \cot \bar{\beta}_2 = \frac{\omega \bar{r}_2}{\bar{v}_{\mu,2}} \tag{4.51}$$

Finally, if $\bar{b}$ is the width of the flow channel in the meridional plane and $\dot{m}$ the mass flow rate, conservation of mass requires that

$$\bar{v}_\mu = \bar{w}_\mu = \frac{\dot{m}}{2\pi \bar{r} \bar{b} \varrho} \tag{4.52}$$

Substitution of (4.52) into (4.50a) results in the expression

$$\frac{1}{\varrho}(\bar{p}_2 - \bar{p}_1) = \frac{1}{2}\left(\bar{v}_1^2 - \frac{\bar{v}_{\mu,2}^2}{\sin^2 \bar{\gamma}_1}\right) + \frac{1}{2}\omega^2 \bar{r}_2^2$$

for the pressure increase from the inlet to the trailing edge of the impeller. Similarly, the pressure increase from the leading edge of the stationary blades to the volute discharge is given by

$$\frac{1}{\varrho}(\bar{p}_3 - \bar{p}_2) = \frac{1}{2}\left(\frac{\bar{v}_{\mu,2}^2}{\sin^2 \bar{\gamma}_1} - \bar{v}_3^2\right) - \omega \bar{r}_2 \bar{v}_{\mu,2} \cot \bar{\gamma}_1 + \frac{1}{2}\omega^2 \bar{r}_2^2$$

Adding the two equations and introducing the additional assumption that the design of the pump is such that the discharge velocity $\bar{v}_3$ equals the inlet velocity $\bar{v}_1$, one obtains the final result for the ideal pressure rise that can be obtained from the pump

$$\frac{1}{\varrho}(\bar{p}_3 - \bar{p}_1) = \omega^2 \bar{r}_2^2 - \omega \bar{r}_2 \bar{v}_{\mu,2} \cot \bar{\gamma}_1 \tag{4.53}$$

High discharge pressures of the propellant pumps are necessary for injecting the propellants into the thrust chamber with a high pressure difference so as to form a spray of small droplets conducive to thorough mixing of the propellants. Also, the thrust chamber itself needs to be at a high pressure so as to avoid overexpansion. On the other hand, the inlet pressures must be chosen by a compromise between the need to mitigate against cavitation and the need to reduce the weight of the tank structure and therefore the tank pressure.

Computational fluid mechanics provides the theoretical means for more accurate calculations of the flow. In particular, the correct geometry of the blade cascade can be taken into account. Results of such calculations reveal strong secondary flows in the space between the blades. These secondary flows are similar to those observed in bent pipes or elbows. Finite Reynolds number effects also can be taken into account.

A requirement common to all propellant feed systems is the need to **start** and **shut down** the rocket motor safely with the least loss of propellant. In systems with gas generators, these are activated before engine start so as to bring the turbo pumps up to speed. During start-up as well as during shutdown, the propellant pumps must operate at off-design conditions. It is important then to avoid conditions that result in massive stall with attendant large pressure fluctuations and vibrations. Some designs therefore incorporate bypass lines and valves that during start and shutdown route parts of the propellant from the pump discharge back to the pump inlet.

Some engine designs allow not only multiple starts and shutdowns but also variations in thrust by **throttling**. Here, too, avoidance of stall of the pumps is an important consideration. The needed control is facilitated if the fuel and oxidizer pumps are driven by separate and separately controlled turbines.

4.14 Thrust Chambers of Liquid-Propellant Motors

An important element of the thrust chamber is the **injector**. In the design illustrated in Fig. 4.22, the oxidizer enters the dome above the **injector plate**, then flows through a large number of small holes in the meridional planes of the injector plate before being injected at high pressure into the thrust chamber. Similarly, the fuel is admitted from an annular manifold through radial holes in the plate. The injection orifices (ports), separate for oxidizer and fuel, cover the entire circular area of the lower side of the injector in the manner of a shower head.

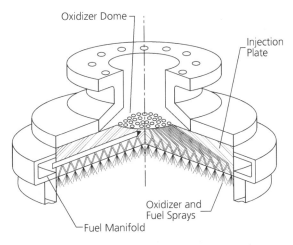

Figure 4.22 Thrust chamber injection plate.

Breaking up the propellant streams into small droplets is aided by the high pressure drop in the orifices. It can be further improved by arranging the angles of the propellant passages in the injector such that the oxidizer jets, and similarly the fuel jets, impinge on their neighbors. Alternatively, the oxidizer jets are made to impinge on the fuel jets and vice versa. In hydrogen–oxygen rocket motors the injection of the gaseous hydrogen, at sonic velocity in the ports, is from openings that surround coaxially the injection ports of the liquid oxygen.

A special problem arises when throttling of the engine over a large range of thrust is required, as is the case in lunar or planetary descent engines. To ensure that the pressure drop at the ports is sufficient to atomize the liquids, movable pintles have been used successfully.

The droplets quickly vaporize by heat transfer from the surrounding combustion products. The gas-phase reaction that follows is aided by the rapid diffusion of the reactants and intermediate combustion products.

Even for propellants that self-ignite on contact, an **igniter** needs to be provided to ensure rapid ignition at engine start. Otherwise, because the pressure in the thrust chamber at the start is low and the reaction may be incomplete, a detonation of unburned propellants in the thrust chamber and beyond could occur.

The minimum needed length of the combustion zone is dictated by the residence time of the propellants that is needed for complete evaporation, mixing, and combustion. Depending on the pressure and type of propellants, this time may vary from 2 to 50 ms. Most of it is required for evaporation and mixing, rather than for the combustion kinetics. Computer models that have attempted to describe in some detail the droplet formation, evaporation, and combustion reaction are useful for extrapolating from known thrust chamber designs to modified designs but otherwise have had only limited success.

The design of thrust chambers requires extensive testing. An important part is the verification that no major, possibly destructive, **combustion instabilities** are present.

Several types of such instabilities have been observed. They have been classified into intrinsic acoustic, injection-coupled acoustic, and low-frequency oscillations.

Intrinsic acoustic oscillations occur at the natural acoustic frequencies characteristic of the chamber geometry. They are therefore associated with sound and shock waves that traverse the chamber at sound speed or higher. Both standing and propagating waves, either longitudinal "organ pipe," radial, or circumferential oscillations have been observed. Potentially the most destructive ones are the circumferential oscillations. Whereas their frequencies can be calculated approximately from standard acoustic theory, much more difficult to predict are the amplitudes, as these can depend on the processes of propellant atomization, evaporation, mixing, and combustion.

Injection-coupled acoustic oscillations depend on the injection velocity, hence on the pressure difference across the injection orifices. Gas pressure fluctuations in the thrust chamber influence the injection pressure difference, hence couple to the propellant injection rate, which in turn induces fluctuations of the rate of propellant burned and therefore of the gas pressure. The stability margins can be improved by injecting at high pressure, at the cost, however, of increased propellant pump power.

Modes of instability in which the combustion chamber pressure is more or less spatially uniform, but fluctuates in time, are referred to as **low-frequency oscillations** or **chugging**. They are caused by the coupling of the chamber pressure with the pressure in the propellant feed system. An important form of low-frequency oscillations, called **pogo** (by analogy with the jumping stick toy), involves, in addition to the thrust chamber and the propellant feed system, the structural modes of vibration, particularly the longitudinal ones, of the entire vehicle. The pogo instability is more amenable to analysis then the high-frequency oscillations that depend intimately on the processes of combustion. It is briefly described in the next section.

Avoidance of dangerous, high-frequency combustion instabilities requires a largely empirical approach. Modifications to the injection pattern, providing baffles, or providing acoustic cavities in the thrust chamber, have been successfully employed.

Small, random pressure fluctuations as high as 5% of the average pressure must be expected even with optimized injection and thrust chamber designs. They are responsible for the very large acoustic output of rocket motors. The resulting sound and shock waves propagate in the atmosphere externally to the launch vehicle but also propagate through the vehicle's structure and cause vibrations and shocks. If unprotected, electronic and other sensitive components either in the vehicle or stationary at the launch site can be damaged.

The sound waves that emanate from the engines and that propagate through the atmosphere at the launch site can even affect components on spacecraft on top of the vehicle, even though they may be partially protected by the payload shroud. The sound pressure levels can reach 140 dB or more, with power spectral densities peaking between about 200 and 2000 Hz.

For vibrations that propagate directly along the vehicle structure, the natural damping by the materials causes the spectra of random vibrations to shift toward lower frequencies, with power spectral densities primarily between 50 and 1000 Hz, depending on the location on the launch vehicle. It is important, for instance in the design of shelves that support electronic hardware on spacecraft, to guard against resonances in this frequency band.

Mention has already been made of the need for **cooling** the thrust chamber. The rate of heat transfer by *convection* from the combustion gas to the thrust chamber walls tends to be highest at the nozzle throat or slightly upstream of it. At this location the combination of the still high gas density and of the already large Mach number (Fig. 4.6) causes a large heat transfer rate.

In comparison with convective heat transfer, *radiative* transfer in liquid propellant motors is smaller. Nevertheless, depending on the type of propellants and size of the motor, radiative transfer can still be significant. At least in very large motors the gas can become partially opaque and in the extreme case would therefore radiate as a blackbody. Two-atom molecules such as hydrogen, oxygen, or nitrogen do not have emission or absorption bands in the wavelength regions that are pertinent for rocket propulsion. On the other hand, heteropolar gases, such as water vapor, carbon monoxide and dioxide, the oxides of nitrogen, and ammonia, have strong bands. Even so, their contribution to the total gas-side heat transfer in liquid propellant thrust chambers rarely exceeds 10%, and less in small motors, where the gas is essentially transparent.

The convective heat transfer rate, $h_c = \dot{q}/\Delta T$ ($\dot{q}$ = heat transferred per unit time and unit area; ΔT = gas-to-wall temperature difference) can be estimated from the slightly modified form of the equation for turbulent flows,

$$\mathrm{Nu} = C \, \mathrm{Re}^{0.8} \, \mathrm{Pr}^{0.34} \quad (4.54)$$

where $\mathrm{Re} = \varrho u d/\mu$ is the Reynolds number referred to the diameter, d, $\mathrm{Pr} = \mu c_p/k$ the Prandtl number, and $\mathrm{Nu} = h_c d/k$ the Nusselt number (u = velocity; ϱ = density; μ = viscosity; k = gas thermal conductivity; c_p = specific heat at constant pressure). The nondimensional constant C depends on geometrical factors. For straight tubes, $C = 0.024$.

ϱ, μ, k, and c_p are obtained from the sum of the mass-weighted amounts of the constituents in the gas mixture. An approximate value for the Prandtl number can also be calculated from the formula

$$\mathrm{Pr} = \frac{4\gamma}{9\gamma - 5}$$

(γ = ratio of the specific heats of the gas mixture) obtained from gas kinetic theory.

For several reasons, (4.54) may underestimate the true heat transfer rate. The addition from radiation has already been mentioned. Recombination at the wall and in the boundary layer of dissociated species will also increase the heat transfer. Semiempirical formulas that are modifications of (4.54) and are based on rocket motor tests have been proposed by Bartz and others.

4.15 Pogo Instability and Prevention

A number of launch vehicles have exhibited an instability referred to as **pogo**. It shows itself as a low-frequency vibration of the entire vehicle, often along its longitudinal axis. The vibration may start some time after launch, grow, and then decay again. Peak accelerations as high as 300 m/s^2 have been observed. The pogo instability and means to suppress it have been studied by Rubin, 1970; Oppenheim and Rubin, 1993 [11]; and others.

The instability arises from the interaction of the vehicle's structural modes of vibration with thrust oscillations. These in turn involve the thrust chamber and the propellant feed system. The analysis is based on the mathematical description of the closed-loop process that includes all of these subsystems.

Simultaneously with the occurrence of pogo, there will also be parametric changes of the vehicle properties, particularly due to propellant depletion and the consequent change in vehicle mass. These, however, are slow compared with the pogo frequencies. The latter can therefore be analyzed as being superposed on a pseudostationary process.

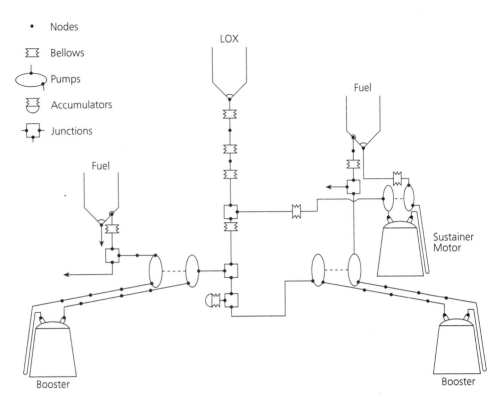

Figure 4.23 Nodes for pogo oscillation analysis. (Adapted from Oppenheim and Rubin [11].)

The analysis starts by defining as discrete nodes all the components that are likely to play a role (Fig. 4.23). These include the thrust chambers, propellant pumps, ducts, their junctions, bellows, tanks and their outlets, hydraulic accumulators, and the principal vehicle structural elements. Each node will receive inputs from, and provide inputs to, some other nodes.

It is important to include the elastic compliance of elements such as tank walls and bellows. More difficult is the proper representation of the contributions from thrust chambers and pumps. These are largely empirical. For instance, the representation of the pumps must include the variability of the flow rate–dependent cavitation in the inducer section and the performance changes caused by variations in the blades' angles of attack. The importance of including among the variables the degree of cavitation has long been recognized [11].

The properties of each node are described by first- and second-order, linear differential equations in the state variables, for which convenient choices are the pressures, flow rates, and structural displacements. This coupled system of equations can be analyzed by classical methods that yield the complex eigenfrequencies and eigenmodes of the system.

It has been demonstrated that the pogo instability can be suppressed by introducing hydraulic accumulators ("pogo suppressors") into the propellant feed system. The accumulators are therefore an essential element in the stability analysis.

4.16 Thrust Vector Control

There are several ways to alter the direction of the thrust for steering the rocket. These methods are jointly referred to as **thrust vector control**.

Most frequently used is a method based on **gimballed thrust chambers** (Fig. 4.24). The gimbals are on two perpendicular axes that allow independent rotations of the thrust chamber. The maximum rotations allowed by the gimbals are typically angles of about ±7°. The rotations, and therefore the vehicle steering, are imposed by actuators, which, for large motors, are usually hydraulic but can also be pneumatic or electric.

To accommodate the motion of the thrust chamber, the propellant ducts must have the necessary flexibility. This is accomplished by a series of bellows, usually made of stainless steel. Bellows are a critical element of gimballed chambers because their skin must be thin enough to offer only limited

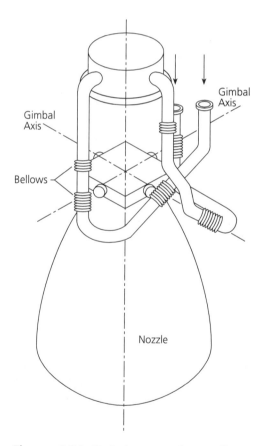

Figure 4.24 Gimbal axes and propellant-line bellows of the LEM engine, NASA. From Ref. 2, Huzel, D. K., et al., "Modern Engineering for the Design of Liquid Propellant Rocket Engines." Courtesy of Rocketdyne Division of Rockwell International. Copyright © 1992, AIAA — reprinted with permission.

resistance to flexing, yet strong enough to withstand the pressure. It is evident that if the motion is limited to, say, ±7° in both directions, the bellows must allow thrust chamber motions of this order of magnitude.

The motions of the hydraulic or pneumatic actuators are controlled by servo-valves, which in turn are controlled by commands from the vehicle's guidance system. Displacement and rate transducers are attached to the actuators. They provide error signals that are the difference between the actual and desired positions and rates. In the control computer, closed-loop control is applied that makes use of the rate signals to ensure stability. In place of rate transducers, digital differentiation of the position signals is also being used.

The figure shows a configuration in which the gimbal axes are placed approximately in the plane of the nozzle throat. Alternatively, they are often placed back of the injection plate and dome (Fig. 4.2).

Figure 4.25 shows the gimbal actuator mounts and articulating propellant ducts in the gimbal plane of the U.S. Space Shuttle main engines.

Another means of thrust vector control is by **lateral injection** into the nozzle. Three such systems are schematically illustrated in Fig. 4.26. In all cases, gas or liquid is injected from the nozzle wall downstream of the throat. A shock front is formed that deflects the main gas stream. The resulting asymmetry of the flow at the nozzle exit plane causes a torque about the center of mass of the vehicle, sufficient to steer it by closed-loop control as commanded by the guidance computer. Four injection ports, each with its servo-valve and spaced 90° apart, are needed. No more than two adjacent injection ports operate at the same time.

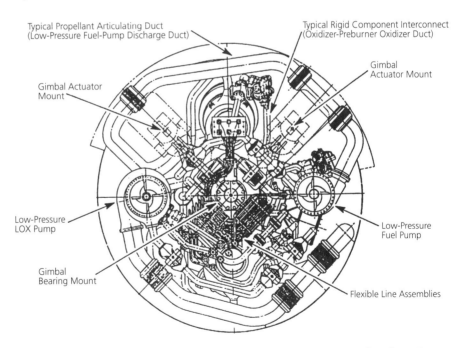

Figure 4.25 Articulating ducts in the gimbal plane of the Space Shuttle main engines, NASA. From Ref. 2, Huzel, D. K., et al., "Modern Engineering for the Design of Liquid Propellant Rocket Engines." Courtesy of Rocketdyne Division of Rockwell International. Copyright © 1992, AIAA—reprinted with permission.

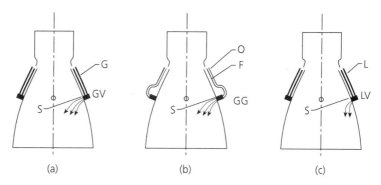

Figure 4.26 Lateral injection systems for thrust vector control: (a) gas chamber tap-off system; (b) bipropellant gas-generator system; (c) liquid system. S, shock front; G, hot gas duct; O, oxidizer; F, fuel duct; L, liquid duct; GV, gas valve; GG, gas generator; LV, liquid injection valve.

In (a), the injected gas is tapped off from the high-pressure region in the combustion chamber. At maximum steering torque, the flow rate of the tap-off gas is typically 1.5 to 2.5% of the primary rocket gas flow. In (b), oxidizer and fuel are combined in gas generators, one for each injection port. A less frequently used scheme in which an inert fluid is injected is illustrated in (c).

Compared with gimballed thrust chambers, a disadvantage of lateral injection is the need to provide ports in a thermally stressed part of the nozzle. This will increase the complexity of providing coolant passages for regenerative cooling. The disadvantage is offset by the simplicity of a fixed mounting of the thrust chamber with rigid propellant ducts.

4.17 Engine Control and Operations

The principal function of the engine control is to ensure the correct flow rates of the propellants. In particular, the flow rates must be such as to result in the optimum mixture ratio for the flight condition at the time ("mixture ratio control"). Also, the flow rates need to be controlled so that toward the end of the firing with the tanks nearly empty, the correct ratio of the remaining fuel and oxidizer is maintained in the tanks ("propellant utilization control").

Many rocket motors are designed for a nominally constant thrust, but others allow the thrust to be varied in response to commands from the flight computer ("thrust control"). The principal components needed for the latter type of engine are indicated in the schematic, Fig. 4.27. Omitted for clarity are various secondary controls such as the tank pressurization controls, safety controls, and the controls needed for start-up and shutdown.

The **propellant utilization control** is based on inputs from transducers that measure the fuel and oxidizer masses in the tanks. Acoustic sensors, capacitance probes, or differential pressure sensors are being used for this purpose. The propellant utilization control is important because there are a number of error sources that can affect the propellant masses actually

4.17 Engine Control and Operations

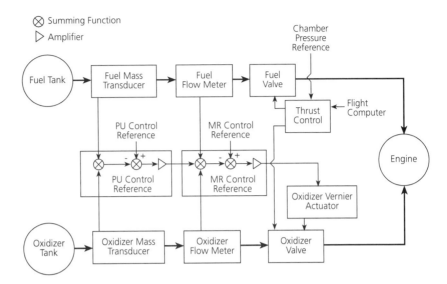

Figure 4.27 Principal components of a typical liquid-propellant control with provision for variable thrust. MR, mixture ratio; PU, propellant utilization.

present in the tanks. These include, for instance, losses of cryogenic propellants from boil-off. Other errors can be present in the initial loading as measured by load cells at the launch site. Without close control, the fuel and oxidizer masses remaining in the tanks toward the end of the firing may significantly deviate from the intended mixture ratio.

The output of the closed-loop propellant utilization control is one of the inputs to the **mixture ratio control**. Additional inputs are those from the flow meters for fuel and oxidizer. The control is designed to maintain a near-optimum mixture ratio, independent of variations of propellant temperature, density, tank pressure, and vehicle acceleration, although modified, if needed, to satisfy the propellant utilization requirement. In particular, the effect of vehicle acceleration on the mixing ratio can be significant, mostly for the propellant in the forward tank because of the long fluid column connecting it to the engine.

Whereas most engine controls are designed to maintain a near-constant mixture ratio, there can be some advantage in purposely modifying the ratio in flight. The heavier propellant is then used at the early parts of the flight at a rate slightly faster than normal and is used slightly less toward the end. The advantage is a small reduction of the gravity loss.

The output from the mixture ratio control is sent to the **vernier actuator**, usually on the oxidizer side. It provides the fine adjustment of the mixture ratio.

Engines that allow throttling also require a **thrust control**. A suitable, although indirect, measurement of the thrust in flight can be obtained from the combustion chamber pressure. A closed-loop control then governs the positions of the main fuel and oxidizer control valves.

Other control schemes are also being used. In all cases is it important to consider the dynamic properties of the components that are being regulated. They can often be characterized by their time delay, for instance, the closing time of valves. Also significant can be the rotational inertia of the turbo pumps and the delay in the pressure buildup of gas generators.

A frequently used type of control is represented by the "proportional–integral–differential" control law

$$y - y_0 = k_1 \varepsilon + \frac{k_2}{\tau_I} \int_0^t \varepsilon \, dt + k_3 \tau_D \frac{d\varepsilon}{dt} \tag{4.55}$$

with y as the control variable. Here, ε is the error term; k_1, k_2, k_3 are gains; τ_I is the integration time; τ_D is the differentiation time; and y_0 is the desired value of the control variable. The addition of the integral term in this equation eliminates the offset inherent in simple proportional controls (but may also cause overshoots). The addition of the differential term provides a faster transient response to rapidly varying conditions.

Typical start-up and thrust cutoff operations are indicated in Fig. 4.28. At **engine start**, depending on the type of propellant and the engine cooling method, either the fuel or the oxidizer flow may lead. Safety considerations are essential elements of engine start and engine shutdown. Usually, the propellant cutoff at **engine shutdown** is such that the condition in the thrust chamber is made fuel rich. Precise thrust cutoff at the time commanded by

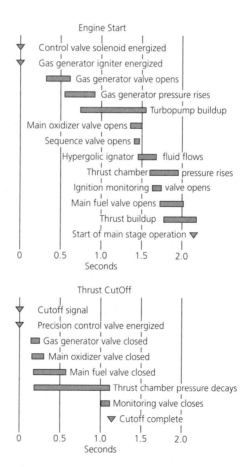

Figure 4.28 Typical sequence of liquid-propellant motor start and cutoff.

the flight computer is particularly important for upper stage motors when the exact instant of shutdown dictates the trajectory after final orbit insertion.

Omitted from the diagram are a number of secondary operations, such as purging and — for cryogenic propellants — chill-down that must precede engine start.

Launch vehicles and upper stage vehicles are often equipped with their own flight computers. It is also possible, and often advantageous, to combine all computers into a single one located in the uppermost stage or even — in case a single spacecraft is being launched — in the spacecraft itself.

4.18 Liquid-Propellant Motors and Thrusters on Spacecraft

Liquid-propellant rocket motors on spacecraft are of two somewhat different types, mostly distinguished by their thrust.

In the first category are the **spacecraft main propulsion motors**. Their applications include final orbit insertion (unless this task is taken over by the final upper stage vehicle), retrofiring prior to a lunar or planetary landing or earth return, and ascent from the surface of a planet or moon.

In a second category are smaller motors, often referred to as **thrusters** or **reaction control motors**. Typical applications are attitude (orientation in space) control of spacecraft and station keeping (keeping the same station on the geosynchronous orbit by correcting for perturbing forces). Other applications are the spin-up and spin-down of spin-stabilized spacecraft and midcourse corrections on deep-space missions.

The boundary between these two categories is fluid. A useful distinction that affects the level of thrust and therefore the size of the engine can be made between maneuvers in a gravitational field and those in a near-weightless environment. In the first case, the time allowed for firing is relatively short, usually counted in tens of seconds or a few minutes. To achieve a desired total impulse, the motor must therefore be relatively large. In the second case, such as in orbit corrections and in several applications to deep-space missions, the period of thrust can be quite long because there will be no gravity loss. Thrust periods of the order of an hour or more are possible. Hence a much smaller motor will be adequate to achieve the needed total impulse.

Except for size, liquid-propellant spacecraft motors are conceptually similar, yet simpler than launch vehicle motors. The propellant flow rates are lower and the tanks are smaller and can be designed for higher pressure. In most cases this obviates the need for turbo pumps. Instead, the simpler system can be used that involves pressurized gas expulsion of the propellants from the tanks.

As an example of a spacecraft **main propulsion motor**, one of the NASA Viking engines is shown in Fig. 4.29. This engine was one of three identical terminal descent engines, spaced equally around the circumference of the craft, and activated shortly before landing on the Mars surface. The propellant load was 85 kg hydrazine, contained in two titanium spheres pressurized with dry nitrogen at an initial pressure of 360 N/cm^2. The deceleration of the

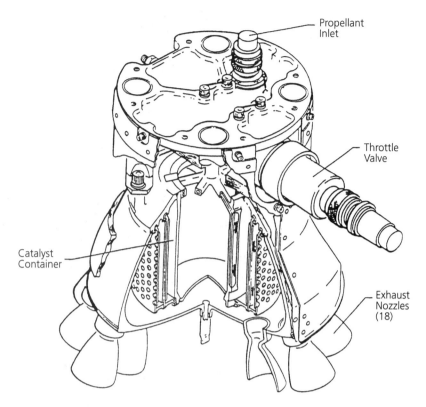

Figure 4.29 Terminal-descent motor for Viking Mission. (From Holmberg, Faust, and Holt, NASA Reference Publication 1027.)

craft was sufficient for the propellant to cover at all times the tank exit ports. Therefore no diaphragms were needed. Thrust variation was provided by a motor-driven throttle valve that controlled the hydrazine flow. The maximum thrust was 2650 N. To avoid generating a dust cloud at the landing site, the exhaust of each engine was divided among six nozzles.

By way of contrast, Fig. 4.30 illustrates a **reaction control motor (thruster)**. Although monopropellant motors are often preferred for their greater simplicity, bipropellant motors, such as the one shown, have a larger specific impulse. The propellant flow rates are controlled by solenoid valves. Cooling is partially by ablation of the carbon–phenolic nozzle and its graphite insert and partially by radiation.

Thrust vector control is also greatly simplified, because the desired result can be achieved by changing the attitude of the entire spacecraft either before or even during the firing of a main motor.

Small **thrusters** are useful for attitude control and station keeping. To allow rotation of the spacecraft about three axes and, independently, acceleration along these axes, a minimum of 12 independently controlled thrusters are needed (four for each axis, since two separate thrusters are needed for oppositely directed thrusts). For redundancy and greater simplicity of the controls, usually more are used.

In most cases these motors need to operate only outside the earth's atmosphere. Their nozzle expansion ratios can therefore be high, without

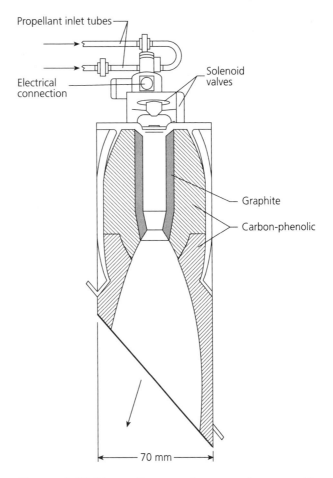

Figure 4.30 Bipropellant reaction control motor with ablation cooling.

problems from overexpansion (Sect. 4.4.4). The larger motors are often capable of restart and throttling, whereas the reaction control thrusters are designed for a large number of low-thrust, short pulses. Because of the dead period between pulses, the cooling requirements are less stringent than would be the case for continuous operation. Sufficient cooling, without the need for active cooling by convection, is obtained from radiation and from the ablation of the nozzle materials, such as carbon–phenolic or carbon–carbon. In some cases, nozzles have also been manufactured from highly refractory metals, either tungsten or columbium.

The pulse duration of reaction control thrusters may vary from a fraction of a second to as much as minutes (although more rarely). Typical thrust levels are between 5 and 100 N. The total number of pulses required during the life of a three-axis stabilized spacecraft may be as much as 1 million or more. Because many spacecraft, particularly communication satellites, require attitude and station keeping control during their entire useful life, which may be 10 years, the initial propellant mass becomes a major fraction of the spacecraft mass — in some cases as much as one-half of the total. To reduce the requirement for propellant, bipropellants with their higher specific impulse are often preferred over the simpler monopropellants. However, when operating in short pulses, the specific impulse is significantly reduced compared with steady-state operation. The loss of enthalpy of the gas to the thrust chamber walls is no longer negligible, not only because of the small volume-to-surface ratio of these motors but also because at the beginning of each pulse the walls are still cold, hence the heat transfer high. Whereas the specific impulse of small hydrazine motors in continuous operation ranges from about 220 to 235 s, it may be only about 150 s when operated in short pulses.

Because spacecraft rocket motors must reliably function even after many years of exposure to the space environment, thermal protection of the tanks, valves, and propellant lines is required. In many cases electric heating must be provided to prevent freezing in deep-space missions or during solar eclipses.

For the propellant in the weightless environment to cover the tank exit ports, the tanks must be equipped with bladders or diaphragms. Because of the small flow rates that are typical of reaction control systems, the need to cover the exit port can also be met by capillary feeds. An example is shown in Fig. 4.31, showing one of the propellant tanks of the NASA Space Shuttle reaction controls.

Figure 4.32 illustrates a bipropellant reaction control system that uses high-pressure helium for the **propellant feed**. Each of the 16 motors is provided with a pair of injection valves, one for the fuel and one for the oxidizer. Together, the motors control the pitch, yaw, and roll of three-axis stabilized spacecraft. Redundancy is obtained by providing two parallel systems. Each can operate independently, yet, by command, the two systems can also be

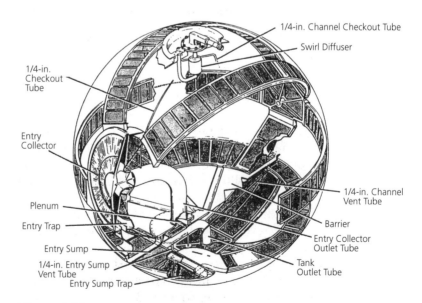

Figure 4.31 Space Shuttle reaction control propellant tank with capillary feed, NASA. (Courtesy of Lockheed Martin, U.S.A.) From Ref. 12, Brown, C. D., "Spacecraft Propulsion," AIAA Education Series. Copyright © 1996, AIAA — reprinted with permission.

4.18 Liquid-Propellant Motors and Thrusters on Spacecraft

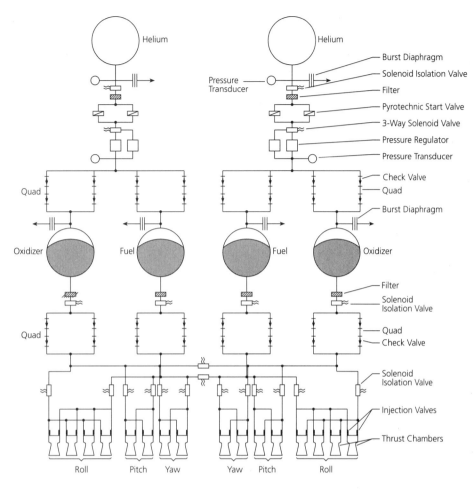

Figure 4.32 Schematic of a bipropellant reaction control system for a three-axis stabilized spacecraft. (Omitted are the fill and purge ports.)

cross-linked so that the propellant of one system can also be used by the motors of the other.

This type of redundancy is analogous to what is common practice in the design of electronic systems. An important consideration is that cross-linking must not introduce new failure modes. In the schematic shown in the figure, this is accomplished by introducing multiple isolation valves. They make it possible, if needed, to separate the two sides of the system and also to isolate from each other the pitch, yaw, and roll motors. The valves are solenoid controlled and are either fully open or fully closed. They are governed by the spacecraft computer in response to the need for thrust or to redirect the propellant flow in case the pressure sensors indicate a failure in a part of the system. Such commands by the spacecraft computer can be overridden by ground command.

Other features include pyrotechnically operated valves. They can be operated only once but have the advantage of being absolutely tight against leaks. They start the flow of the pressurization gas by bursting their diaphragm. This type of valve will stay leak free for many years, a feature that is particularly important when the system will not be needed until much later in a long-duration mission, as is the case with planetary descent or ascent vehicles. To guard against a pyrotechnic valve not opening on command, two parallel valves are provided on each side of the system.

Because the helium pressure decreases as the propellants are being forced from their tanks, pressure regulators, again duplicated for redundancy, are located between the helium and propellant tanks. They reduce the pressure in the propellant tanks to the prescribed motor injection pressure.

Backflow of helium or of propellant is prevented by check valves. These are arranged as quadruplets in an arrangement that is both in parallel and in series. (If one of the valves is stuck closed, the parallel string will still permit flow in the proper direction; if one does not close in spite of backflow, the valve that is in series will do so.)

Burst diaphragms are placed on the helium side. They serve to prevent more serious damage if a pressure regulator should fail or if errors are committed in ground operations.

4.19 Components of Solid-Propellant Rocket Motors

In solid-propellant motors the fuel and oxidizer, together with a binder and other additives, are premixed and contained as a solid mass in the motor case. The specific impulse is lower than that of most liquid-propellant motors, but this disadvantage is compensated by greater simplicity and lower cost.

Solid-propellant motors are particularly well suited for boosters slung to the sides of launch vehicles and for rocket upper stages. When used as boosters, their lower specific impulse is acceptable because by their added thrust they reduce the gravity loss otherwise typical of massive launch vehicles.

4.19.1 Propellants

So-called **double-base** propellants are a homogeneous mixture, typically a nitrocellulose explosive dissolved in nitroglycerin plus some small amounts of additives. Each of these components is both a fuel and an oxidizer (therefore the designation "double base"). They are explosive both separately and together.

Composite propellants are mixtures of oxidizer crystals, often ammonium perchlorate, and a fuel in powder form, mixed with a binder such as polybutadiene. The processing is somewhat less hazardous than it is with double-base propellants.

Also in use are combinations of these two types. They are then called **composite double base**. They consist of a crystalline oxidizer and a powdered fuel in a matrix of nitrocellulose–nitroglycerin.

Among the additives are **binders** (which often also participate as fuels in the combustion), curing agents, and catalysts. The latter are important for improving the rheological properties of the mixture and for controlling the burn rate.

At slightly elevated temperatures the premixed propellants turn into highly viscous fluids that can be cast into the motor case. After curing, and back at normal temperature, the propellants form a solid, rubber-like mass. Long-term storage requires a temperature-controlled, low-humidity environment.

At low temperature, as may be encountered after an extended period in space, solid propellants may become brittle. Cracks or separation from the motor case may occur, probably with catastrophic consequences at the time of firing. The propellants therefore must be protected against the large temperature swings and the low temperatures frequently encountered after launch.

The hazards encountered with solid propellants are fires and explosions, particularly in the processing and casting operations. These must be carried out with stringent controls. Once cast and cooled to normal temperature, the principal hazard is from electric sparks. As a precaution, the motor case and tools in contact with the propellant need to be electrically grounded.

Oxidizers

The most frequently used oxidizer is **ammonium perchlorate** in the form of small, white crystals. It is chemically compatible with most fuels. Combined with high-energy fuels, it results in a high specific impulse. Like all other perchlorates, it produces hydrogen chloride in the combustion gas, rendering it toxic and corrosive.

Ammonium nitrate and other nitrates have the advantage of producing combustion gases that are smokeless and relatively nontoxic but result in a lower specific impulse in comparison with ammonium perchlorate. Ammonium nitrate is often used in the gas generators that drive the turbo pumps of large liquid-propellant rocket motors.

A high-performance oxidizer is **sodium perchloride**. Its use as an oxidizer for rocket propulsion is limited by its hygroscopic nature.

Fuels and Binders

Powdered **aluminum** has become an important fuel. Typically it constitutes up to 20% of the weight of the propellant. Upon combustion, it forms a slag of aluminum oxide, which is liquid at the motor temperature. Some fraction of it is retained in the motor case, particularly when reentrant nozzles are used. Retained slag does not contribute to the thrust, hence results in a (small) decrease of the specific impulse.

Beryllium is much lighter than aluminum and, by comparison, can increase the specific impulse by about 15 s. Its use has been impeded by concerns about human health, because inhalation of beryllium powder is highly toxic. **Beryllium hydride** could increase the specific impulse still further by about 25 s. However, it deteriorates quickly through chemical reaction during storage and is therefore rarely used.

Binders are most often polymers, such as several types of polybutadiene or plastisol binders such as polyvinyl chloride. They are essential to the safety and reliability of the motor. In the process of curing they determine the rheological properties of the final product by their cross-linking and branch chaining. The binders participate as a fuel in the combustion process and thereby add to the enthalpy of the combustion gas.

4.19.2 Burn Rate and Ignition

The burn proceeds along the entire exposed surface, essentially at a spatially uniform rate. Given the initial propellant ("grain") surface, the burn surface at a later time into the burn can be found by applying at each time step a uniform recession distance perpendicular to the preceding surface.

The **burn rate**, b, which is the rate at which the burn surface recedes, depends primarily on the gas pressure, p_{ch}, in the chamber ("bore"). Because the Mach number of the flow is low, in most cases p_{ch} is essentially the same as the stagnation pressure. In turn, p_{ch} is dictated by the instantaneous area of the burn surface relative to the area of the nozzle throat.

The burn rate increases with pressure. Empirically it is found that it can be expressed by an exponential law of the form

$$b = c\,(p_{ch}/\bar{p}_{ch})^n \tag{4.56}$$

where n is referred to as the **burn rate pressure exponent**. Depending on the propellant, n can vary from as low as 0.2 to as high as 0.8. Propellants with n close to 1 are hazardous, because the steep increase of the burn rate with pressure can lead to a runaway condition of gas generation and pressure.

It is convenient to choose for $\bar{p}_{ch}$ in (4.56) a constant pressure, such as 100 atm, which is in the midrange of rocket motor pressures. (Different choices merely affect the factor c, not the burn rate pressure exponent.)

The factor c depends strongly on the propellant. Adding a catalyst or reducing the oxidizer particle size can greatly increase c and therefore the burn rate. In addition, c depends somewhat on the initial propellant temperature. It is lower when the propellant is cold, say $-20°$C, higher when warm, say $40°$C. This dependence on temperature needs to be considered particularly when the motor has been exposed for some time to the space environment, resulting in a cold propellant. As an example, for a composite ammonium nitrate propellant the burn rate at $p_{ch} = 50$ atm is 1.7 mm/s at $-20°$C and 2.1 mm/s at $40°$C. Figure 4.33, adapted from Sutton and Ross [1], provides additional data.

It follows from the definition of the burn rate that the gas mass flow rate, $\tilde{m}$, is given by

$$\tilde{m} = A_b \varrho_{pr} b \tag{4.57}$$

where A_b is the burn surface area at some specified time and ϱ_{pr} the propellant density.

It is important to note that if there are cracks in the propellant or if the propellant mass has partially separated ("debonded") from the case wall, the effective burn area can be greatly increased over the intended area. Because the nozzle throat area is fixed, the increase in mass flow rate will increase the pressure in the motor case, possibly leading to a catastrophic failure. Cracks and separations are more likely to occur when the propellant in a spacecraft motor is very cold. To prevent this, the propellant temperature

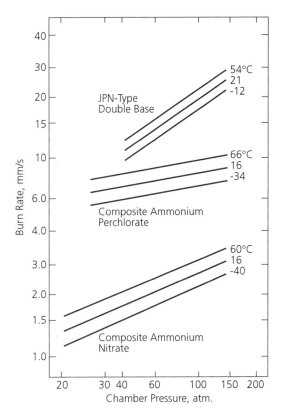

Figure 4.33 Burn rate versus chamber pressure for several solid propellants and initial propellant temperatures. (Adapted from Sutton and Ross [1].)

needs to be controlled, either by thermal blankets or combined with electrical heating. Cracks and debonding can also occur after long-term storage in a less than perfectly controlled environment. Ultrasonic and x-ray probes are useful for prelaunch checks.

Even in the absence of cracks there will be some unavoidable combustion instabilities. The use of powdered aluminum as a fuel additive has largely eliminated severe transverse oscillations. Slots or perforations in the propellant grain also tend to dampen the pressure oscillations.

It is often desirable to design the propellant configuration such that the pressure in the chamber stays approximately constant during the entire burn. The advantage is that the pressure can stay close to the maximum that is tolerated by the motor case. The result will be that the thrust is also nearly constant and a maximum for the case and nozzle design. In comparison to a thrust level that changes with time, yet has the same peak value, the gravity loss of the vehicle will be reduced.

If the bore surface were simply cylindrical and the pressure constant, the burn area would increase linearly with time. Because the mass flow rate would increase, the pressure would in fact increase, resulting in a still more rapid increase of burn area, pressure, and thrust. This effect can be alleviated by designing the initial propellant inner surface to have, for instance, a

174 CHAPTER 4 *Chemical Rocket Propulsion*

star-shaped cross section, thereby increasing the initial burn surface without increasing the final one near burnout. For the same purpose, perforations or radial slots in the propellant can be provided (at the cost of a more complicated mandrel).

To meet varying mission requirements, solid-propellant motors can be **offloaded**. This means that the motor is loaded with less propellant than could be accommodated by the case. Hence the motor is lighter and provides a smaller total impulse, as may be required for some missions.

Ignition is initiated by **squibs**, which are small cylinders containing an explosive charge fired by an electric filament ("bridge wire"). For redundancy, some squibs have more than one filament. The squid starts an **explosive transfer train** that is similar to but much faster than the type of powder cord used, for instance, in mining. In turn, the transfer assembly ignites the propellant.

Every possible precaution must be taken to prevent premature ignition. A typical arrangement is that in the **safe position** the explosive transfer train is physically separated from the squib. **Arming** then consists of eliminating this separation by the mechanical action provided by a small electric motor. All electrical parts that form part of the ignition system are designed such that at least two independent failures would have to occur before a premature ignition becomes possible.

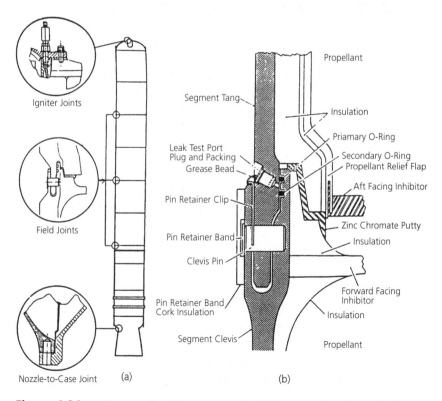

Figure 4.34 Solid-propellant boosters on the U.S. Space Shuttle, early design. (a) Stack of the segments, indicating locations of field joints; (b) design of field joints prior to Challenger accident.

4.19.3 Case Design

Usually, the motor case is of approximately cylindrical or spherical shape. Smaller, cylindrical cases can be fabricated as a single piece ("monolithic design"). Steel or composites are commonly used.

For very large motors, such as the solid-propellant boosters attached to major launch vehicles, the difficulty of casting the propellant and transporting the motor to the launch site requires a buildup of a stack consisting of separate segments. The case walls of the segments are connected by bolts or by tang and clevis held in position by pins. An example of the latter design is the boosters on the U.S. Space Shuttle. Figure 4.34 shows an earlier version, now superseded.

The figure indicates some of the precautions needed to mate the individual segments properly. Insulation and flame-inhibiting materials are used to prevent the flame from spreading into the interface between adjoining segments. To ensure good bonding to the case, a layer of inert, organic material is used. The same material also acts as a thermal insulator to the case when the propellant is close to being consumed at any one location. The joints between segments are designed to be leak tight by the use of rubber O-rings in series, protected from the hot gas by a zinc chromate putty.

4.20 Hybrid-Propellant Rocket Motors

In these motors the fuel is solid, in a configuration that is similar to that of solid-propellant motors. In contrast, the oxidizer is liquid, stored in a tank and pumped to impinge on the burn surface of the fuel.

In terms of simplicity of the design and usually also in terms of the specific impulse, hybrids are intermediate between solid- and liquid-propellant motors. They have the ability to be throttled, to be extinguished, and to be restarted.

An example is provided by a class of motors that use hydrogen-terminated-polybutadiene (HTPB) as the fuel and N_2O as the oxidizer. The turbine that drives the oxidizer pump is provided with gas from a gas generator that makes use of the same oxidizer and of an azide-based polymer solid fuel.

Nomenclature

a	speed of sound
b	burn rate (Eq. 4.56)
c_p, c_v	specific heats
e	internal energy
g	gravitational acceleration; g_0: standard gravitational acceleration
h	enthalpy
m	mass
$\dot{m}$	mass flow rate

n_i	number of moles of the ith constituent
p	pressure
r	radius
r_m	mixing ratio
s	entropy
u, v, w	velocities
x	coordinate
x_i	mole fraction of ith constituent
y	control variable
A	area
F_t	thrust
I_{sp}	specific impulse
$K(T)$	equilibrium constant (Eq. 4.41)
M, Nu, Pr, Re	Mach, Nusselt, Prandtl, Reynolds numbers
R	gas constant
T	absolute temperature
α	constant in Summerfield criterion (Eq. 4.25)
β_{ex}	divergence angle
ε	staging residual mass fraction; also degree of reaction
λ	Lagrange multiplier
μ	Mach angle; also viscosity
μ_j	molecular weight of jth constituent
ν_i	stoichiometric coefficient of ith constituent
$\nu(M)$	Prandtl–Meyer function
ϱ	density
ψ	sequential mass fraction; Ψ: payload mass fraction
ω	angular velocity
$(\ )_a$	ambient
$(\ )_\theta$	azimuthal
$(\ )_{ch}$	chamber
$(\ )_{ex}$	nozzle exit plane
$(\ )_\mu$	meridional
$(\ )_{pl}$	payload
$(\ )_{pr}$	propellant
$(\ )_{rs}$	residual
$(\ )_{vac}$	vacuum condition
$(\ )^*$	condition at throat

Problems

(1) Consider the nozzle represented in Fig. 4.35. An axisymmetric bell nozzle with zero divergence angle and the dimensions (in mm) indicated in the figure is assumed. There is zero ambient pressure at the nozzle exit.

    The calculation can be based on the approximation of a steady, one-dimensional, isentropic flow of a thermally and calorically perfect gas with an average ratio of 1.30 of the specific heats and a gas constant of 330 m²/(s² K). The stagnation temperature can be taken equal to the flame

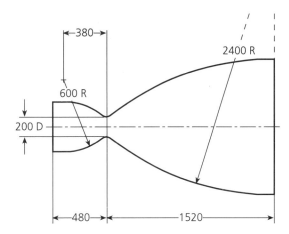

Figure 4.35 Axisymmetric bell nozzle. (Dimensions are in mm.)

temperature of 3200 K in the combustion chamber. The stagnation pressure, which can be taken equal to the chamber pressure, is 800 N/cm².

(a) Find the Mach number at the nozzle exit.
(b) Find the pressure, density, temperature, and velocity at the nozzle throat.
(c) Find the mass flow rate, the theoretical thrust (including the "pressure thrust"), and the specific impulse.
(d) Graph as functions of the axial coordinate the Mach number, the temperature, and the pressure.

(2) Consider a solid-propellant motor with the grain cross section indicated in Fig. 4.36. This cross section can be assumed to be uniform over the length of 1500 mm of the grain. The density of the propellant is 1700 kg/m³. The nozzle throat area is 4000 mm², the nozzle exit area 0.300 m². The motor operates in vacuum. The flame temperature in the gas passage perforation is 3300 K and can be taken equal to the stagnation temperature of the flow.

A perfect gas is assumed with ratio of the specific heats of 1.25 and a gas constant of 280 m²/(s² K).

(a) Given a constant burn rate of 3.5 mm/s of the propellant, show graphically the inner contour of the propellant at 0, 20, 40, 60, and 80 seconds after ignition and find the mass flow rate at these times.
(b) Find the stagnation pressure (= pressure in the center perforation) at these times.
(c) Find the thrust (including the "pressure thrust") at these times.
(d) From the burn rate of 3.5 mm/s at ignition and a burn rate pressure exponent of 0.6, find improved values for the burn rates at 20, 40, 60, and 80 seconds by using the results obtained in (b). With these improved numbers, recalculate the thrust at these times.

(3) For the rocket motor defined in Problem 1 and using the results obtained there, find

(a) The ambient pressure at which there will be a normal shock at the nozzle exit plane.

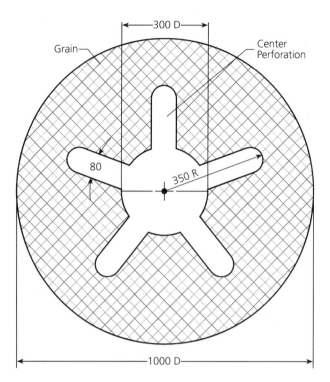

Figure 4.36 Cross section of a solid-propellant grain. (Dimensions are in mm.)

(b) The ambient pressure above which flow separation in the nozzle is likely to occur. [Make use of the Summerfield criterion, Eq. (4.25).]

(4)* Consider a propellant pump designed to pump liquid oxygen (density 1.19 10^3 kg/m^3 at a mass flow rate of 35.0 kg/s.

The pump is of the centrifugal type, similar to the one shown in Fig. 4.20. The flow enters the pump in the axial direction, then flows radially outward, driven by the impeller blades. Finally, the flow is redirected by the stationary blades in the volute such as to leave the pump in a conduit without swirl.

To simplify the calculation, an infinite number of blades are assumed, so that the flow is steady and independent of the azimuthal coordinate. To further simplify, the entire flow is represented by the flow tangential to the mean flow surface (defined as the axisymmetric surface for which one-half of the mass flow is on the inside, one-half on the outside). Losses will be neglected.

At the pump entrance, the hub radius is 30 mm, the blade tip radius 100 mm. The impeller exit radius is 150 mm. The widths of the flow passages are such that the throughput velocity (i.e., the velocity component in the meridional planes) is the same everywhere. The impeller blades are curved and turn the flow relative to the impeller through an angle of 60°, measured along the mean flow surface. The rate of rotation of the impeller is 7000 rpm.

The oxygen upstream of the pump is at a temperature of −220°C and a pressure of 2.00 atm, which corresponds to a boiling temperature of −176°C.

(a) Compute the pressure at the impeller exit and the pressure at the pump exit.

(b) Compute the theoretical (no loss) power required to drive the pump.

(5)* A space vehicle that, in addition to the payload, has two rocket stages is assumed to accelerate in gravity-free space along a rectilinear path.

The specific impulse for the lower stage is 350 s, for the upper stage 440 s. For both stages, the stage residual mass fraction is 0.12. The ratio of payload mass to total takeoff mass of the vehicle is 0.10.

Using the method of Lagrange multipliers, compute the optimum ratio of the first-stage mass (including its propellant) to the takeoff mass. Similarly, compute the ratio of the second-stage mass to the takeoff mass. (Answers: 0.522 and 0.377.)

(6)* Consider the chemical reaction

$$3S_{-2} + S_{-1} \longleftrightarrow 2S_1 + S_2$$

where the S_i designate the participating species. The degree of reaction will be designated by ε (referred to the product species S_1), the ratio of the pressure to a standard pressure by p/p_0, and the equilibrium constant by $K(T)$.

Write the mass action law in the form

$$f(\varepsilon, p/p_0) = \log K(T)$$

by expressing the function f explicitly in terms of ε and (p/p_0).

(7)* Assume a rocket motor with a fission reactor and hydrogen as the propellant. The hydrogen cools the reactor and enters a nozzle for expansion to a supersonic velocity. The hydrogen will be partially dissociated.

For an assumed temperature of the hydrogen at the nozzle entrance of 3000 K and a pressure of 30 atm, compute the degree of reaction pertaining to the dissociation to atomic hydrogen. (The equilibrium constant for this reaction, at 3000 K, is 2.838 10^{-2}; the reference pressure is 1 atm.)

References

1. Sutton, G. P. and Ross, D. M., "Rocket Propulsion Elements," 5th ed, John Wiley & Sons, New York, 1986.
2. Huzel, D. K. and Huang, D. H., eds., "Design of Liquid-Propellant Rocket Engines," American Institute of Aeronautics and Astronautics, Washington, DC, 1992.
3. Vertregt, M., "A Method for Calculating the Mass Ratios of Step Rockets," *Journal of the British Interplanetary Society*, Vol. 15, pp. 95–97, 1956.
4. Liepmann, H. W. and Roshko, A., "Elements of Gasdynamics," Galcit Aeronautical Series, John Wiley & Sons, New York, 1957.
5. Eyring, H., Lin, S. H. and Lin S. M., "Basic Chemical Kinetics," John Wiley & Sons, New York, 1980.
6. Hansen, C. F., "Rate Processes in the Gas Phase," NASA Reference Publication 1090, NASA, Washington, DC, 1983.

7. Wilkins, R. L., "Theoretical Evaluation of Chemical Propellants," Prentice-Hall, Englewood Cliffs, NJ, 1963.
8. Glassman, I. and Sawyer, R. F., "The Performance of Chemical Propellants," NATO Advisory Group for Aerospace Research and Development, AGARDograph No. 129, Technivision Services, Slough, England, 1969.
9. Vance, J. M., "Rotordynamics of Turbomachinery," John Wiley & Sons, New York, 1988.
10. Brennen, C. E., "Hydrodynamics of Pumps," Oxford University Press, Oxford, 1994.
11. Oppenheim, B. W., and Rubin, S., "Advanced Pogo Stability Analysis for Liquid Rockets," *Journal of Spacecraft and Rockets*, Vol. 30, No. 3, pp. 360–373, 1993.
12. Brown, C. D., "Spacecraft Propulsion," AIAA Education Series, American Institute of Aeronautics and Astronautics, Washington, DC, 1996.

5

Orbital Maneuvers

In this chapter, we will consider some special flight paths that are of particular interest for a number of space missions, as described in Refs. 1 to 4. As before, attention will be restricted to flight paths, or segments of such paths, where the gravitational field is a central, inverse-square force field. Perturbations, such as those that are produced by the oblateness of the earth, will therefore be neglected.

In some cases, the spacecraft center of mass motion that is considered here will be one of merely coasting (i.e., in free fall) without thrust. Other cases that are considered are flight paths where intervals of coasting are separated by short, quasi-instantaneous bursts of thrust. Here, the thrust vector, pointing in a fixed direction, is represented as a delta function of time. In many cases, this turns out to be a good approximation because the thrusting time (the so-called burn) of chemical rocket motors is usually just a few minutes or less, much shorter than orbit periods.

Whereas most examples in the present chapter will apply to artificial satellites and other spacecraft, in a later section, "gravity turns," maneuvers typical for many launch vehicles, will be considered.

5.1 Minimum Energy Paths

In the planning of space vehicle trajectories, it is often of interest to find the **minimum energy path** that leads from a prescribed point of departure to a prescribed target point. Both points, as well as the center of attraction, are assumed to be fixed in the same inertial reference system, for instance, the heliocentric system.

Because by (3.12) the energy is directly related to the semimajor axis, the problem becomes a purely geometrical one. It is illustrated in Fig. 5.1 for the case of elliptic orbits.

Given a point of departure P_1 and target point P_2, there are infinitely many arcs of conic sections that connect the two points and are in the plane containing the center, designated by F, of the gravitational attraction.

For a fixed trajectory plane and a fixed center of attraction in it, the number of parameters needed to determine the trajectory would be three: the semimajor axis a, the eccentricity e, and an angle to fix the inclination of the major axis. If, in addition, two distinct points, such as P_1 and P_2, through which the trajectory must pass are specified, the number of free parameters, in the general case, is reduced to one.

In the figure, two of the infinitely many possible ellipses through P_1 and P_2 with center of attraction F are shown. The two ellipses have different

182 CHAPTER 5 *Orbital Maneuvers*

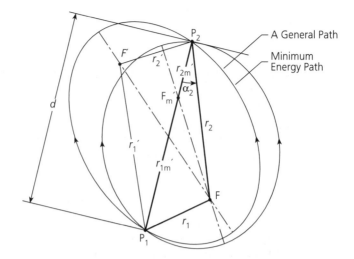

Figure 5.1 Elliptic trajectories connecting two fixed points P_1 and P_2 in an inverse-square gravitational field.

secondary foci, designated by F' and F'_m, respectively. The triangle P_1–P_2–F is given by the data, but the secondary foci are as yet unknown.

For each ellipse, there are two complementary arcs that connect the point of departure P_1 to the target point P_2. The energy, designated by w (potential plus kinetic energy) per unit mass of the space vehicle, is a constant of motion, the same for both arcs. However, in general, the flight times will differ.

From a well-known geometrical property of ellipses, the radii indicated in the figure must satisfy the relations

$$r_1 + r'_1 = r_2 + r'_2 = 2a \tag{5.1}$$

It follows from $r'_1 = 2a - r_1$ and $r'_2 = 2a - r_2$ that such elliptic arcs always exist irrespective of the choice of P_1, P_2, and F, provided only that the semimajor axis is chosen large enough. For then the two circles of radius r'_1 about P_1 and radius r'_2 about P_2 will intersect.

Let d designate the distance P_1–P_2 and $s = (1/2)(r_1 + r_2 + d)$, hence one-half of the perimeter of the triangle P_1–P_2–F. From the triangular inequality $r'_1 + r'_2 \geq d$ and (5.1) follows $2a \geq s$. Therefore, for a prescribed triple of points P_1, P_2, and F, the minimum value, a_m, that the semimajor axis can have is

$$a_m = \tfrac{1}{2} s \tag{5.2}$$

The radii r'_{1m} and r'_{2m} associated with this minimal ellipse become

$$r'_{1m} = s - r_1, \qquad r'_{2m} = s - r_2 \tag{5.3}$$

and therefore

$$r'_{1m} + r'_{2m} = d \tag{5.4}$$

This shows that the secondary focus F'_m of the minimal ellipse lies on the line P_1–P_2, with the distance d divided according to (5.4).

5.1 Minimum Energy Paths

For the minimal ellipse, the energy w_m per unit mass from (3.12) is

$$w_m = -\mu/s \qquad (5.5)$$

Since the energy decreases with decreasing semimajor axis, the minimal ellipse is also the path from P_1 to P_2 that requires the least energy.

It also follows that the minimal ellipse corresponds to the smallest kinetic energy at the point of departure, hence to the *minimum initial velocity*. One concludes from (3.14) that this velocity (in a nonrotating reference frame attached to the center of attraction) is

$$v_{1m} = \sqrt{2\mu\left(\frac{1}{r_1} - \frac{1}{s}\right)} \qquad (5.6)$$

It should be recalled, however, that the preceding development applies only to cases where the points P_1 and P_2 are fixed in the nonrotating reference frame attached to the center of attraction. This is the case, for example, for the exoatmospheric path of a rocket or for the aircraft-launched suborbital flight of a vehicle, but not for a mission from the earth to a planet, because in this last case neither the point of departure nor the target is fixed. Nevertheless, even in this case, the preceding equations are often useful for estimating purposes.

The eccentricity, e_m, of the minimal ellipse can be found as follows. Applying to the triangle P_1–P_2–F the identity that relates for any triangle one of its half-angles to its perimeter and the lengths of its sides, one finds

$$\cos\frac{\alpha_2}{2} = \sqrt{\frac{s(s-r_1)}{r_2 d}}$$

where α_2 is the angle formed by r_2 and r'_{2m}.

The distance between the foci F and F'_m is $2a_m e_m = s e_m$. Therefore from the triangle F'_m–P_2–F

$$\begin{aligned}
s^2 e_m^2 &= (r'_{2m})^2 + r_2^2 - 2 r'_{2m} r_2 \cos\alpha_2 \\
&= (s-r_2)^2 + r_2^2 - 2(s-r_2)r_2\left[\frac{2s(s-r_1)}{r_2 d} - 1\right] \\
&= s^2 - \frac{4s(s-r_1)(s-r_2)}{d}
\end{aligned}$$

hence

$$e_m = \sqrt{1 - \frac{4(s-r_1)(s-r_2)}{sd}} \qquad (5.7)$$

Having found the semimajor axis and eccentricity of the minimum energy path, the flight time between the point of departure and target can be calculated from Kepler's equation (3.21). Although straightforward, the calculation is algebraically cumbersome unless **Lambert's theorem**, discussed in the next section, is used.

Results analogous to those derived for elliptic paths can also be found for hyperbolic paths. The energy, w, in this case is always positive, hence

5.2 Lambert's Theorem

Lambert (1728–1777) established the surprising fact that in a central inverse-square force field the flight time between two points (fixed, as before, in the nonrotating reference frame attached to the center of attraction) is proportional to the Kepler orbit period, with a proportionality constant that depends only on the straight line distance between the points and the sum of the distances of the points from the center of attraction.

Figure 5.2 illustrates the theorem and indicates the notation that will be used. The proof of Lambert's theorem makes frequent application of the identities that relate sums of trigonometric functions to their products and vice versa. The theorem is valid for all conic sections but will be derived here only for elliptic paths.

The time of flight from point P_1 to point P_2 as given by Kepler's equation (3.21) is

$$t_2 - t_1 = (P/2\pi)[E_2 - E_1 - e(\sin E_2 - \sin E_1)]$$
$$= (P/2\pi)[E_2 - E_1 - 2e\cos\tfrac{1}{2}(E_1 + E_2)\sin\tfrac{1}{2}(E_2 - E_1)] \quad (5.8)$$

where P is the period of the orbit. This form is useful because it shows that the flight time depends, other than on the period, only on the sum and difference of the two eccentric anomalies E_1 and E_2.

In place of E_1 and E_2 one specifies new variables f and g by the two equations

$$\begin{aligned} f + g &= 2\cos^{-1}\left(e\cos\tfrac{1}{2}(E_1 + E_2)\right) \\ f - g &= E_2 - E_1 \end{aligned} \quad (5.9)$$

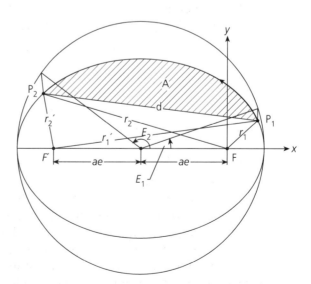

Figure 5.2 Illustration of Lambert's theorem.

(We postpone until later the consequences of the multivaluedness of the arc cos function.) From substitution into (5.8),

$$t_2 - t_1 = (P/2\pi)[f - \sin f - (g - \sin g)] \tag{5.10}$$

The same functions f and g can also be used to express the distance d between P_1 and P_2. For, if (x_1, y_1) and (x_2, y_2) are the Cartesian coordinates, as shown in the figure, of P_1 and P_2, respectively,

$$\begin{aligned}d^2 &= (x_2 - x_1)^2 + (y_2 - y_1)^2 = a^2(\cos E_2 - \cos E_1)^2 \\&\quad + a^2(1 - e^2)(\sin E_2 - \sin E_1)^2 \\&= 4a^2 \sin^2 \tfrac{1}{2}(E_2 - E_1)\left[\sin^2 \tfrac{1}{2}(E_1 + E_2) + (1 - e^2)\cos^2 \tfrac{1}{2}(E_1 + E_2)\right] \\&= 4a^2 \sin^2 \tfrac{1}{2}(f - g) \cdot \sin^2 \tfrac{1}{2}(f + g)\end{aligned}$$

from (3.19) and the substitutions from (5.9). From the matching trigonometric identity it follows therefore that

$$d = a(\cos g - \cos f) \tag{5.11}$$

An equally simple expression can also be obtained for the sum $r_1 + r_2$ of the two radii. Using the equation following (3.19) and applying twice the trigonometric identities relating sums and products,

$$\begin{aligned}r_1 + r_2 &= 2a[1 - \tfrac{1}{2}e(\cos E_1 + \cos E_2)] \\&= 2a\left[1 - \cos\frac{f+g}{2}\cos\frac{f-g}{2}\right] \\&= a[2 - \cos f - \cos g] \tag{5.12}\end{aligned}$$

Finally, solving (5.11) and (5.12) for $\cos f$ and $\cos g$, one obtains

$$\begin{aligned}\cos f &= 1 - (2a)^{-1}(r_1 + r_2 + d) \\\cos g &= 1 - (2a)^{-1}(r_1 + r_2 - d)\end{aligned} \tag{5.13}$$

The ambiguity, noted earlier, in solving for f and g can be resolved as follows: We define f^* and g^* as the principal values of f and g, so that

$$f = \pm f^* + 2\pi k, \qquad g = \pm g^* + 2\pi l \tag{5.14}$$

$k, l = \cdots -1, 0, 1, 2, \ldots$. Since $d \geq 0$, it then follows from (5.13) that

$$f^* \geq g^* \tag{5.15}$$

Also, (5.10) can be rewritten as

$$t_2 - t_1 = (P/2\pi)[\pm(f^* - \sin f^*) \mp (g^* - \sin g^*) + 2\pi m] \tag{5.16}$$

where $m = 0$ or 1 and the choices of the algebraic signs of the first and second summands are as yet independent of each other. (We exclude orbits consisting of more than one revolution; their flight times are obtained trivially by adding integer multiples of the Kepler orbit period.)

Four cases, represented in Fig. 5.3, need to be distinguished. They depend on whether the area A (to the left in moving from P_1 to P_2) bounded by the arc P_1–P_2 and the line P_1–P_2 does or does not contain the focal points F and F'.

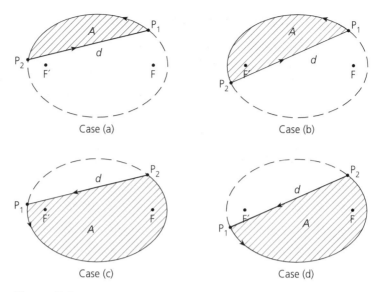

Figure 5.3 The four cases of Lambert's theorem.

Case (a): Here neither F nor F' is in A. The value of m and the algebraic signs in (5.16) can most easily be found from the limiting case when P_1 and P_2 coincide, that is, when $d = 0$, $t_2 - t_1 = 0$. Clearly, $r_1 = r_2 < 2a$. Hence from (5.13) $\cos f = \cos g \neq -1$ and $f^* < \pi$, $g^* < \pi$. By continuity, also in the general case, $m = 0$ and

$$t_2 - t_1 = (P/2\pi)[+(f^* - \sin f^*) - (g^* - \sin g^*)] \qquad \text{if } F \notin A, F' \notin A \quad \textbf{(5.17a)}$$

since, by assumption, $t_2 - t_1 \geq 0$ and $f^* - \sin f^* \geq g^* - \sin g^*$.

The limit of case (a) is reached when F' is on the line P_1–P_2. In that case, $d = r_1' + r_2'$. From $r_1 + r_1' = r_2 + r_2' = 2a$ and (5.13) follows that $\cos f = -1$, hence $f^* - \sin f^* = \pi$ at the transition from case (a) to case (b).

Case (b) occurs when F', but not F, is in A, **case (c)** when both F and F' are in A, and **case (d)** when F, but not F', is in A. Finding m and the algebraic signs in (5.16) in each of these cases proceeds analogously to case (a). The results can be summarized as follows:

$$t_2 - t_1 = (P/2\pi)[2\pi - (f^* - \sin f^*) - (g^* - \sin g^*)] \qquad \text{if } F \notin A, F' \in A$$
(5.17b)

$$t_2 - t_1 = (P/2\pi)[2\pi - (f^* - \sin f^*) + (g^* - \sin g^*)] \qquad \text{if } F \in A, F' \in A$$
(5.17c)

$$t_2 - t_1 = (P/2\pi)[+(f^* - \sin f^*) + (g^* - \sin g^*)] \qquad \text{if } F \in A, F' \notin A$$
(5.17d)

Lambert's theorem, which has just been derived, is similar to Kepler's equation insofar as it allows one to compute from orbital data the elapsed time of flight between two points. The inverse problem, the calculation of the location of a space object at a given time, usually requires a numerical inversion or else a series expansion as discussed in Sects. 3.3.1 and 3.3.2.

5.2.1 Application to Suborbital Flights

Lambert's theorem provides a particularly convenient way to calculate the exoatmospheric flight time of suborbital vehicles between two points. We neglect here for simplicity the corrections introduced by aerodynamic drag, earth rotation, and gravity anomalies. Assuming that the chosen path is a minimum energy trajectory (Section 5.1), the secondary focus, F′, coincides with the line connecting the two points. The limiting case of the transition from case (a) to case (b) therefore applies, with $f^* - \sin f^* = \pi$. (The complementary path would lie in the earth's interior.) From (5.13), defining the constant δ by

$$\delta = \frac{2(r_1 + r_2 - d)}{r_1 + r_2 + d} \quad (5.18)$$

follows

$$\cos f = -1, \quad \cos g = 1 - \delta$$

Therefore the elapsed flight time for the minimum energy path becomes, from (5.17a) or from (5.17.b),

$$t_2 - t_1 = (P/2\pi)\left[\pi - \cos^{-1}(1-\delta) + \sqrt{1-(1-\delta)^2}\right] \quad \textbf{(5.19)}$$

where the arc cos and the square root functions are evaluated on their principal branches. The period, P, of the complete orbit, from (5.2) is

$$P = \frac{\pi}{4}\sqrt{\frac{(r_1 + r_2 + d)^3}{\mu}} \quad (5.20)$$

5.2.2 Application to Planetary Missions

Spacecraft paths from the earth to the planets or to the moon, or between planetary flybys, are calculated by highly accurate numerical methods. These are outside the scope of this book. The methods outlined in Sects. 5.1 and 5.2, if iterated, are useful, however, for purposes of preliminary flight planning.

We assume here that the spacecraft is already outside the sphere of influence of the earth and has a velocity sufficient to reach the planet on a minimum energy path without further use of thrust.

In the heliocentric reference frame, an iterative scheme can be developed as follows: Starting with the earth's position (point P_1) at the time of departure of the spacecraft and a rough estimate of the planet's position (point P_2) at the time of arrival of the spacecraft, an estimate for the radii r_1, r_2 and the distance d between P_1 and P_2 is obtained. Lambert's theorem then allows one to calculate a first estimate of the flight time for a minimum energy path from P_1 to P_2. From this, a second, improved, estimate of the planet's position at the time of arrival can be obtained. In turn, from Lambert's theorem, a second approximation is obtained for the flight time. Still better results will generally be obtained by continuing the iteration.

5.3 Maneuvers with Impulsive Thrust

Injection of a spacecraft into a final orbit or hyperbolic trajectory almost always consists of one or more lengthy coasting intervals separated by much shorter intervals of thrust.

Chemical propulsion lends itself to short thrust durations ("burns"). They can be of the order of minutes or less. For small, tactical missiles with high acceleration, the thrust duration may be of the order of only seconds.

For the spacecraft maneuvers considered in this section, good approximations can be obtained by assuming that the thrust has a uniform direction and is **impulsive**, that is, represented by a delta function in time. The result is a change in speed

$$\Delta v = \lim \int_{t_1}^{t_2} \frac{F(t)}{m(t)} dt \qquad (5.21)$$

where $F_t(t)$ is the thrust, $m(t)$ the vehicle mass, and the intended limit is obtained by letting $t_2 - t_1 \to 0$, $F_t \to \infty$.

The maneuver is carried out by first orienting the vehicle so that its thrust axis is in the direction of the desired velocity change.

Sometimes the objective of the maneuver is to produce a change in the orbital plane. At other times, the objective may be to increase the magnitude of the velocity. Often, these are combined by the application of a single, impulsively applied thrust.

5.3.1 Plane Changes

Figure 5.4 illustrates a simple plane change from one circular orbit to another such orbit. The two orbital planes are defined by their inclinations i_1 and i_2 to the equatorial plane and the right ascensions Ω_1 and Ω_2 of the ascending nodes. The speed, v, in this simple case is assumed to be

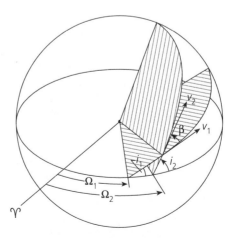

Figure 5.4 Plane change by impulsive thrust from a circular orbit to another circular orbit.

unchanged. With $\mathbf{v}_1$ designating the velocity immediately preceding and $\mathbf{v}_2$ the velocity immediately following the thrust, there is $v_1 = v_2 = v$.

The turning angle β is obtained from

$$\cos\beta = \cos i_1 \cos i_2 + \sin i_1 \sin i_2 \cos(\Omega_2 - \Omega_1) \quad \text{(pure plane change)} \tag{5.22}$$

as follows easily from the trigonometric relation relating the angles i_1, $\pi - i_2$, and $\Omega_2 - \Omega_1$ in the spherical triangle shown in the figure.

In this case, the two velocity vectors form an isosceles triangle. Hence the magnitude Δv of the $\Delta\mathbf{v}$ vector that must be produced by the thrust is

$$\Delta v = 2v \sin\left(\tfrac{1}{2}\beta\right) \quad \text{(pure plane change)} \tag{5.23}$$

Plane changes are needed, for instance, when injecting a geostationary satellite or other equatorial satellite from a launch site not on the equator. As indicated in Fig. 5.5, the ground trace of the satellite follows at first a great circle through the launch site. A plane change can be made at either point I or at point I′ in the diagram, that is, on the line of intersection of the initial and final orbit planes.

In the example, case (a), shown in the figure, a launch is assumed from the U.S. launch complex at Cape Canaveral (latitude 28.7° north) with a launch direction approximately east (to take advantage of the earth's rotation and to ensure reentry of the launch boosters over the ocean). Considered here is the plane change from a nonequatorial, circular orbit into an equatorial circular orbit of the same radius and equatorial crossings at points I and I' located at longitudes 90° and 270° beyond the longitude of the launch site. As is easily seen from the right angle spherical triangle shown in the figure, the turning angle β is just equal to the latitude δ_L of the launch site. Therefore, from (5.23), $\Delta v/v = 2\sin(\delta_L/2) = 0.496$ for Cape Canaveral. For the European Space Agency, French-operated launch site (latitude approximately 5°) in Guiana, the corresponding number is 0.087.

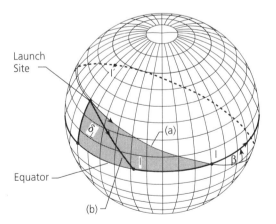

Figure 5.5 Ground traces for launch from Cape Canaveral, Florida ($\delta = 28.3°$) for insertion into equatorial orbits: (a) from a circular orbit and (b) from a suborbital trajectory.

Another insertion Δv can be obtained by first launching into a suborbital trajectory with a heading in a more southerly direction, case (b) in the figure. The transition into the final, circular, equatorial orbit in general will then be a combination of plane change and speed increase.

For a given turning angle β, Δv is less at lower speed v, indicating that pure turning is advantageously carried out for circular orbits at higher altitude and for elliptic orbits at the apoapsis.

5.3.2 Repeated Thrusts at Periapsis

For a given initial velocity $\mathbf{v}_1$ and a required final velocity $\mathbf{v}_2$, the needed velocity increment Δv, relative to v_1 is

$$\Delta v / v_1 = \sqrt{1 + (v_2/v_1)^2 - 2(v_2/v_1) \cos \beta} \tag{5.24}$$

where β is the angle between $\mathbf{v}_1$ and $\mathbf{v}_2$. If $\beta = 0$, this ratio is a minimum.

As an example, we consider the effect of repeated impulsive thrusts at the periapsis of elliptic orbits. Figure 5.6 illustrates the effect of separate firings, at the same point, of upper stage motors, typically solid-propellant motors, of a multistage vehicle. The motor firings occur at periapsis, therefore where the velocity is highest for each elliptic path. Before each firing, the attitude of the spacecraft is readjusted to ensure that the thrust at the periapsis is in the

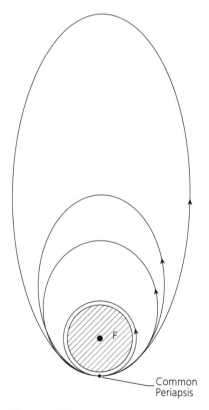

Figure 5.6 Repeated upper stage firings at a common periapsis.

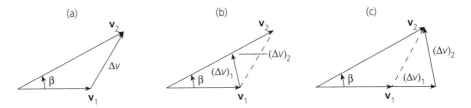

Figure 5.7 Comparison of one versus two impulsive applications of thrust.

direction of the path. Each elliptic path returns to the same periapsis height, with a velocity that is larger than for the preceding path. The elapsed time for each elliptic path allows enough time for stage separation and attitude adjustment for the next firing.

5.3.3 Plane Change and Speed Change Combined

It is often advantageous to combine plane changes with speed changes. Figure 5.7 represents the plane that contains the initial velocity $\mathbf{v}_1$ and final velocity $\mathbf{v}_2$. It shows three different modes for comparison.

Often, the twin objectives of plane change and speed increase can be achieved with a single motor firing, case (a) of the figure. It is of interest to compare this with the two other cases: case (b) to make the plane change first, to be followed (still at essentially the same point on the trajectory, i.e., at the same potential energy) by increasing the speed, and case (c) to increase the speed first and then make the plane change.

It follows from the triangular inequality applied to cases (b) and (c) that the required Δv is smallest in case (a), that is, when the plane change and speed increase are combined in a single firing.

5.3.4 Multiple Stages with Impulsively Applied Thrust

In practice, these theoretically optimal combinations for achieving a minimum Δv for a given objective can serve only as a general guide. This is because mission designs are necessarily almost always based on the use of already developed and available rocket motors. Whereas some liquid-propellant motors are capable of multiple starts and terminations, this is not the case for solid-propellant motors, the class of motors (possibly off-loaded) that is most frequently used for maneuvers as described here. The mission designer therefore must typically choose from a relatively small set of existing, or at least planned, motors and from several possible choices a near-optimal solution.

In what follows, we assume a two-stage vehicle. The motors are fired in quick succession, so that the gravitational potential energy is unchanged. The choice of rocket motors and stage characteristics then usually consists of maximizing the increase in the vehicle's total energy obtained by firing the two stages. (Extensions to more than two stages, and to a delayed firing of the second stage, hence at a different potential energy, are immediate.)

Let

m_1 = mass of stage 1 before burn
m_2 = mass of stage 2 before burn
m_3 = spacecraft mass
μ_1 = stage 1 ratio of residual mass to mass m_1
μ_2 = stage 2 ratio of residual mass to mass m_2
v_0 = velocity before stage 1 burn
$(\Delta W)_3$ = energy increase of the spacecraft, resulting from the stage firings

The residual masses referred to comprise such motor components as the void motor case, nozzle, igniter, and thrust cone of solid-propellant motors. Typically, the residual mass is no more than about 10% of the total stage mass before firing.

The increase in the spacecraft's energy, $(\Delta W)_3$, is

$$(\Delta W)_3 = \tfrac{1}{2} m_3 (v_0 + (\Delta v)_1 + (\Delta v)_2)^2 \quad (5.25)$$

where $(\Delta v)_1$ and $(\Delta v)_2$ are the velocity increments provided by stages 1 and 2, respectively. From the "rocket equation" (2.28), they are

$$(\Delta v)_1 = g_0 I_{\text{sp},1} \ln \frac{m_1 + m_2 + m_3}{\mu_1 m_1 + m_2 + m_3} \quad (5.26a)$$

$$(\Delta v)_2 = g_0 I_{\text{sp},2} \ln \frac{m_2 + m_3}{\mu_2 m_2 + m_3} \quad (5.26b)$$

where $I_{\text{sp},1}$ and $I_{\text{sp},2}$ denote the specific impulse of the stage 1 and 2 motors.

As far as propellant consumption is concerned, there is no difference whether the residual of the last stage remains attached to the spacecraft or is separated from it. Separation may be required to prevent contamination of the spacecraft from outgassing that occurs from motors that are still hot.

The mission designer's task is therefore to select a combination of rocket motors that will result, at least approximately, in the smallest combined mass of the two stages for the required spacecraft energy increment ΔW_3.

5.4 Hohmann Transfers

A frequently executed maneuver is to raise a spacecraft from an initial, often circular, orbit into a higher, again often circular, orbit. The two orbit planes may or may nor coincide. This maneuver can be accomplished by two impulsive thrusts, one on the initial orbit, the other ("orbit insertion") immediately preceding the final orbit. Such a maneuver is called a **Hohmann transfer**.

In what follows, we consider the transfer of a spacecraft between two circular orbits, not necessarily in the same plane. One instance of this would be the transfer from a low earth orbit, with its orbital plane through the launch site, into a geostationary orbit. The maneuver is illustrated in Fig. 5.8.

The initial orbit has radius r_1 and velocity v_1. Similarly, the final orbit has radius r_2 and velocity v_2. The initial and final orbit planes and the transfer

5.4 Hohmann Transfers

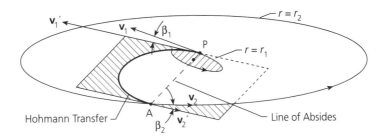

Figure 5.8 Hohmann transfer between two circular orbits; P, periapsis; A, apoapsis.

orbit plane are assumed to intersect all in a common line, referred to as the **line of absides**. The first thrust changes the velocity v_1 to v_1' through the angle β_1 so that the new velocity is in the Hohmann transfer plane. Similarly, the second thrust changes the velocity v_2' to v_2 through the angle β_2 so that the new velocity is in the final orbit plane. The transfer orbit therefore is a semiellipse with its periapsis on the initial orbit and its apoapsis on the final orbit.

It follows that the semimajor axis, a', of the transfer orbit is $a' = (1/2)(r_1 + r_2)$. Therefore, from (3.14),

$$v_1 = \sqrt{\mu/r_1}, \qquad v_2 = \sqrt{\mu/r_2}$$
$$v_1' = \sqrt{\frac{2\mu r_2}{r_1(r_1 + r_2)}}, \qquad v_2' = \sqrt{\frac{2\mu r_1}{r_2(r_1 + r_2)}} \qquad (5.27)$$

With the definitions $\Delta v_1 = v_1' - v_1$ and $\Delta v_2 = v_2 - v_2'$, the magnitudes of the velocity increments that must be delivered by the two thrusts are

$$\Delta v_1 = \sqrt{v_1^2 + v_1'^2 - 2v_1 v_1' \cos \beta_1}, \qquad \Delta v_2 = \sqrt{v_2^2 + v_2'^2 - 2v_2 v_2' \cos \beta_2} \qquad (5.28)$$

as follows from the cosine law applied to the triangles formed by the velocities.

The transit time $t_2 - t_1$ from the initial to the final orbit is one-half of the period of the transfer ellipse, hence

$$t_2 - t_1 = \pi \sqrt{\frac{1}{\mu}\left(\frac{r_1 + r_2}{2}\right)^3} \qquad (5.29)$$

Important for estimating the total propulsion requirement for the maneuver is the sum $\Delta v = \Delta v_1 + \Delta v_2$. It is easily shown from (5.27) and (5.28) that in the case of no plane change and for a fixed initial orbit radius, Δv at first increases with increasing radius r_2, then decreases again. The maximum required Δv occurs at a radius ratio r_2/r_1 of 15.58, where $\Delta v/v_1 = 0.536$. In the limit, as $r_2/r_1 \to \infty$, $\Delta v/v_1 \to \sqrt{2} - 1$.

In the more often encountered case in which there is a plane change, the sum of the angles β_1 and β_2 will be known. The split between the plane change at periapsis and the one at apoapsis is open to the mission designer. Usually, the approach used is to minimize the sum Δv of Δv_1 and Δv_2.

Since $v_1 v_1'/(v_2 v_2') = (r_2/r_1)^{3/2}$, the factor of $\cos \beta_1$ in (5.28) is larger (for $r_2 > r_1$) than the corresponding term

$$v_2 v_2' = \frac{\mu}{r_2} \sqrt{\frac{2r_1}{r_1 + r_2}}$$

For instance, for a geostationary orbit, to be reached from a low earth orbit, the ratio $(r_2/r_1)^{3/2}$ is about 17. Reducing β_1 at the expense of β_2 will reduce Δv and therefore also the propellant consumption. The optimal split between the two plane changes will depend on the radius ratio r_2/r_1.

In the important case of the geostationary final orbit, starting from a low-altitude parking orbit, the split will be optimal if almost the entire plane change is done at the apogee. The benefit derived from introducing a partial plane change also at perigee (where the velocity is higher) is no longer significant.

The Hohmann transfer orbit satisfies the condition that the second (empty) focal point of the ellipse lies on the line connecting the transfer's initial and final points. It then follows from the discussion in Sect. 5.2 that the Hohmann transfer is the minimum energy path between the Hohmann periapsis and apoapsis.

5.5 Other Transfer Trajectories

Transfer trajectories other than Hohmann transfers are frequently used. The choice is often dictated by the need to match the impulse of solid-propellant motors to the mission requirement. These motors must burn to completion and may have excess impulse for the intended spacecraft mass. (Sometimes, the motors are "offloaded," that is, filled with less propellant than the capacity of the motor case would allow; offloading, however, requires a new mandrel for the casting process.)

The type of transfer orbit that is optimal will depend significantly on the radius ratio or semimajor axis ratio of the final orbit relative to the initial one. For transfers from an initial circular orbit to a final circular orbit and a radius ratio larger than about 15, it may happen that the sum of the Δv's for non-Hohmann transfers is slightly less, at best about 4%, than for the Hohmann transfer. (Because the point at which the transfer path meets the final orbit is allowed to vary, this does not contradict the earlier statement that the Hohmann trajectory is the path of least energy between the Hohmann periapsis and apoapsis points.)

Often it is required not only that a final orbit be achieved but also that the *time phasing* be correct, that is, that the spacecraft arrive in its final orbit at the required place and time. In particular, this is the case in rendezvous maneuvers where the spacecraft must meet, and possibly dock, with another spacecraft. The transfer time will generally be longer than would be the case in the absence of such a requirement.

In what follows in the present section, it will be assumed that the initial and final orbits are *circular* and that their planes and also the transfer plane are *coplanar*. In Fig. 5.9 three different types of transfer orbits

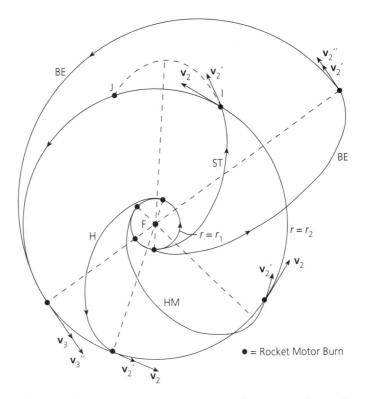

Figure 5.9 Comparison of four transfer trajectories: H, Hohmann; HM, Hohmann modified; BE, bielliptic; ST, semi-tangential.

(modified Hohmann, bielliptic, and semitangential), all for the same initial and final orbits, are shown and compared with the Hohmann transfer.

For these orbits, such data as the velocity increments and the time needed for the transfer can all be calculated by using the algebraic equations derived in Chap. 3. However, transfers between *elliptic* orbits can be calculated in this way only in some special cases. In all other cases, recourse must be made to numerical methods.

The **modified Hohmann transfer** is a two-impulse path similar to the Hohmann transfer but has a somewhat higher apoapsis. As the transfer orbit meets the final orbit, a small change of the flight angle, in addition to the velocity increment, is made. This maneuver is frequently advantageous when solid-propellant motors are used. The basic requirement is that the propellant be fully consumed when the spacecraft has the correct, or very nearly correct, velocity for the desired final orbit.

To avoid needless repetition, we indicate simply the sequence of calculations by their equation numbers in Chap. 3. Given μ, r_1, r_2, and an assumed Δv_1, the sequence used to find Δv_2 and the transfer time is as follows:

From μ and r_1 follows v_1. Adding to this Δv_1 gives v_1' and the angular momentum h' of the transfer path. From (3.10), with $\theta = 0$, follows the transfer path's eccentricity e'. From (3.11) follows the energy, w', and hence from (3.12) the semimajor axis. From r_2 and (3.10) follows the true anomaly θ_2 at the (second) intersection of the transfer path with the final orbit. The radial and azimuthal components of $\mathbf{v}_2'$ follow from (3.14). The required $\Delta \mathbf{v}_2 = \mathbf{v}_2 - \mathbf{v}_2'$, needed to start the final orbit, is then immediate. Finally, the eccentric anomaly E_2' at θ_2 is found from (3.20), and the transfer time, $t_2 - t_1$, is calculated from (3.21).

The **bielliptic** transfer, for instance, between two circular, coplanar orbits (Fig. 5.9), uses three impulsive thrusts. The path consists of two semiellipses, the first one tangential to the initial orbit, the second one to the final orbit. There are no changes in flight angle. Prior to orbit insertion, the spacecraft reaches an altitude higher than the altitude of the final orbit. At the point of termination of the second semiellipse, thrust is applied tangentially to the path but in the *reverse* direction ("retro-thrust"). The calculation can proceed in the same manner as just outlined.

The **semitangential** transfer is qualitatively the same as the modified Hohmann transfer but usually goes to a higher altitude. Depending on the choice of the semimajor axis, a favorable time phasing, important for instance for some rendezvous maneuvers, can be obtained. Insertion can take place at either point I or J, as may be closest for matching the rendezvous requirement.

5.6 On-Orbit Drift

Once a spacecraft is in its intended orbit, a relatively modest expenditure of propellant will suffice to change the time phasing of the spacecraft.

The necessary maneuver consists of two short burns of a low-thrust motor. The thrusts are nearly tangential to the orbit. The first thrust imparts a small incremental velocity, either in the direction of the orbital motion or opposite to it. The result is a relative drift motion, which is stopped by a second thrust in the direction opposite to the first.

This maneuver makes it possible, for instance, to change the longitude of a geostationary satellite. If several satellites, with one or more perhaps serving as spares, have been placed in the same orbit, their relative positions can be changed and a satellite made to assume the function of another, perhaps disabled, one.

There are still other applications, for instance, in rendezvous maneuvers. Prior to docking, when the two spacecraft are already on the same orbit, although possibly still widely separated, the drift velocity imparted to one of them will, over time, close the distance.

Prior to final orbit insertion, or before the start of a hyperbolic trajectory, spacecraft are very often put into a parking orbit about the earth. The usual purpose is to check out the spacecraft functions, as they may have been affected by the launch. The choice of the parking orbit and time phasing is dictated by the communications requirements imposed by the location of the ground station that does the checkout. After completion of the checks, which may take several weeks, the on-orbit drift maneuver can be used to place the spacecraft into its operational position and phasing or into a new position for further acceleration.

Assuming for simplicity a circular orbit, the amount of propellant consumed is easily estimated. (The slight increase in eccentricity that follows the first thrust can be neglected for this purpose.) If r is the radius of the orbit, m_s the mass of the spacecraft, I_{sp} and $\tilde{m}$ the specific impulse and mass flow rate of the motor, and Δt_{pr} the duration of the burn, the impulse given to the spacecraft is $g_0 I_{sp} \tilde{m} \Delta t_{pr}$. Hence the drift velocity, Δv, imparted by the first thrust is

$$\Delta v = \frac{g_0 I_{sp} \Delta m_{pr}}{m_s}$$

where Δm_{pr} is the mass of propellant consumed by each of the two thrusts. If the drift is to advance (or retard) the spacecraft by a polar angle (true anomaly) difference $\Delta \theta$ in comparison with the unperturbed motion, and if Δt is the time allowed for the drift, the amount of propellant consumed by the two thrusts is

$$2\Delta m_{pr} = \frac{2\Delta \theta r m_s}{g_0 I_{sp} \Delta t} \qquad (5.30)$$

Taking as an example a geostationary spacecraft ($r = 42\,164$ km), $m_s = 1000$ kg, $I_{sp} = 200$ s, $\Delta \theta = 90°$, and $\Delta t = 60$ days, the total propellant consumption calculated from this equation is $2\Delta m_{pr} = 13.0$ kg.

5.7 Launch Windows

The final orbit plane of a satellite, or the plane of the trajectory of a spacecraft on a deep-space mission, is determined by the plane's inclination to the equatorial plane and by the right ascension of the ascending node (Sect. 3.4). Because of the need to place the spacecraft trajectory in the selected plane and taking into account the earth's or planetary rotation — and with it the motion of the launch site — the acceptable launch times are restricted.

Also, because of various operational requirements, margins in the time of launch must be provided for. Providing these margins requires that the flight and ground computer software be designed to initiate the necessary adjustments of the flight path whenever the actual launch time differs from the nominal one.

Time intervals acceptable for a launch are referred to as **launch windows**. Other things being equal, the launch window for a launch from the earth repeats itself approximately every 24 h.

Depending largely on the required inclination of the final orbit, launch windows may vary in duration from a few minutes to several hours. As far as astrodynamic considerations are concerned, the time of launch into *equatorial* orbits, in principle, is not restricted.

On the other hand, *near-polar* orbits necessitate very short windows. Otherwise, excessive propellant consumption might result from the plane changes that would be needed if the actual and nominal times of launch differed by more than a few minutes.

In addition to the requirement of placing the spacecraft into the selected orbital plane, it may be necessary to control the time at which the spacecraft passes its ascending node, narrowing the available window further.

Launch windows can be widened if there is excess propellant either in the launch vehicle or in the upper stages. Liquid-propellant systems always carry a small amount — sometimes as little as 1% — of excess propellant to cope with uncertainties, for instance, in the propellant mixing ratio or in the initial propellant loading. If the nominal flight path and the available motors are such that excess propellant above the amount needed for a margin of safety can be carried, the launch window can be widened. The excess propellant then becomes available for plane changes and other flight path adjustments.

Other than the restrictions imposed by astrodynamics, operational considerations also influence the choice of an acceptable time of launch. Such operational aspects include the weather at the launch site, particularly high-altitude winds, the sun angle on the spacecraft before and after deployment of solar panels, the need sometimes for sufficient daylight visibility of the launch, limitations caused by the boil-off of cryogenic propellants, and the availability of all of the needed communications links.

Figure 5.10 is a graphic representation of a launch window for a particular mission. It applies to the placement of a payload into orbit by means of the U.S. Space Shuttle and two upper stages launched from Cape Canaveral. The Space Shuttle mean altitude was 278 km. The final orbit of the payload was circular, at an altitude of 20,187 km (a 12-h orbit), with an inclination to the equator of 55.0°. The two upper stages released by the Space Shuttle provided two additional velocity increments of 2107 m/s by the lower and 1888 m/s by the upper stage. Both solid-propellant motors of these

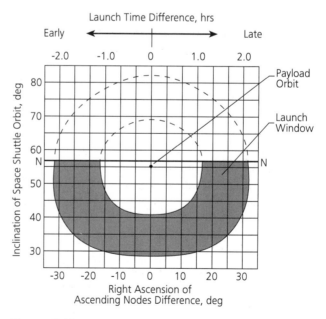

Figure 5.10 Launch window for a Cape Canaveral Space Shuttle launch for a mission with two upper stage firings. N–N, maximum inclination allowed by range safety. Adapted from Chobotov, V. A., "Orbital Mechnics," 2nd ed., Ref. 4.

upper stages had excess propellant. The chosen flight path therefore was a modification, by raising the apogee and a plane change, of a Hohmann transfer.

In the figure, the ordinate represents possible inclinations of the flight path of the Space Shuttle when launched from Cape Canaveral. The right ascension of the Space Shuttle's path is the abscissa in the figure and is shown as the difference from the right ascension of the payload orbit. The abscissa therefore also represents the delay, or advance, of the true launch time (e.g., a 15° increase of the right ascension corresponds to a delay of 1 hour). Also represented in the diagram is the payload orbit.

The shaded area represents the Space Shuttle's launch window that is compatible with the payload's final orbit and with the amount of propellant in the upper stage motors. If it is launched at the nominal time, the launch inclination, because of the excess propellant available in the upper stages, could be anywhere from about 28° to 41°. The window's boundaries are limited, however, by the maximum launch inclination (about 57°, corresponding to a launch azimuth of 35° north) allowed by range safety considerations at this particular launch site.

5.8 Injection Errors and Their Corrections

The injection — for instance, by a Hohmann transfer — into an intended orbit will unavoidably result in some random errors in the orbital parameters. These must be corrected if the high precision that is usually required for the final orbit is to be met. The corrections are typically made by first reorienting the spacecraft in the direction needed for the thrust, followed by a short burn of an upper stage motor.

The following discussion is limited to the case of final orbits that are *circular*. The extension to elliptic orbits is straightforward, however.

If the final, intended orbit is circular, the approximate orbit, that is, the orbit resulting from a Hohmann or similar transfer, can have at most three sources of error. They are an error, designated by Δi, of the inclination of the orbital plane; an error, designated by Δa, of the semimajor axis; and an error, designated by Δe, of the eccentricity. It is convenient to discuss these error sources separately. The total error and the required corrective maneuvers can then be obtained by superposition.

5.8.1 Inclination Error

Figure 5.11 shows the geometrical relations represented on the unit sphere. Using primed symbols for the approximate and unprimed ones for the final orbit, there are

$$\Delta i = i' - i, \qquad a' = a, \qquad e' = e = 0$$

where $|\Delta i| \ll 1$ by assumption. Let $\Delta \delta(t)$ be the spacecraft's latitude and $\lambda'(t)$ the spacecraft's longitude before the correction. Both latitude and longitude here refer to the final orbital plane. The origin of time is taken at the moment

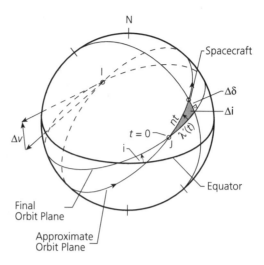

Figure 5.11 Orbit plane inclination error and correction.

when the spacecraft on the approximate orbit crosses the final orbital plane. Since the semimajor axes (here the radii) are, by assumption, the same for both paths, the "mean angular velocities" (Sect. 3.2)

$$n' = n = \sqrt{\frac{\mu}{a^3}}$$

are also the same.

From the law of sines applied to the spherical, right angle triangle shown in the figure,

$$\sin \Delta\delta = \sin \Delta i \sin(nt)$$

Expanding $\sin \Delta\delta$ and $\sin \Delta i$ into their power series and omitting third-order terms,

$$\Delta\delta = \Delta i \sin(nt), \qquad (a'=a, e'=e=0) \tag{5.31}$$

The spacecraft on the approximate orbit is therefore seen to change its latitude relative to the final orbit sinusoidally in time, with amplitude Δi.

Let the longitude difference $\Delta\lambda = \lambda' - \lambda$, where $\lambda = nt$. Then, from the same spherical triangle,

$$\cos(nt) = \cos\lambda' \cos\Delta\delta = [\cos(nt)\cos\Delta\lambda - \sin(nt)\sin\Delta\lambda]\cos\Delta\delta$$

Expanding in power series the trigonometric functions of the perturbation terms,

$$\cos(nt) = \left[\cos(nt)\left(1 - \tfrac{1}{2}\Delta\lambda^2 + \cdots\right) - \sin(nt)\left(\Delta\lambda - \tfrac{1}{3}\Delta\lambda^3 + \cdots\right)\right]$$
$$\times \left(1 - \tfrac{1}{2}\Delta\delta^2 + \cdots\right)$$

After dropping the higher order terms and substituting $\Delta\delta$ from (5.31), the final result for $\Delta\lambda$ to the lowest significant order becomes

$$\Delta\lambda = -\tfrac{1}{4}\sin(2nt)\Delta i^2, \qquad (a'=a, e'=e=0) \tag{5.32}$$

5.8 Injection Errors and Their Corrections

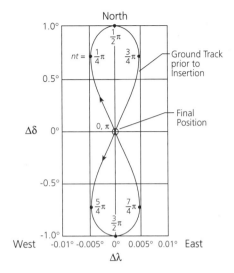

Figure 5.12 Ground track of geostationary satellite prior to correction of 1° inclination error.

The longitude difference is therefore only of second order in Δi, with a frequency twice that of the latitude difference.

The error in inclination can be corrected by applying thrust perpendicular to the plane of motion, at either point I or point J in the diagram, where the two planes intersect. The needed velocity increment Δv is seen to be $\Delta v = \Delta i\, v$, where the velocity v is from (3.14), so that

$$\Delta v = \sqrt{\mu/a}\, \Delta i \qquad (a' = a, e' = e = 0) \tag{5.33}$$

The geographic latitude and longitude are obtained from $\Delta\delta$ and $\Delta\lambda$ by a straightforward coordinate transformation.

A particularly simple case occurs for *geostationary* satellites. Shown in Fig. 5.12 is the ground track, that is, the radial projection of the geostationary satellite's path onto the earth surface. An inclination Δi of 1.0° is assumed. The mean angular velocity in this case is $n = 2\pi/d_{si}$ (d_{si} = length of sidereal day). As given by (5.31) and (5.32), the satellite in its approximate orbit before the corrective maneuver is seen to describes a figure eight, with a much smaller variation in longitude than in latitude.

5.8.2 Semimajor Axis Error

Let $\Delta a = a' - a$ and

$$\Delta n = n' - n = \sqrt{\frac{\mu}{(a + \Delta a)^3}} - \sqrt{\frac{\mu}{a^3}}$$

Expanding by means of the binomial theorem and retaining only the lowest significant order terms in $\Delta a/a$ gives,

$$\Delta n = -\tfrac{3}{2} n\, \Delta a/a \qquad (\Delta i = 0, e' = e = 0) \tag{5.34}$$

This shows that for a positive (negative) Δa, prior to the satellite's injection into the final orbit, there is a slow westward (eastward) drift relative to the final motion.

The corrective maneuver consists of a Hohmann transfer between the two coplanar circular orbits (Sect. 5.4). Therefore two impulsive thrusts are needed. The two required velocity changes Δv_1 and Δv_2, obtained from (5.27), are, to lowest order in $\Delta a/a$,

$$\Delta v_1 = \sqrt{\frac{\mu}{a+\Delta a}} - \sqrt{\frac{2\mu a}{(a+\Delta a)(2a+\Delta a)}} = \frac{1}{4}\sqrt{\frac{\mu}{a^3}}\Delta a \quad (5.35a)$$

$$\Delta v_2 = \sqrt{\frac{2\mu(a+\Delta a)}{a(2a+\Delta a)}} - \sqrt{\frac{\mu}{a}} = \frac{1}{4}\sqrt{\frac{\mu}{a^3}}\Delta a \quad (5.35b)$$

hence

$$\Delta v_1 = \Delta v_2, \quad (\Delta i = 0, e' = e = 0) \quad (5.35c)$$

5.8.3 Eccentricity Error

Let $\Delta e = e' - e = e'$ since, by assumption, $e = 0$. The eccentric anomaly, E', for the orbit prior to the correction is, from the Fourier–Bessel series (3.30),

$$E' = M' + 2\sum_{k=1}^{\infty}\frac{1}{k}J_k(k\Delta e)\sin(kM')$$

where $M' = n'(t - t_p) = n(t - t_p)$ since the mean angular velocity, n, depends only on the semimajor axis. The time of periapsis passage is designated by t_p. Expressing the Bessel functions by their Taylor series and dropping terms of third order in Δe and higher, the final result, expressed by the difference $\Delta E = E' - E$ between the eccentric anomalies before and after the corrective maneuver, is therefore

$$\Delta E = \Delta e \sin(n(t - t_p)) + \tfrac{1}{2}(\Delta e)^2 \sin(2n(t - t_p)) \quad (\Delta i = 0, a' = a) \quad (5.36)$$

In place of the difference of the eccentric anomalies, it may be more convenient in some applications to compute the difference of the true anomalies. They, as well as the radial difference, can be obtained, for instance, from (3.20).

The corrective maneuver can take place at either the periapsis or apoapsis. Computing the needed velocity change is straightforward.

5.9 On-Orbit Phase Changes

It is sometimes necessary to change the position of a spacecraft on its orbit without changing the orbit itself. What is being altered, therefore, is the **time phasing** of the spacecraft motion.

For instance, because of changing patterns in communications traffic, there may be a need to move a geostationary communications satellite along its orbit to a different longitude. Another such instance may occur when a

5.9 On-Orbit Phase Changes

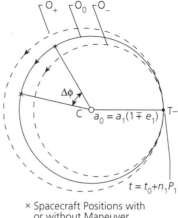

× Spacecraft Positions with or without Maneuver

Figure 5.13 Change of phase of a spacecraft on a circular orbit. O_0, basic orbit; O_+ and O_-, transfer orbits; C, center of attraction; T, location of thrust.

constellation of satellites, all on the same orbit, needs to be rearranged because one of the spacecraft has reached the end of its useful life.

In what follows, we assume, for simplicity, a circular orbit, designated by O_0 in Fig. 5.13. The extension to elliptic orbits is straightforward, however.

To accomplish the change, two impulsive thrusts, both at point T in the figure, are applied, either in the direction of motion or opposite to it. The two thrusts are assumed to be separated by a time difference Δt. If the first thrust is in the direction of motion, the spacecraft will follow an elliptic transfer orbit O_+ which has an increased semimajor axis, hence longer period. This orbit will lead the spacecraft back to point T. After one or, in more common applications, many revolutions, a second, short thrust in the *opposite* direction will place the satellite back on its original orbit. Correspondingly, if the first thrust is in the retro-direction, the spacecraft will at first follow the transfer orbit O_-, before a second thrust will bring it back to O_0.

The point T is the periapsis of O_+ and the apoapsis of O_-. The result of following O+, with its longer period, will be a satellite position *behind* the one without the maneuver. Correspondingly, by following O_-, with its shorter period, the satellite will be *ahead*.

To minimize the expenditure of propellant, the two impulses are kept small, resulting in transfer orbits that differ little from the original orbit. Unavoidably, the time needed for completion of the maneuver then becomes long, sometimes as much as several months.

Let a_0 be the radius of the circular orbit O_0 and a_1 the semimajor radius of a transfer orbit. It follows from the geometry of ellipses that

$$a_0 = a_1(1 \mp e_1) \tag{5.37}$$

where e_1 is the eccentricity of a transfer orbit and the upper sign here and in what follows applies to orbits of the type O_+, the lower sign to orbits O_-.

The periods P_0 of the original (circular) orbit and P_1 of the transfer orbits are to lowest significant order

$$P_0 = 2\pi \sqrt{\frac{a_0^3}{\mu}}, \qquad P_1 = 2\pi \sqrt{\frac{1}{\mu}\left(\frac{a_0}{1 \mp e_1}\right)^3} = P_0(1 \mp e_1)^{-3/2} \qquad (5.38)$$

If n_1 is the number of revolutions on a transfer orbit, the time between thrusts, hence the time needed to execute the maneuver, is $\Delta t = n_1 P_1$. In this time, the spacecraft moves through a polar angle (true anomaly) of $2\pi \Delta t/P_1$. Without the maneuver, the angle would be $2\pi \Delta t/P_0$. Therefore the difference

$$\Delta\phi = 2\pi \Delta t\left(\frac{1}{P_1} - \frac{1}{P_0}\right)$$

is the ultimate angular position of the satellite after carrying out the maneuver relative to what the position would have been without the maneuver. For transfer orbits illustrated by O_+ in the figure $\Delta\phi$ is negative; that is, the spacecraft is *delayed* by the maneuver. Correspondingly, for transfer orbits illustrated by O_-, the spacecraft is *advanced*.

Substituting for the periods from (5.38), expanding by the binomial series, and retaining only the lowest order terms in e_1 result in the approximation

$$\Delta\phi = \mp\frac{3}{2}\sqrt{\frac{\mu}{a_0^3}} e_1 \Delta t \qquad (5.39)$$

Let v_0 be the velocity of the spacecraft before the first and again after the second thrust. At the periapsis of O_+ (apoapsis of O_-) the velocity v_1, from (3.14) and (5.37), is

$$v_1 = \sqrt{\mu\left(\frac{2}{a_0} - \frac{1}{a_1}\right)} = \sqrt{\frac{\mu}{a_0}(1 \pm e_1)}$$

so that Δv_1, the change in velocity by the first thrust, and Δv_2, by the second thrust, are

$$\Delta v_1 = -\Delta v_2 = v_1 - v_0 = \sqrt{\frac{\mu}{a_0}(1 \pm e_1)} - \sqrt{\frac{\mu}{a_0}}$$

$$= \pm\frac{1}{2}\sqrt{\frac{\mu}{a_0}} e_1 \qquad (5.40)$$

where again the binomial expansion for small e_1 was used. Combining (5.39) and (5.40),

$$\Delta v_1 \Delta t = -\tfrac{1}{3} a_0 \Delta\phi \qquad \textbf{(5.41)}$$

valid for either orbit. This shows that for a given displacement $\Delta\phi$ the quantity $2|\Delta v| = 2|\Delta v_1|$ that characterizes the expenditure of propellant by the maneuver can be made arbitrarily small, although only at the cost of a longer delay time Δt. For advancing (retarding) the spacecraft by an angle between 0 and 180°, a smaller $2|\Delta v|$ will result when the thrusts occur at the apoapsis (periapsis) of the transfer orbit.

For instance, if it is necessary to *increase* the geographical longitude (i.e., a displacement toward the east, hence in the (assumed) direction of the motion of the satellite) by an angle less than 180°, a velocity *decrease* by the first thrust and a velocity *increase* by the second thrust are required. The reverse will apply for a displacement toward the west.

To consider another example: A navigational system may have eight equally spaced satellites on each of its (circular) orbits. Replacement of one of the satellites by one of its neighbors will therefore require an angular displacement of 45°. For a radius $a_0 = 20{,}000$ km and allowing a time interval Δt of 30 days, $|\Delta v_1| = |\Delta v_2|$ from (5.41) becomes 2.02 m/s.

Because of the lack of exact sphericity of the earth, geostationary satellites tend to drift slowly from their intended station. The drift in the north–south direction resulting from the J_2 term in the gravitational potential (Sect. 2.1) is considerably larger than the drift in the east–west direction. Common practice is to correct the drifts by firing small thrusters on the spacecraft, perhaps every few days or every day.

The need for frequent north–south corrections can be reduced if the plane of the orbit is allowed to have a small inclination relative to the equatorial plane. This will save satellite propellant. For communication satellites, a complication, however, is that spot beams intended for downlinks to small receiving areas need to be steered, either mechanically or electronically.

5.10 Rendezvous Maneuvers

For the purposes of a rendezvous and, possibly, docking of two space vehicles, not only their orbits but also their positions on the orbit at a time must be matched. For docking, the relative angular orientation ("attitude") of the two vehicles must also be controlled. In the present section, we consider only the motions, and particularly the relative motions, of the centers of gravity of the two vehicles.

Rendezvous of spacecraft are useful for such purposes as the resupply of a spacecraft or space station by a cargo vehicle, for moving a spacecraft to a new orbit by means of a "space tug," or for a vehicle ascending from a planetary surface to meet an orbiting vehicle.

It is useful to distinguish several phases in the rendezvous maneuver. The ones that we consider here are

1. The transfer of one of the vehicles, here called the "active vehicle," from a parking orbit to the approximate orbit and time phasing of the second, or "passive vehicle." This phase of the motion is formulated in the geocentric (or planetocentric or heliocentric, depending on the application) reference frame, which for the present purposes is a very close approximation of a true inertial frame.

2. The closing maneuver that brings the two vehicles together and starts from what was at the end of the first phase only an approximate vicinity. To achieve the required precision, this phase is formulated in a reference frame that is fixed in the active vehicle. In general, this will not be an inertial frame.

5.10.1 Rendezvous Maneuver, First Phase

As illustrated in Fig. 5.14, it is assumed that the active spacecraft is already on a parking orbit, which is taken here to be circular. For simplicity, the passive vehicle is assumed to be on a coplanar, also circular, orbit. The two orbits are connected by a Hohmann transfer semiellipse. The maneuver therefore requires two impulsive thrusts by the active vehicle, one in leaving the parking orbit and a second one at the end of the transfer so that the positions, speeds, and directions of motion of the two vehicles can be matched.

Let r_1 and r_2 ($r_2 > r_1$) be the radii of the two orbits. The corresponding periods are, from (3.16),

$$P_1 = 2\pi \mu^{-1/2} r_1^{3/2}, \qquad P_2 = 2\pi \mu^{-1/2} r_2^{3/2} \tag{5.42}$$

The semimajor axis of the Hohmann transfer orbit is $(1/2)(r_1 + r_2)$. If t_1 is the time of the initiation of the Hohmann transfer by the active vehicle, t_2 the time of the rendezvous, then

$$\begin{aligned} t_2 - t_1 &= \pi \mu^{-1/2} \left(\frac{r_1 + r_2}{2} \right)^{3/2} \\ &= 2^{-5/2} P_1 \left(1 + \frac{r_2}{r_1} \right)^{3/2} \end{aligned} \tag{5.43}$$

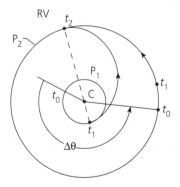

Figure 5.14 Rendezvous maneuver, first phase, with Hohmann transfer. Circular orbits with periods P_1 (active spacecraft) and P_2 (passive spacecraft). RV, rendezvous point; C, center of attraction.

If at some arbitrary initial time $t = t_0$, the angular separation of the two vehicles is $\Delta\Theta$, where

$$\Delta\Theta = 2\pi n + \Delta\theta$$

where $0 \leq \Delta\theta < 2\pi$. Here, $n = 0, 1, 2, \ldots$ indicates the number of additional full revolutions on the circular orbit that the active vehicle may have to make to meet the rendezvous condition.

The angular separation (true anomaly difference) $\Delta\theta$ between the two vehicles on their circular orbits is repeated regularly after each interval of time, P_{sn}, called the **synoptic period**. (The term is borrowed from astronomy, where it designates the time interval from one approximate alignment of sun, earth, and a planet to the next such alignment.) It follows from equating the polar angles of the two vehicles that

$$2\pi P_{\text{sn}}/P_1 = 2\pi (P_{\text{sn}}/P_2 + 1)$$

therefore

$$P_{\text{sn}} = P_1 P_2 / (P_2 - P_1) \tag{5.44}$$

The condition for the two vehicles to meet at the same point in space and time therefore becomes

$$\frac{2\pi (t_1 - t_0)}{P_1} + \pi = \frac{2\pi (t_2 - t_0)}{P_2} + \Delta\Theta$$

or

$$(P_2 - P_1)(t_1 - t_0) = P_1 P_2 \left(\Delta\Theta/2\pi - \tfrac{1}{2}\right) + \pi P_1 \mu^{-1/2} \left(\frac{r_1 + r_2}{2}\right)^{3/2}$$

from which

$$t_1 - t_0 = \left(\frac{\Delta\Theta}{2\pi} - \frac{1}{2}\right) P_{\text{sn}} + \frac{2^{-5/2} P_1 ((r_2/r_1) + 1)^{3/2}}{(r_2/r_1)^{3/2} - 1} \tag{5.45}$$

from (5.42) and (5.43). This time difference is the needed delay before the active vehicle initiates the Hohmann transfer.

To illustrate the result by numerical data: For a parking orbit at 300 km altitude above the earth and a geostationary satellite, the two periods P_1 and P_2 are 90.5 and 1436 minutes, respectively. The synoptic period, P_{sn}, from (5.44), is 96.6 minutes. The time $t_2 - t_1$ needed for the Hohmann transfer, from (5.43), is 316.5 minutes. If Θ is assumed to be 180°, the delay time $t_1 - t_0$ before initiating the transfer orbit, from (5.45), is 21.3 minutes.

5.10.2 Rendezvous Maneuver, Second Phase

The first phase of the maneuver will bring the active vehicle into the vicinity of the passive vehicle. But the reference frame and the two separate Kepler orbits for the two vehicles that have been used in the preceding section are not accurate enough to formulate the close-in final approach of the two vehicles.

Instead, a suitable reference frame for describing this second phase is one that is fixed in the *active* vehicle. In general, this will not be an inertial frame. This frame also has the advantage of being directly applicable to the radar and optical sensors on the active vehicle. These sensors image the

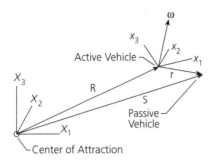

Figure 5.15 Docking maneuver: illustration of Eq. (5.47).

passive vehicle and are needed in the performance of a close-in maneuver to the necessary precision. In what follows we will be concerned only with the relative motion of the centers of mass of the vehicles, not with their relative angular orientation, which, of course, must also be closely controlled before docking.

Referring to Fig. 5.15, X_i, i = 1, 2, 3, are the Cartesian coordinate axes of the (nonrotating) frame of reference with origin in the mass center of the planet. Fixed to the active vehicle, with the origin in its center of mass, the x_i axes define the second reference frame that will be used. We designate the X_i components of the radius vector **R** from the center of the planet to the center of mass of the active vehicle by R_i, and correspondingly the components of the radius **S** to the passive vehicle by S_i. The components of the vector $\mathbf{r} = \mathbf{S} - \mathbf{R}$ in the x_i reference frame are designated by r_i. The angular velocity (relative to an inertial reference) of the active vehicle is designated by ω, with components ω_i in the x_i reference frame.

We designate by m the mass, by $\mathbf{F}_g$ the force of gravity, and by $\mathbf{F}_t$ the thrust, all referring to the active vehicle. Time differentiations are indicated by $d(\ )/dt$ in the X_i reference frame and by $(^\bullet)$ in the x_i frame.

In indicial notation, the equation of motion of the active vehicle in the planetocentric reference frame X_i, from (2.1), is

$$\frac{d^2 R_i}{dt^2} = -\frac{\mu R_i}{R^3} + \frac{F_{t,i}}{m} \quad (i = 1, 2, 3) \tag{5.46}$$

To express the equation of motion of the *passive* vehicle in the frame x_i that is fixed in the active vehicle, we make use of the symbol

$$e_{ijk} \begin{cases} = 1 & \text{if } (i, j, k) \text{ is an even permutation of } (1, 2, 3) \\ = -1 & \text{if } (i, j, k) \text{ is an odd permutation of } (1, 2, 3) \\ = 0 & \text{otherwise (i.e., the same index appears at least twice)} \end{cases}$$

The Cartesian components of vector products, such as the ith component of $\mathbf{a} \times \mathbf{b}$ are therefore

$$(\mathbf{a} \times \mathbf{b})_i = \sum_{j,k} e_{ijk} a_j b_k$$

In this notation the equation (1.4) for transformations between two reference frames in relative motion to each other, when applied to the passive vehicle, becomes

$$\ddot{r}_i = -\frac{\mu S_i}{S^3} - \frac{d^2 R_i}{dt^2} + \sum_j (\omega_j^2 r_i - \omega_i \omega_j r_j) - \sum_{j,k} e_{ijk}(2\omega_j \dot{r}_k + r_k \dot{\omega}_j)$$

The first term on the right can be expanded by the binomial series. Retaining only first-order terms in $|r_i/R| \ll 1$,

$$S^2 = \sum_j (R_j + r_j)^2 = \sum_j R_j^2 \left(1 + \frac{2r_j}{R_j}\right) = R^2 \left[1 + \frac{2}{R^2} \sum_j R_j r_j\right]$$

Using the binomial expansion once more results in

$$\frac{S_i}{S^3} = \frac{R_i(1 + r_i/R_i)}{R^3 \left(1 + \frac{2}{R^2} \sum_j R_j r_j\right)^{3/2}} = \frac{R_i}{R^3} \left[1 + \frac{r_i}{R_i} - \frac{3}{R^2} \sum_j R_j r_j\right]$$

Combining this with (5.46), the zero-order terms are seen to cancel so that only first-order terms in r_i and its derivatives remain.

When solved for the components of the control thrust $\mathbf{F}_t$, the final result obtained is

$$\frac{F_{t,i}}{m} = -\ddot{r}_i - \mu \frac{R_i}{R^3} \left[\frac{r_i}{R_i} - \frac{3}{R^2} \sum_j R_j r_j\right]$$
$$+ \sum_j (r_i \omega_j^2 - \omega_i \omega_j r_j) \sum_{j,k} e_{ijk}(2\omega_j \dot{r}_k + r_k \dot{\omega}_j) \qquad (i = 1, 2, 3) \quad \textbf{(5.47)}$$

This equation is basic to the development of the software for computing the thrust that is needed to control the active vehicle on a prescribed approach path. The vehicle's angular velocity ω and its derivative are obtained from inertial measurement units or similar devices based on the use of mechanical or ring-laser gyroscopes. At least three gyros and three rate gyros are needed. The distance vector $\mathbf{r}$ and the range rate are computed from data that are provided by radar and optical sensors. In this manner the required thrust as a function of time can be obtained by a closed-loop feedback control that compares at each instant the desired and actual distance vectors and initiates the needed corrective thrust.

5.11 Gravity Turn

Large launch vehicles normally take off from the launch site nearly vertically. Similarly, space probes, after having landed on a planet or moon, take off along a vertical. To make the transition to an orbit, the trajectory must then turn away from the vertical.

Because of their stringent weight limitations, launch vehicles cannot be designed structurally to withstand appreciable aerodynamic forces and moments. For this reason, the attitude of the vehicle, while in the denser part of the atmosphere, needs to be controlled to keep the angle of attack within at most a few degrees. Upper atmosphere winds and wind shear aggravate the structural and steering problems.

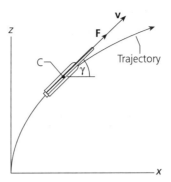

Figure 5.16 Gravity turn, schematic. C, center of mass of launch vehicle.

A trajectory that turns from a near-vertical path, yet minimizes the angle of attack, is call a **gravity turn** [5]. Once a slight deviation from the vertical (the "kick angle") has been initiated by steering, gravity alone will deflect the trajectory from the near vertical toward a more horizontal flight path (Figs. 5.16 and 5.17).

The turning of the trajectory will generally take place over a distance small compared with the radius of the earth or planet. Therefore the curvature of the planet's surface can be neglected, and so can the change with altitude of the gravitational acceleration.

As indicated in Fig. 5.16, let $x(t)$ be the horizontal, $z(t)$ the vertical position of the center of mass of the launch vehicle, $\mathbf{v}(t)$ the velocity, and $\gamma(t)$ the flight angle. $\mathbf{F}_t$ is the thrust, $\mathbf{F}_d$ the drag force acting on the vehicle. Both forces are nominally parallel to $\mathbf{v}$. The magnitude of the thrust will be assumed to be constant, whereas the drag will change with altitude and velocity. We write $\mathbf{F} = \mathbf{F}_t + \mathbf{F}_d$.

In coordinates fixed to the planet,

$$m(t)\, dv_x/dt = F(t)v_x/v$$
$$m(t)\, dv_z/dt = F(t)v_z/v - gm(t) \tag{5.48}$$

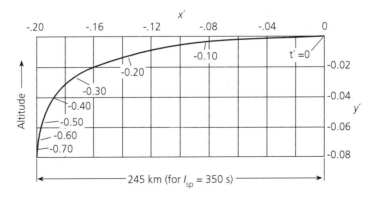

Figure 5.17 Solution of the equations (5.55) for a gravity turn. For $g = g_0$, $v'(t' = 0) = 1.00$, $k = 4.00$. Drag neglected.

where $m(t)$ is the vehicle's mass and $g =$ const. the gravitational acceleration. Multiplying the first equation by v_z, the second by v_x, and subtracting

$$v_z\, dv_x/dt - v_x\, dv_z/dt = gv_x \qquad (5.49)$$

From

$$\frac{d}{dt}(\tan\gamma) = \frac{d}{dt}\left(\frac{v_z}{v_x}\right) = \frac{1}{v_x^2}\left(v_z\frac{dv_x}{dt} - v_x\frac{dv_z}{dt}\right) = -\frac{g}{v_x}$$

follows

$$\frac{d\gamma}{dt} = -\frac{g}{v}\cos\gamma \qquad (5.50)$$

The temporal derivative of the speed is given by

$$\frac{dv}{dt} = \frac{F}{m} - g\sin\gamma \qquad (5.51)$$

the change in altitude by

$$\frac{dz}{dt} = v\sin\gamma \qquad (5.52)$$

It is convenient to use γ and v as the dependent variables.

In addition to this (nonlinear) system of differential equations, there are the auxiliary relations

$$m(t) = m_0 - \frac{F_t}{g_0 I_{sp}} t \qquad (5.53)$$

and

$$F_d = AC_d(M)\tfrac{1}{2}\rho_a(z)v^2 \qquad (5.54)$$

The former follows from (2.24) and

$$F_t = g_0 I_{sp} \tilde{m} = -g_0 I_{sp}\, dm/dt$$

Here, $\tilde{m}$ is the propellant mass flow rate, assumed to be constant, m_0 the vehicle mass at $t = 0$, A the vehicle maximum cross-sectional area, $C_d(M)$ the drag coefficient, and $\rho_a(z)$ the atmospheric density. The drag coefficient depends principally on the Mach number, M.

The solution of the system of equations (5.50) to (5.54) is obtained numerically. Neglecting the aerodynamic drag can serve as a useful first approximation. An example is shown in Fig. 5.17.

It is convenient in this case to introduce the nondimensional quantities

$$t' = t/I_{sp}, \quad m' = m/m_0, \quad v' = v/(g_0 I_{sp}), \quad k = F_t/(g_0 m_0)$$
$$x' = x/(g_0 I_{sp}^2), \quad y' = y/(g_0 I_{sp}^2)$$

where k is a constant. If, as a first approximation, aerodynamic drag is neglected, the equations of motion in terms of the nondimensional quantities become

$$\left.\begin{aligned}\frac{d\gamma}{dt'} &= -\frac{g}{g_0}\frac{\cos\gamma}{v'} \\ \frac{dv'}{dt'} &= \frac{k}{1-kt'} - \frac{g}{g_0}\sin\gamma\end{aligned}\right\} \quad (5.55)$$

To avoid the singularity that arises when the flight angle approaches 90°, it is advantageous to integrate the equations backward in time, starting, for instance, at the point where the flight path would become horizontal.

An example of the numerical solution of these equations is shown in Fig. 5.17. The calculation applies to a launch from the surface of the earth ($g = g_0$). The origin of time is taken at the point where the flight path would have become horizontal. The parameters chosen are $v'(t' = 0) = 1.00$, $k = 4.00$. For an assumed I_{sp} of 350 s, the horizontal distance traveled from the point of launch to the point of zero flight angle becomes 245 km. The velocity at this latter point becomes 3430 m/s.

Nomenclature

a	semimajor axis
e	eccentricity
E	eccentric anomaly
F	force
f, g	variables in Lambert's theorem, Eqs. (5.9)
g	gravitational acceleration; g_0: standard gravitational acceleration
i	inclination (Fig. 5.4)
I_{sp}	specific impulse
m	mass
n	mean angular velocity; number of complete revolutions
P	orbital period
r, R	radius vectors
S	radius vector (Fig. 5.15)
v	velocity
w	energy per unit mass
ΔW	energy increment resulting from impulsive thrust [(Eq. 5.25)]
β	turning angle (Fig. 5.7)
γ	flight angle (Fig. 5.16)
δ	latitude
Θ	true anomaly, including integer multiples of 2π
λ	longitude
μ	gravitational parameter
$\Delta\phi$	relative angular position
ω	angular velocity vector
$()_m$	minimum
$()_p$	periapsis
$()_{pr}$	propellant

( )$_s$ spacecraft
( )$_{sn}$ synoptic

Problems

(1) A spacecraft is on a 8000 km radius, circular orbit about Mars. A short-duration, impulsive thrust in the direction of motion is applied to increase the spacecraft's velocity further. Find numerically the minimum velocity increment that is needed to cause the spacecraft to escape from the Mars gravitational field. (The gravitational parameter of Mars is 42.81 10^3 km^3/s^2.)

(2) To avoid an accumulation of space debris in the geostationary orbit, a spacecraft, having served out its usefulness, is being propelled into a new orbit. The orbit change consists of two short-duration, impulsive rocket motor burns. The thrust vectors are chosen to be either parallel or antiparallel to the direction of motion at the time. The first of the two thrusts, which is tangent to the geostationary orbit, is hence at the perigee (or apogee, if the thrust is retrograde) of the resulting elliptic orbit. The second thrust is at the apogee (perigee) of this orbit.

The intent of the maneuver is to ensure that the final orbit is at a minimum distance of 300 km from the geostationary orbit. The final orbit can be either entirely on the outside or entirely on the inside of the geostationary orbit. The propellant consumption for removing the spacecraft will be proportional to the sum of the magnitudes of the two velocity changes Δv_1 and Δv_2.

Determine the magnitudes and directions of the two velocity changes for the case in which the sum $|\Delta v_1| + |\Delta v_2|$ of the magnitudes, hence the propellant consumption, is a minimum. (The earth's gravitational parameter is 3.9860 10^5 km^3/s^2.)

(3) Consider a **Hohmann transfer** from a circular orbit of radius r_1 to a second, coplanar circular orbit of radius r_2. Let Δv the sum of the two velocity increments needed for the maneuver.

Show that for a fixed gravitational parameter and radius r_1 the maximum required Δv occurs for $r_2 = 15.58 r_1$.

(4)* Consider a spacecraft that coasts on a **minimum energy** elliptic path in the gravitational field of the sun. The path starts at a point P_1 in the vicinity of Jupiter. P_1 is at 75.00° ecliptic longitude, 1.00° ecliptic latitude, and at a radius from the sun's center of 5.150 AU. It ends at point P_2 in the vicinity of Pluto. P_2 is at 335.00° ecliptic longitude, $-12.00°$ ecliptic latitude and at a radius of 38.00 AU. (1 AU = 1.495979 10^8 km; gravitational parameter of the sun = 1.32712 10^{11} km^3/s^2.)

  (a) Compute the semi major axis and eccentricity of the path.
  (b) Compute the time required to travel from P_1 to P_2.

(5)* A space vehicle is launched from the earth at sea level. The takeoff is vertical, followed by a maneuver, assumed instantaneous, that results in a flight angle of 88.0° (i.e., 2.0° from the vertical, the so-called "kick angle"). After this, the vehicle is to follow a **gravity turn** trajectory. The thrust and specific impulse, hence also the mass flow rate, are assumed constant throughout.

Let $(\ )_0$ designate quantities at the time of initiation of the gravity turn. The following numerical data are assumed:

Mass	m_0	$= 400\,000$ kg
Thrust	F_t	$= 7.00\ 10^6$ N
Specific impulse	I_{sp}	$= 350$ s
Flight angle	γ_0	$= 88.0$ deg.
Velocity	v_0	$= 300$ m/s

(a) Write a computer program for the gravity turn trajectory in Cartesian coordinates. A reference frame fixed to the earth is assumed. The effect of the earth's rotation, earth surface curvature over the horizontal distance of the trajectory, and change of gravity with altitude are neglected. Continue the calculation to the point where the trajectory would be horizontal.

(b) Modify the program so as to include aerodynamic drag. The atmospheric density, ρ_a, can be approximated by

$$\rho_a = \rho_{sl} \exp(-h/h_0)$$

where h is the altitude, $h_0 = 9295$ m the scale height of the atmosphere (near sea level), and $\rho_{sl} = 1.226$ kg/m^3 the standard atmospheric density at sea level. The maximum cross-sectional area of the vehicle is 18 m^2. The drag coefficient is assumed to be 1.0 throughout the gravity turn trajectory. Compare the result with that in (a).

(c) Modify program (a) by formulating it in the more exact inertial frame represented by polar coordinates in an earth-centered, nonrotating reference frame. Launch in the equatorial plane is assumed. Include the change in gravity with altitude for a spherical earth. Compare the result with that in (a).

References

1. Kaplan, M. H., "Modern Spacecraft Dynamics and Control," John Wiley & Sons, New York, 1976.

2. Brown, C. D., "Spacecraft Mission Design," AIAA Education Series, American Institute of Aeronautics and Astronautics, Washington, DC, 1992.

3. Prussing, J. E. and Conway, B. A., "Orbital Mechanics," Oxford University Press, New York, 1993.

4. Chobotov, V. A., "Orbital Mechanics," 2nd ed., American Institute of Aeronautics and Astronautics, Washington, DC, 1996.

5. Culler, G. J. and Fried, B. D., "Universal Gravity Turn Trajectories," *Journal of Applied Physics*, Vol. 28, No. 6, pp. 672–676, 1957.

6
Attitude Control

The need for close control of the **attitude** (i.e., orientation in space relative to some frame to be defined in each case) of *spacecraft* follows from such requirements as the need to point antennas and sensors toward the earth or other astronomical objects. The required accuracy greatly depends on the particular application. It can be relatively low for broadcasting satellites. In some scientific applications the required accuracy of the attitude control may be 1 arc minute or less.

Attitude control is also needed in the case of *launch vehicles* or *upper stage vehicles*, particularly because their orientation will influence the direction of thrust, hence the flight path.

A useful distinction can be drawn between external and internal disturbances that trigger the attitude control. An example of an **external disturbance** is the effect that is produced on a spacecraft by solar radiation pressure. Usually the spacecraft configuration is such that it is asymmetric with respect to the sun line; consequently, there will be a torque.

Such external torques can also be produced by gas leaking from a spacecraft, by aerodynamic torques in the upper atmosphere, and by the torque produced by the gravity gradient. In the case of a spacecraft that is not fully degaussed, the vehicle's permanent magnetic dipole, by interacting with a planetary magnetic field, will in general produce a small torque.

In Fig. 6.1, rough estimates of the more important external torques acting on typical earth-orbiting spacecraft are shown. Above about 10,000 km altitude, the predominant external perturbation is usually caused by the solar radiation pressure. At a distance of 1 AU from the sun, this pressure is $4.4 \ 10^{-6} \ N/m^2$. For small spacecraft, or if the spacecraft has approximate symmetry, the solar radiation torque at this distance from the sun can be as small as 10^{-5} Nm or even less.

External torques perturb the total angular momentum of the vehicle. Ultimately, this must be corrected by a countertorque, such as can be provided by pairs of small thrusters.

An example of an **internal disturbance** is the oscillating torque acting on the propellant tank, hence on the remainder of the vehicle, by fluid sloshing. Although this will not alter the vehicle's total angular momentum, it will disturb the orientation of sensors, particularly attitude control sensors, that are mounted on the vehicle's shell.

Particularly critical is also the attitude control of *reentry vehicles* and of vehicles that make use of aerobraking in a planetary atmosphere.

Because the stream of propellant gas at the nozzle exit of *rocket motors* is highly turbulent, perturbation torques, much larger than those shown in Fig. 6.1 for spacecraft, are produced each time such a motor is fired. On the

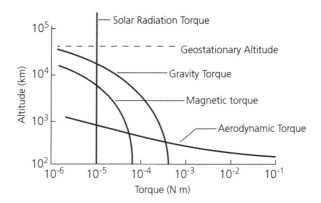

Figure 6.1 Orders of magnitude of the principal disturbance torques affecting earth-orbiting spacecraft. (Adapted from DeBra, D. B. and Cannon, R. H., "Momentum Vector Consideration in Wheeel-Jet Satellite Control," Guidance, Control and Navigation Conference, American Rocket Society, Stanford University, 1961.)

other hand, the needed accuracy of control, typically of the order of 1 or 2 degrees, can be less in this case.

In the case of *launch vehicles*, the main concern is usually the aerodynamic torque, which may be caused either by steering or by upper atmosphere wind shear. This effect can be critical because launch vehicles, particularly very large ones for which weight saving is at a premium, are not designed to tolerate substantial bending moments.

In this chapter, reference will be made mostly to *spacecraft*. Most results, however, translate easily to other vehicle types.

Concerning attitude control, many spacecraft can be classified into one of the two types represented in Fig. 6.2: spin-stabilized spacecraft or three-axis stabilized ones.

Spin-stabilized spacecraft (Fig. 6.2a) maintain their attitude by the gyroscopic effect resulting from spinning the cylindrically shaped portion of the spacecraft about its longitudinal axis. For full stability, the gyroscopic effect must be supplemented by so-called nutation dampers, passive devices that dissipate energy.

The cylindrical portion is usually the more massive one because it typically contains such heavy items as the propellant needed for attitude control, the storage batteries, and often the empty case of the solid-propellant motor that was used for the final orbit insertion. Angular velocities of the order of one revolution per second are quite common. Spin-up is achieved by pairs of thrusters firing in a plane perpendicular to the nominal spin axis.

The bore sights of the antennas and of some of the sensors need to be pointing in a desired direction, for instance, toward the earth. They are therefore mounted on a **despun platform**, which is supported by a shaft and bearings in the cylindrical section. The designation "despun" here refers to a reference frame corotating with the earth, not with inertial space. Since the angular velocity, relative to inertial space, of the platform is very low

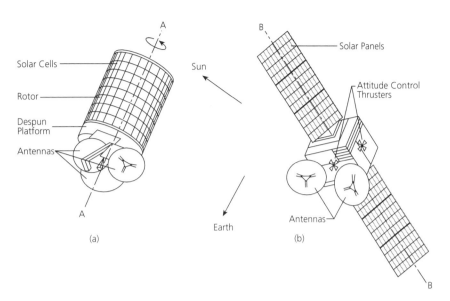

Figure 6.2 (a) Spin-stabilized spacecraft; (b) three-axis stabilized spacecraft. A–A, spin axis; B–B, solar panel rotation axis.

(for a circular orbit 2π times the inverse of the orbital period), the despun section does not appreciably contribute to or subtract in this case from the gyroscopic effect of the spinning section.

Electric power and signal circuits connect the two sections by means of slip rings. The bearing and slip ring friction is compensated for by small electric motors (as otherwise the two sections would start to share their angular momenta, speeding up the despun platform and slowing down the spinning section).

For orbiting spacecraft, the nominal spin axis is perpendicular to the orbit plane. This allows the antennas on the despun section to keep on pointing in the desired direction as the spacecraft moves along its orbit.

Except during maneuvers, spin-stabilized spacecraft do not require active sensors for their attitude control. The control system is therefore substantially simpler than that needed for three-axis stabilized spacecraft. On the other hand, because at any time only approximately one-half of the solar cells are illuminated, the maximum available solar electric power is more limited. Another reason for this is that the drum that carries the solar cells of spin-stabilized spacecraft is restricted in size by the launch vehicle payload space, whereas three-axis stabilized spacecraft can make use of fold-out solar panels, the size of which is not so limited.

To carry out the maneuvers needed to reach final orbit, the entire vehicle is usually despun prior to its reorientation to match the intended thrust direction. This can be accomplished by firing a pair of small thrusters or by the yo–yo mechanism described in Sect. 1.3. Having acquired the new orientation, the vehicle is spun up again. This serves to keep the attitude steady during the firing of the motor.

Three-axis stabilized spacecraft (Fig. 6.2b) use actuators, such as thrusters, to maintain the desired orientation. For instance, in the case of geostationary communication satellites, the desired orientation is such that the communication antennas remain pointed toward the ground stations. Relative to inertial space, this requires one full rotation of the spacecraft once every 24 h.

In the case of deep-space missions, the normal orientation of the vehicle is such that the high-gain antenna is directed toward the earth.

An additional consideration is that the deployed solar panels should be close to being perpendicular to the sun line, so as to produce the maximum solar electric power. For this purpose, they are mounted on a shaft, often on a common shaft when there are two solar panels on opposite sides of the spacecraft. Relative to the spacecraft, this shaft is slowly rotated by an electric motor. For an orbiting spacecraft, the rate of rotation will correspond to the orbital period. With optimal orientation of the spacecraft, the maximum possible solar incidence angle on the panels will be 90° at the time of spring and fall equinox and 66.5° (90° less the 23.5° inclination of the ecliptic) at solstice. Figure 7.7 illustrates the astronomical and geometrical constraints.

The twin requirements of pointing the antenna beams and at the same time the solar panels require that the orientation of all three spacecraft axes be controlled. This is accomplished by a control system that receives inputs from sensors and commands actuators. For orbiting spacecraft, the most important sensors are so-called horizon sensors that scan periodically the astronomical body that is being orbited. Several types of actuators are in use. Common ones are small thrusters arranged such that in various combinations they can rotate the spacecraft about any of its axes.

Because they are peculiar to space technology, the emphasis in this chapter is placed on **sensors** and **actuators**. They are an essential part of all space vehicles. By contrast, the electronic aspects and the underlying **control theory** differ only in detail from their very broad applications in many other fields of technology. The goals, of course, are to achieve stability, quick response, and relative insensitivity to parameter changes. Control theory has become a discipline all by itself, usually taught in separate courses. It is therefore largely passed over in this book.

The development of the **software** needed for attitude control is a major task. It is usually based on the description of the spacecraft rotations by Euler angles and their transformations. Because quaternions are mathematically the most natural way to describe rotations, it has also become customary to use them in software developments. This also goes beyond the scope of this book.

The treatment in this chapter follows in major parts that of Kane, Likins, and Levinson [1] and that of Bryson [2]. Different treatments and additional material can be found in Refs. 3 to 6.

6.1 Principal Axes and Moments of Inertia of Spacecraft

As a preliminary to attitude control analysis it is necessary to determine the principal axes and the principal moments of inertia of the spacecraft, which, for this purpose, is assumed to be a rigid body. The importance of

6.1 Principal Axes and Moments of Inertia of Spacecraft

these concepts derives from their connection to the angular momentum vector, say **L**, about the center of mass of the spacecraft. If **r** is the position vector extended from the center of mass, and $\omega = \omega(t)$ the angular velocity of the spacecraft relative to inertial space,

$$\mathbf{L} = \int \mathbf{r} \times (\omega \times \mathbf{r})\, dm \qquad (6.1)$$

where the integration is extended over the total mass of the spacecraft. Making use of the vector identity

$$\mathbf{r} \times (\omega \times \mathbf{r}) = r^2 \omega - (\omega \cdot \mathbf{r})\mathbf{r}$$

a useful alternative expression for **L** is

$$\mathbf{L} = \omega \int r^2\, dm - \int (\omega \cdot \mathbf{r})\mathbf{r}\, dm$$

We next introduce a set of mutually perpendicular axes with their origin at the center of mass and fixed to the spacecraft, hence corating with it. Let L'_i, x'_i, ω'_i ($i = 1, 2, 3$) denote the components of **L**, **r**, ω in this set of axes. Therefore

$$\left.\begin{array}{l} L'_1 = I'_{11}\omega'_1 + I'_{12}\omega'_2 + I'_{13}\omega'_3 \\ L'_2 = I'_{21}\omega'_1 + I'_{22}\omega'_2 + I'_{23}\omega'_3 \\ L'_3 = I'_{31}\omega'_1 + I'_{32}\omega'_2 + I'_{33}\omega'_3 \end{array}\right\} \qquad (6.2)$$

where

$$\left.\begin{array}{l} I'_{11} = \int (x'^2_2 + x'^2_3)\, dm, \quad I'_{22} = \int (x'^2_3 + x'^2_1)\, dm, \quad I'_{33} = \int (x'^2_1 + x'^2_2)\, dm \\ I'_{ij} = I'_{ji} = -\int x'_i x'_j\, dm \quad (i \neq j) \end{array}\right\} \qquad (6.3)$$

More compactly, where δ_{ij} is the Kronecker delta,

$$I'_{ij} = \int (\delta_{ij} x'_k x'_k - x'_i x'_j)\, dm \qquad (6.3')$$

The terms I'_{11}, I'_{12}, etc. are the components of the **inertia tensor**, say **I**. (That it is a Cartesian tensor, i.e., that it satisfies the transformation equations defining such tensors, requires a proof, not given here.) As is seen from (6.3), **I** is symmetric and has the matrix representation

$$\begin{bmatrix} I'_{11} & I'_{12} & I'_{13} \\ I'_{12} & I'_{22} & I'_{23} \\ I'_{13} & I'_{23} & I'_{33} \end{bmatrix} \qquad (6.4)$$

Equation (6.2) can be written more compactly as

$$\mathbf{L} = \mathbf{I}\omega \qquad (6.5)$$

Finding the **principal axes** amounts to determining the eigenvectors of the inertia tensor. By way of mathematical background, the reader will recall that the eigenvectors of a tensor **T** and the associated eigenvalues are obtained by considering the vector equation

$$\mathbf{T}\mathbf{x} = \lambda \mathbf{x} \tag{6.6}$$

If a vector $\mathbf{x}$, other than $\mathbf{x} = 0$, satisfies this equation, the vector is called an eigenvector, say $\mathbf{e}$, of $\mathbf{T}$. Associated with each eigenvector $\mathbf{e}$ is its eigenvalue λ. Because the equation is homogeneous in $\mathbf{e}$, there is no loss in generality by assuming that all eigenvectors are of magnitude one ("unit eigenvectors"). Equivalently, the equation can be expressed by a set of homogeneous, scalar equations for the components. For a nontrivial solution to exist, the determinant formed from the coefficients in these equations must vanish, which gives rise to the so-called characteristic (or "secular") equation for the eigenvalues. Furthermore, if, as is the case for the inertia tensor, $\mathbf{T}$ is real and symmetric, it can be shown that all eigenvalues are real and that the eigenvectors can be chosen to be mutually orthogonal.

Applying this to the inertia tensor, there will be three orthonormal eigenvectors $\mathbf{e}$, each with its eigenvalue λ. The Cartesian components of any chosen eigenvector in the coordinate system (x_1', x_2', x_3') will be designated by e_1', e_2', e_3'. The three scalar equations corresponding to (6.6) for this eigenvector are therefore

$$\left.\begin{array}{r}(I_{11}' - \lambda)e_1' + I_{12}'e_2' + I_{13}'e_3' = 0 \\ I_{12}'e_1' + (I_{22}' - \lambda)e_2' + I_{23}'e_3' = 0 \\ I_{13}'e_1' + I_{23}'e_2' + (I_{33}' - \lambda)e_3' = 0\end{array}\right\} \tag{6.7}$$

The condition that there exist solutions other than the trivial one $\mathbf{e} = 0$ therefore leads to the characteristic equation

$$\begin{vmatrix} I_{11}' - \lambda & I_{12}' & I_{13}' \\ I_{12}' & I_{22}' - \lambda & I_{23}' \\ I_{13}' & I_{23}' & I_{33}' - \lambda \end{vmatrix} = 0 \tag{6.8}$$

hence a cubic in λ. As a consequence of the symmetry of the inertia tensor, all three solutions are real. They will be designated by $\lambda_1, \lambda_2, \lambda_3$.

Having determined the eigenvalues from the characteristic equation, the components of each unit eigenvector in the (x_1', x_2', x_3') coordinate system are found successively from (6.7), together with the normalization condition

$$e_1'^2 + e_2'^2 + e_3'^2 = 1$$

If no eigenvalues are equal, we can choose a new Cartesian coordinate system (x_1, x_2, x_3), with origin at the center of mass, so that the unit eigenvectors just calculated are the base vectors of a new (right-handed) Cartesian coordinate system. The new axes are called **principal axes**. As is easily shown, in this coordinate system the products of inertia vanish, so that the matrix representation of the inertia tensor in the principal axes system is

$$\begin{bmatrix} I_1 & 0 & 0 \\ 0 & I_2 & 0 \\ 0 & 0 & I_3 \end{bmatrix} \tag{6.9}$$

where

$$I_1 = \int \left(x_2^2 + x_3^2\right) dm, \qquad I_2 = \int \left(x_3^2 + x_1^2\right) dm, \qquad I_3 = \int \left(x_1^2 + x_2^2\right) dm \tag{6.10}$$

I_1, I_2, I_3 are called the **principal moments of inertia**. (A single subscript is used to distinguish them from the moments of inertia in the general case.) It follows easily that

$$I_1 = \lambda_1, \qquad I_2 = \lambda_2, \qquad I_3 = \lambda_3 \tag{6.11}$$

If two of the eigenvalues — and therefore also two of the principal moments of inertia — are equal, the eigenvector associated with the third eigenvalue determines one of the principal axes. The other two principal axes are perpendicular to this axis and to each other but otherwise can be chosen arbitrarily. A circular cylinder with uniform density is an example. More generally, a geometrically irregular body, possibly with nonuniform density, can still be an example if two of the eigenvalues are equal. One then speaks of an "inertially symmetric" body.

To simplify the attitude control system, many spacecraft are designed to be inertially symmetric. Often, this symmetry can be obtained solely by a judicious arrangement of the components within the spacecraft envelope. Quite common are component distributions that result in off-diagonal elements of the inertia matrix that are no more than about 1% of the principal moments of inertia.

If all three eigenvalues are equal, the orientation of the (mutually orthogonal) principal axes can be chosen arbitrarily, and all moments of inertia are equal. The inertia tensor is said to be isotropic.

6.1.1 Application to Spacecraft Calculations

In computing the principal axes and moments of inertia of a spacecraft, the most time-consuming task is the calculation of the integrals in (6.3) for an *a priori* specified coordinate reference system (x_1', x_2', x_3').

For this purpose, the spacecraft is segmented into a large number of parts whose contributions to the integrals are separately evaluated and then summed. Software exists that facilitates the segmentation and also has available the principal axes and principal moments of inertia of many geometrical shapes with constant or variable density, such as prismatic bodies, circular cones, or spherical shells. The spacecraft designer then need only specify the dimensions, the density, the location of the center of mass in the spacecraft reference coordinates, and the direction cosines of the part's reference axes with respect to the spacecraft reference axes.

Two such segments, one of general shape and one prismatic, are illustrated in Fig. 6.3, together with their reference axes.

Generally, the centers of mass of the segmented part and of the spacecraft will be different. So will the orientations of the respective reference axes. Therefore a transformation will be needed to make the segment's inertial tensor that is contained in the software useful for the calculation of the spacecraft inertial tensor. This transformation can be viewed as a two-step

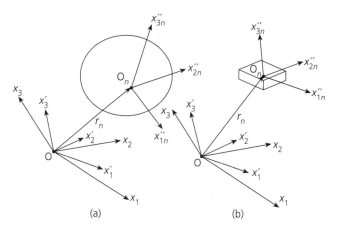

Figure 6.3 Computation of the spacecraft inertia tensor from the inertia tensors of its components. Examples: (a) propellant tank; (b) electronic box. O, center of mass of spacecraft; O_n of nth component.

process: (1) a parallel translation that brings the two centers of mass into coincidence and (2) a rotation that causes the two sets of reference axes to coincide.

Let (x'_1, x'_2, x'_3) designate the **spacecraft reference coordinates** of a point on the spacecraft, with their origin at the spacecraft center of mass. For convenience, one of the axes is usually chosen to be parallel to the nominal pointing direction of the principal sensor or of the high-gain antenna, or in the direction of the thrust axis. In general, the spacecraft reference axes will not be principal axes. (At the start of the design process, these are not known.)

Further, let $(x''_{1,n}, x''_{2,n}, x''_{3,n})$ designate the **component reference coordinates** of a point in the nth segmented part. The origin is chosen to coincide with its center of mass. Analogous to (6.3'), the (i, j)th component $J''_{ij,n}$ of the inertia matrix of the part is

$$J''_{ij,n} = \int \left[\delta_{ij} x''_k x''_k - x''_i x''_j\right]_n dm \tag{6.12}$$

If $\mathbf{r}_n$ is the radius vector from the spacecraft center of mass to the center of mass of the nth part, the (i, j)th component $I''_{ij,n}$ of the inertia matrix, after the parallel displacement of the part's reference axes so that their origin coincides with the spacecraft center of mass, is

$$I''_{ij,n} = \int \left[\delta_{ij}(r'_k + x''_k)(r'_k + x''_k) - (r'_i + x''_i)(r'_j + x''_j)\right]_n dm$$

The mixed terms that result from the multiplication vanish as a consequence of the definition of the center of mass, so that, with m_n as the mass of the nth part

$$I''_{ij,n} = J''_{ij,n} + m_n\left[\delta_{ij} r'_k r'_k - r'_i r'_j\right]_n \tag{6.13}$$

Let $l'_{ij,n} = \cos(x''_{i,n}, x'_{j,n})$ be the direction cosines of the reference axes of the segmented part with those of the spacecraft. The (i, j)th element of the inertia matrix for the entire spacecraft, when expressed in the spacecraft reference system, is therefore

$$I'_{ij} = \sum_n l'_{ki,n} l'_{lj,n} \left(J''_{kl,n} + m_n [\delta_{kl} r'_m r'_m - r'_k r'_l]_n \right) \quad (i, j, k, l, m = 1, 2, 3)$$

(6.14)

This is the same quantity that was expressed earlier by (6.3), except now more detailed as needed for practical calculations. The computation required for finding the eigenvalues and eigenvectors of the spacecraft proceeds as in (6.8) and (6.7).

As a simple example we may consider a prismatic body of constant density. This is often a useful approximation for **electronic boxes** on spacecraft (Fig. 6.3b). In this example it is convenient to define the reference axes of the part to coincide with its principal axes. If the length, width, and height are b_1, b_2, and b_3, the corresponding moments of inertia are easily found to be

$$J''_{ij} = \begin{cases} (m_b/12)(b_1^2 + b_2^2 + b_3^2 - b_i^2) & \text{if } i = j \\ 0 & \text{if } i \neq j \end{cases}$$

where m_b is the mass of the box.

Hence from (6.13)

$$I''_{ij} = \begin{cases} m_b \left[\dfrac{b_1^2 + b_2^2 + b_3^2 - b_i^2}{12} + r_1'^2 + r_2'^2 + r_3'^2 - r_i'^2 \right] & \text{if } i = j \\ -m_b r'_i r'_j & \text{if } i \neq j \end{cases}$$

The contribution by the box to the spacecraft inertia tensor, expressed in the spacecraft reference system, then follows from (6.14).

6.2 The Euler Equations for Time-Dependent Moments of Inertia

In analyzing the attitude control of spacecraft, it can usually be assumed that the inertial properties are time independent. Although the properties will change as propellants are expended, this change is usually very slow compared with the attitude control frequencies and can be neglected. The attitude control of such spacecraft can then be described by means of the classical Euler's equations for the rotational motion of rigid bodies.

In some other, more rare, instances the temporal change of the moments of inertia needs to be taken into account. This can occur with small, rapidly spinning vehicles with fast-burning solid propellants at the time of the main motor firing. As propellant is expended, the center of mass moves relative to the rest of the vehicle, and — more important — the moments of inertia change. As will be assumed in the following, in these applications the orientation of the principal axes relative to the vehicle will not change.

In what follows, we derive the modifications of Euler's equations that are needed when the time rates of change of the moments of inertia are comparable to the rate of rotation of spinning vehicles. In practice, the case of interest is the inertially symmetric vehicle with nominal spin axis, thrust axis, and one of its principal axes all coinciding. This axis will be designated by x_3.

The angular momentum equation in the form most useful for applications to spacecraft refers to the total angular momentum **L** of the spacecraft about its (instantaneous) center of mass. If **M** is the resultant moment (such as may result from the disturbance torques shown in Fig. 6.1) about the center of mass of the spacecraft and caused by external forces acting on it,

$$d\mathbf{L}/dt = \mathbf{M} \tag{6.15}$$

Here, and in what follows in this chapter, $d(\ )/dt$ refers to the time derivatives in inertial space, whereas $(\dot{\ })$ refers to time derivatives in the space fixed to the spacecraft. (In addition to external moments, there can also be internal moments, such as those caused by the sloshing of liquid propellants, that disturb the attitude; these do not affect the total angular momentum.)

From (1.1) for the connection between the two time derivatives,

$$d\mathbf{L}/dt = \dot{\mathbf{L}} + \omega \times \mathbf{L}$$

where ω is the instantaneous angular velocity of the spacecraft relative to inertial space. In component form, expressed in the spacecraft principal axes coordinate system (x_1, x_2, x_3),

$$L_1 = I_1 \omega_1, \qquad L_2 = I_2 \omega_2, \qquad L_3 = I_3 \omega_3$$

as follows from (6.2) when applied to the principal axes. From this follows the final result

$$\left.\begin{array}{l} I_1 \dot{\omega}_1 + \omega_1 \dot{I}_1 + (I_3 - I_2)\omega_2 \omega_3 = M_1 \\ I_2 \dot{\omega}_2 + \omega_2 \dot{I}_2 + (I_1 - I_3)\omega_3 \omega_1 = M_2 \\ I_3 \dot{\omega}_3 + \omega_3 \dot{I}_3 + (I_2 - I_1)\omega_1 \omega_2 = M_3 \end{array}\right\} \tag{6.16}$$

It may be noted that the vector ω is special in the sense that there is no distinction between its derivative in the two reference spaces. For it follows immediately from (1.1) that $d\omega/dt = \dot{\omega}$. Strictly speaking, the components $\omega_1, \omega_2, \omega_3$ are not simply time derivatives of certain angles. Instead, they are nonintegrable combinations of time derivatives of the angular displacements.

The second term in (6.16) represents the effect caused by the rate of change of the principal moments of inertia. When these rates can be neglected, the well-known Euler equations

$$\left.\begin{array}{l} I_1 \dot{\omega}_1 + (I_3 - I_2)\omega_2 \omega_3 = M_1 \\ I_2 \dot{\omega}_2 + (I_1 - I_3)\omega_3 \omega_1 = M_2 \\ I_3 \dot{\omega}_3 + (I_2 - I_1)\omega_1 \omega_2 = M_3 \end{array}\right\} \tag{6.16'}$$

for the rotational motion of solid bodies result. When the rates of change of the inertial terms are small compared with the attitude control frequencies, as happens in the majority of attitude control problems, the inertial terms can be treated as quasi-constant. That is, Eq. (6.16') can be applied at each instant, although the constants change on the (longer) time scale that characterizes the rate of propellant depletion.

In applications, most often the moments of inertia and the external torques are given, and the task is to find the angular velocity. Although there are important special cases in which the equations are linear and simple to solve, in the general case Euler's equations are nonlinear.

6.3 The Torque-Free Spinning Body

As an example of the application of the Euler equations (6.16') we consider the rotational motion of a rigid, spinning body. This topic is treated in most textbooks in mechanics. Because it has numerous applications to spacecraft, it is also included here.

6.3.1 The Torque-Free, Inertially Symmetric, Spinning Body

Spin-stabilized spacecraft are usually designed so that two of the principal moments of inertia, say I_1 and I_2, are very close to being equal. Such spacecraft are said to be **inertially symmetric**. In practice, the agreement may be 1% or even better. This can be achieved by judiciously placing within the envelope of the spacecraft its various components. Because most spacecraft are more weight than volume limited, the designer has considerable freedom in placing the components to achieve this symmetry. The attitude control of inertially symmetric, spinning spacecraft is substantially simpler than for asymmetric vehicles.

Let x_1, x_2, x_3 be the body's principal axes and ω_1, ω_2, ω_3 the corresponding of the angular velocity components relative to inertial space. Rigid bodies for which one of the principal moments of inertia, say I_3, is smaller than the two (equal) others are called **prolate** (Latin *latus*: side, flank; literally "protruding flank," i.e., pencil-like). If larger, the body is called **oblate** (literally "shortened flank," i.e., disklike). The axis x_3, with moment of inertia I_3, is usually the nominal spin axis. The two cases are illustrated in Fig. 6.4.

Assuming that $I_1 = I_2$ and that there are no external torques, Euler's rigid body equations simplify to

$$\left. \begin{array}{r} \dot{\omega}_1 + \left(\dfrac{I_3}{I_1} - 1\right) \omega_2 \omega_3 = 0 \\[6pt] \dot{\omega}_2 - \left(\dfrac{I_3}{I_1} - 1\right) \omega_1 \omega_3 = 0 \\[6pt] \dot{\omega}_3 = 0 \end{array} \right\} \quad (6.17)$$

As indicated by the last of these equations, ω_3 is constant. The first two equations therefore are a pair of linear differential equations for ω_1 and ω_2. Let

$$\Omega = \left(\frac{I_3}{I_1} - 1\right) \omega_3 \quad (6.18)$$

therefore also a constant. With this definition, the general solution of (6.17) is seen to be

$$\left. \begin{array}{l} \omega_1 = \omega_{12} \, \cos(\Omega t - \chi) \\ \omega_2 = \omega_{12} \, \sin(\Omega t - \chi) \end{array} \right\} \quad (6.19)$$

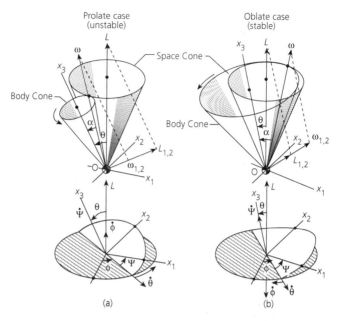

Figure 6.4 Torque-free, inertially symmetric, spinning body. (a) Prolate body ($I_1 = I_2 > I_3$, unstable motion); (b) oblate body ($I_1 = I_2 < I_3$, stable motion). O, center of mass; L, angular momentum; ω, angular velocity; x_1, x_2, x_3, body-fixed, principal axes; θ, nutation angle.

where χ is an arbitrary phase angle and ω_{12} the amplitude. It follows that

$$\omega_1^2 + \omega_2^2 = \omega_{12}^2 = \text{const.}$$

which shows that ω_{12} is the projection of $\boldsymbol{\omega}$ on the (x_1, x_2) plane. The projected vector, $\boldsymbol{\omega}_{12}$, is seen to rotate in time around the x_3 axis at the rate Ω. Also, since ω_{12} and ω_3 are both constant, the magnitude, ω, of the angular velocity remains constant. For the same reason, the angle, say α, between the angular velocity vector and the x_3 axis stays constant.

Because, by assumption, there is no external torque, the angular momentum, **L**, is constant relative to inertial space (although not relative to a body-fixed reference). From (6.2), when specialized to principal axes, and with $I_1 = I_2$, it follows that

$$L_1 = I_1\omega_1, \qquad L_2 = I_1\omega_2, \qquad L_3 = I_3\omega_3$$

The projection $\mathbf{L}_{12}$ on the (x_1, x_2) plane of **L** is therefore collinear with the projection of $\boldsymbol{\omega}$. The angular momentum, the angular velocity, and the x_3 principal axis of the body are therefore all in the same plane.

Given that initially the x_3 axis includes an angle θ with **L**, the x_3 axis will precess about **L** at a constant rate and constant θ. This angle is often called the **nutation angle** (nutation = nodding motion; the expression stems from the more general case in which the three principal moments all differ and θ, contrary to the inertially symmetric case, can be oscillating, hence the "nodding").

There exists a simple relation between the two angles α and θ. Since, as is readily seen, $\tan \alpha = \omega_{12}/\omega_3$ and $\tan \theta = L_{12}/L_3$, it follows that

$$\tan \theta = (I_1/I_3) \tan \alpha \qquad (6.20)$$

Hence in the prolate case $\alpha < \theta$, in the oblate case $\alpha > \theta$.

As illustrated in the figure, the angular velocity vector can be visualized as being coincident with the contact line between two circular cones, one centered on the x_3 axis and fixed to the body. It is called the **body cone**. The other cone is centered on **L**, is fixed in inertial space, and is called the **space cone**. It is easily shown that the two cones roll on each other without sliding. In the prolate case, the two cones are exterior to each other; in the oblate case, the body cone is exterior to the space cone. Different initial conditions merely change the half-angles of the two cones.

The same system of axes (x_1, x_2, x_3) is also shown in the lower part of the figure, together with the Euler angles and their time derivatives.

The angular velocity vector, in place of decomposing it into its components $\omega_1, \omega_2, \omega_3$, can also be represented as the vector sum of a vector **p** parallel to the angular momentum vector **L** and a vector **s** parallel to the body symmetry axis, x_3. The first is referred to as the **precession**, the second as the **spin**. The magnitude of **p** is the rate of precession and measures the rate at which the plane that contains **L**, ω, and the x_3 axis rotates about **L**.

Given the moments of inertia, there are *two* independent variables, for instance, ω_3 and θ. All other quantities of interest can be expressed in terms of these. For instance, from $p \sin \theta = \omega_3 \tan \alpha$, together with (6.20), follows

$$p = \frac{I_3}{I_1} \frac{\omega_3}{\cos \theta} \qquad (6.21)$$

6.3.2 Stability of the Motion

Even in the absence of external torques, spinning spacecraft can slowly lose kinetic energy of rotation. Propellant sloshing and friction in the bearings between the spinning part and the despun platform can convert some of the kinetic energy into thermal one. The total angular momentum will be unaffected, but the nutation angle can slowly change.

This consideration leads naturally to the question of **stability** of the gyroscopic motion. In an analysis that is at least qualitatively correct, one assumes that at each instant the motion can be described as in Sect. 6.3.1 and that there is merely a slow change of the parameters. (The notion of folding all dissipative effects into a single scalar quantity that acts to diminish the rotational kinetic energy of the spacecraft is being used frequently. It appears to be impossible to establish this entirely from first principles. Indeed, some conditions, although pathological in practice, have been discovered where this breaks down. In realistic problems of spacecraft motions no such anomalies have ever been found.)

The kinetic energy of rotation is

$$T = \frac{1}{2} \int (\omega \times \mathbf{r})^2 \, dm \qquad (6.22)$$

where **r** is the position vector extended from the center of mass. Evaluating the vector product and using the definitions (6.3) of the moments and products of inertia result in

$$T = \tfrac{1}{2}\omega_i' \omega_j' I_{ij}', \qquad i, j = 1, 2, 3$$

Writing (6.2) for the angular momentum more compactly,

$$L_i' = I_{ij}' \omega_j', \qquad i, j = 1, 2, 3$$

Hence

$$T = \tfrac{1}{2}\boldsymbol{\omega} \cdot \mathbf{L} \tag{6.23}$$

As is readily seen from Fig. 6.4, the projection of the angular momentum on the x_3 *principal* axis is $L_3 = L\cos\theta$. Also $L_3 = I_3\omega_3 = I_3\omega\cos\alpha$. Therefore

$$\omega = \frac{L\cos\theta}{I_3 \cos\alpha}$$

Writing in (6.23) the scalar product in terms of the *principal* axes components and using the expression just found for ω,

$$T = \frac{L^2}{2I_3} \cdot \frac{\cos\theta \cos(\theta - \alpha)}{\cos\alpha} \tag{6.24}$$

The factor containing L and I_3 is constant. The second factor containing the cosines, when expanded in a power series about $\theta = 0$, results in

$$T = \frac{L^2}{2I_3}\left[1 - \left(1 - \frac{I_3}{I_1}\right)\theta^2\right] \tag{6.25}$$

valid approximately for small nutation angles.

The factor $1 - I_3/I_1$ is positive for prolate, negative for oblate bodies. As the rotational kinetic energy decreases because of internal dissipative effects, θ is seen to increase in the prolate, decrease in the oblate case. A **prolate** body, if initially spun approximately about its inertial axis of symmetry, will tend to increase its nutation angle, hence is **unstable**. It will finally

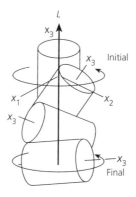

Figure 6.5 Unstable motion of spinning body with $I_1 = I_2 > I_3$.

go into a flat spin, as illustrated in Fig. 6.5. On the other hand, an **oblate** body will tend to return to a spin about its inertial axis of symmetry, hence be **stable**.

The rate of dissipation of kinetic energy in spacecraft is low. The growth of the instability of prolate vehicles is therefore correspondingly slow. In practice, it may take many minutes before it reaches amplitudes that could be of concern. **Nutation dampers**, discussed in Sect. 6.6.2, are used to eliminate entirely the instability of spinning, prolate vehicles.

For aerodynamic and structural reasons, the diameter of the shroud that contains the spacecraft on the launch vehicle is typically smaller than its length. Therefore this often calls for prolate spacecraft. Nutation dampers therefore find frequent applications.

6.3.3 General Case of the Torque-Free, Spinning Body

The analysis of the general case of a torque-free, spinning body with principal moments of inertia that all differ from each other is considerably more complicated than the inertially symmetric case. One reason is that the simple reduction of the solid-body Euler equations to a linear set is no longer possible.

This general case forms a classical topic in mechanics. It was studied by Poinsot (1777–1859), Klein (1849–1925), and others. The solution can be expressed in terms of elliptic integrals. But with the advent of high-speed computers it is much more practical to integrate Euler's equations directly for the parameter values of interest to the spacecraft designer.

The motion, relative to inertial space, of such a body can be viewed as consisting of three parts: (1) a rotation about one of the principal axes; (2) a precession of this axis about the (constant) angular momentum vector; (3) a nutation (nodding) of the axis relative to the angular momentum vector.

The motion turns out to be **stable** if the body is initially spun about the principal axis that corresponds to the *maximum* moment of inertia (or, more generally, if the angle between the angular velocity and the axis with the maximum moment of inertia is less than 90°).

Although spin stabilization of inertially nonsymmetric spacecraft is feasible, few such vehicle designs have been considered. When physical and geometric requirements imposed by the payload dictate a nonsymmetric design, three-axis stabilized (nonspinning) spacecraft are usually preferred.

It has happened on occasion that because of a faulty deployment of a spacecraft component, or because of an error in the attitude control system, a spacecraft was sent into an **uncontrolled tumbling motion**. Often, the control can be recovered and an otherwise catastrophic event prevented by first deducing from telemetry the character and magnitude of the tumbling motion, followed by computer modeling of the event, using the full Euler equations. This allows one to examine the consequences of proposed corrective actions. The final step is then to initiate the selected remedial actions by firing the correct thrusters in the correct sequence and for the correct duration.

6.4 Attitude Control Sensors

To control the attitude of spacecraft, reliance is placed on sensors of various types. This is particularly the case for three-axis stabilized vehicles but also applies to spin-stabilized vehicles.

Sensors that are directed toward a star or toward the earth's horizon depend on sources *external* to the spacecraft. Others, such as gyroscopes, are *internal*. In all cases, an error signal is generated that then is used by the attitude control system to initiate the needed correction, for instance, by firing the appropriate attitude control thrusters.

Usually, the attitude of a spacecraft is defined relative to inertial space, that is, relative to axes that do not rotate. For orbiting spacecraft, it is sometimes preferred to relate the vehicle attitude to the type of reference frame familiar from aircraft attitude control by introducing the *pitch* (relative to the local horizontal), *roll*, and *yaw* angles. It should be noted, however, that as the spacecraft moves on its orbit, the local horizontal rotates. Hence allowance must be made for the fact that the aircraft-type reference is not an inertial reference.

6.4.1 Star Sensors

In deep-space missions, as opposed to earth or planet orbiting missions, the spacecraft attitude is usually determined by observations of stars by small, optical telescopes. The attitude is then specified by the orientation of a set of spacecraft-fixed reference axes relative to the three mutually orthogonal axes that can be defined by the ecliptic plane and the vernal equinox line (Fig. 1.1). For this purpose, the declination and right ascension (Fig. 1.4) of selected guide stars are programmed into a star catalog contained in the memory of the flight computer.

To fix the attitude, a minimum of two bright stars ("guide stars") need to be sighted. For high accuracy, the angular separation between the two stars should be not much less than 45°. To avoid locking on the wrong star, the immediate vicinity of the selected guide stars should be void of other bright stars.

A frequently selected star is Capella. It satisfies these conditions and also has the advantage of a high declination. For deep-space missions close to the ecliptic plane, as most such missions are, a high declination avoids optical interference from the sun or from the planets.

In addition to the declination and right ascension of the guide stars, the flight computer should contain in its memory the positions of other bright stars. This greatly facilitates the reacquisition of the guide stars in case lock-on has temporarily been lost.

The attitude of the spacecraft is therefore determined by the lines of sight to two guide stars relative to spacecraft-fixed reference axes. The geometric relation between the selected astronomical reference (e.g., based on the ecliptic plane together with the vernal equinox line) and the spacecraft reference axes can then be specified either by the direction cosines between the two systems of axes or by their Euler angles.

6.4.2 V-Slot Sun Sensors

Sun sensors are used to find the declination, say δ, of the sensor axis relative to the heliocentric reference. The right ascension cannot be determined from a sun sensor. Therefore, to determine the spacecraft attitude completely, additional information, for instance from a star sensor or from a gyrocompass, is needed.

Most sun sensors are made to rotate continuously by an electric motor. Alternatively, an oscillating mirror can be used to scan the sun. In the case of spin-stabilized vehicles, the needed rotation can be provided by attaching the sensor directly to the spinning part.

The principle of operation is explained in Fig. 6.6. Figure 6.6a is a view parallel to the sensor axis of rotation. Figure 6.6b is the projection on the plane containing the sensor axis of rotation and the perpendicular to the ecliptic plane.

The functioning of the sensor can best be explained by basing it on the geometry of a sphere with center C on the axis of rotation. For this reason, the sensor is depicted in the figure as a sphere of unit radius. In practice, the sensor outer surface could just as well be a flat surface.

There are two slots, S_1 and S_2. When the sun is in the plane C–S_1, a portion of the light will strike the photocell P_1, which is located on the line C–A_1 approximately through the midpoint of S_1. Hence the photocell is

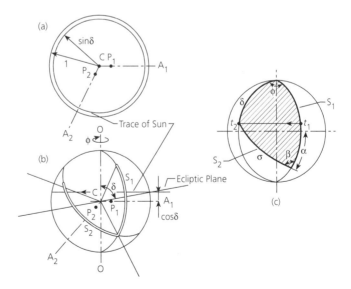

Figure 6.6 Sun sensor represented by unit sphere. (a) Top view; (b) elevation; (c) spherical triangle used for calculation.

pulsed once on every revolution of the sensor. The slot S_2 and the photocell P_2 function analogously.

The two slots are arranged in the form of a letter V. Slot S_1, but not S_2, is in a plane that contains the axis of rotation.

As the sensor rotates, the path of the sun as seen by the unit sphere is incident on it on a plane parallel to the sphere's equator. The declination, δ, of the sensor axis of rotation in the heliocentric reference is the angle in the diagram between the axis of rotation and the intersection of the plane of incidence with the unit sphere.

The period, say P, of the sun sensor, relative to inertial space, is equal to the time interval between two successive pulses of photocell P_1. By averaging over many rotations, P is determined to high precision. If t_1 is an instant of time when P_1 is triggered, and t_2 the instant following t_1 when P_2 is triggered, then the azimuthal angle, ϕ, traversed by the sensor during the interval $t_2 - t_1$ is

$$\phi = (2\pi/P)(t_2 - t_1) \qquad (6.26)$$

The angle ϕ will be needed to find δ. This is seen from considering the spherical triangle shown in Fig. 6.6c. The angles α and β are geometrical properties of the sensor, hence are known. Also known, from (6.26), is ϕ. The remaining side, σ, of the spherical triangle can be eliminated as shown next:

From the two Napier formulas of spherical trigonometry,

$$\tan\frac{\delta + \sigma}{2} = \frac{\cos(\phi - \beta)/2}{\cos(\phi + \beta)/2} \tan\frac{\alpha + \pi/2}{2}$$

and

$$\tan\frac{\delta - \sigma}{2} = -\frac{\sin(\phi - \beta)/2}{\sin(\phi + \beta)/2} \tan\frac{\alpha + \pi/2}{2}$$

The desired result for the declination δ of the sensor axis follows from applying the inverse tangent function to both sides of each equation and adding, with the result that

$$\delta = \tan^{-1}\left[\frac{\sin(\phi - \beta)/2}{\sin(\phi + \beta)/2} \cotan\alpha\right] - \tan^{-1}\left[\frac{\cos(\phi - \beta)/2}{\cos(\phi + \beta)/2} \cotan\alpha\right] \qquad (6.27)$$

The declination can therefore be found from the measured time intervals between photocell pulses and the application of software that is programmed to express Eqs. (6.26) and (6.27).

6.4.3 Horizon Scanning Sensors

Orbiting spacecraft often use as an input to their attitude control system observations of the **horizon** of the orbited astronomical body. The required sensors usually operate in a scanning mode in which the sensors' lines of sight are swept repeatedly over the astronomical body such that they cross its horizon.

In the case of *three-axis stabilized* spacecraft, the scanning motion is achieved either by a continuous conical sweep of the sensors' lines of sight

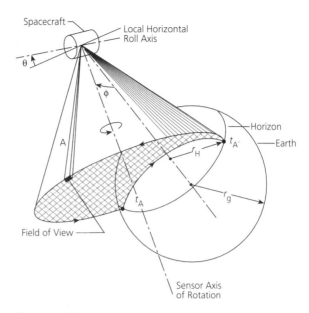

Figure 6.7 Horizon scanning sensor (only a single line-of-sight cone is shown). t_A, start; t'_A, end of intersection with earth; θ, pitch angle; ϕ, roll angle.

or by motor-driven mirrors that oscillate back and forth through some finite angle. In *spin-stabilized* spacecraft the sensors can be rigidly attached to the spinning portion of the vehicle. In this case the vehicle itself provides the sweeping motion of the lines of sight.

The basic principle of operation of such sensors is illustrated in Fig. 6.7. In this figure the earth is taken as an example. The line of sight moves on a circular cone that intersects the earth during part of the time. As is illustrated in the figure, the intersection starts and terminates on the horizon as seen from the spacecraft. On a given sweep the intersection starts at some time t_A and terminates at t'_A. It is the difference between these two times that provides the information needed for the attitude control.

In the usual arrangement, illustrated in Fig. 6.8, there are two lines of sight, A and B, provided either by a single sensor or by two separate but synchronized sensors. The vehicle's pitch angle (relative to the local horizontal) is designated, in conformity with the usual aircraft terminology, by θ, the roll angle by ϕ, and the yaw angle (not shown) by ψ. It is evident that because of the spherical symmetry of the earth, ψ cannot be determined by horizon sensors. To obtain it requires information derived from gyroscopes. (More precisely, the outputs of two rate gyros, one for the roll rate and one for the yaw rate, are combined with the Kalman filtered roll angle obtained from the horizon sensors.)

The photodetectors and optics of horizon sensors are designed to operate in the near-infrared, usually in the wavelength range 10 to 20 μm. The advantage is that in this spectral range the earth's atmosphere is more sharply defined. Also, there is less variation of the signal's amplitude between a sunlit and a dark spacecraft horizon.

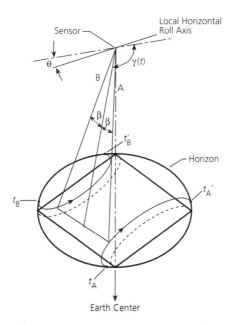

Figure 6.8 Horizon scanning sensor with dual lines of sight A and B. Dashed lines; earth trace at zero spacecraft roll; solid lines; general case.

Let r_g be the earth's radius (averaged for the orbit and including a correction for the effective height of the atmosphere in the near-infrared region) and r the distance of the spacecraft from the earth's center. The radius, r_H, of the horizon seen from the spacecraft is therefore

$$r_H = r_g \sqrt{1 - r_g^2/r^2} \qquad (6.28)$$

Figure 6.8 shows the ground traces of the two lines of sight A and B for a given (here positive) roll angle ϕ. Shown by dashed lines for reference are the ground traces for zero roll angle. These are symmetric with respect to the plane defined by the roll axis and the local vertical.

The angle, say 2β, between the two sensor telescopes with the lines of sight A and B in principle can be chosen arbitrarily as long as A and B sometimes intersect the earth. For comparable sensitivity of the pitch and roll angle determinations it is, however, advantageous to choose β such that for zero roll angle the four points of tangency with the earth form approximately a square. This is also shown in the figure. It then follows that the preferred choice for β is

$$\beta = \tan^{-1}\left(1 + 2r^2 \, r_H^2/r_g^4\right)^{-1/2} \qquad (6.29)$$

Let $\gamma(t)$ be the (positive) angle between the spacecraft roll axis (i.e., the reference axis for the pitch) and the plane containing the two lines of sight. This angle can be determined by a shaft encoder on the axis of rotation of the sensor.

As the plane is swept over the earth, there will be an instant, say t_v, when this plane is vertical. This will occur at the time

$$t_v = \tfrac{1}{2}(t_A + t'_A) = \tfrac{1}{2}(t_B + t'_B) \tag{6.30}$$

The value of γ at this time is sufficient to determine the **pitch angle** θ of the spacecraft, since evidently

$$\theta = \gamma(t = t_v) - 90° \tag{6.31}$$

The period of the sweep is of the order of a few seconds, which is short compared with the time for appreciable changes of the spacecraft's pitch. The measurements of γ at the midpoints between t_A and t'_A and similarly between t_B and t'_B, averaged for greater precision as indicated in (6.30), in effect provide a quasi-continuous reading of the pitch angle.

The **roll angle** can be determined from the time intervals $t'_A - t_A$ and $t'_B - t_B$. This is seen from Fig. 6.8 as follows: For roll angles $\phi \neq 0$, the lines of sight, when just touching the horizon, become displaced by comparison with the case $\phi = 0$, resulting in a lengthening or shortening of the ground traces, hence of the time intervals. If $(\Delta t)_0$ is the value of the time intervals for zero roll angle, it follows from the geometrical relations shown in the figure that

$$t'_A - t_A = (\Delta t)_0 + \frac{2(1 + 2r^2/r_g^2)}{1 + 2r^2 r_H^2/r_g^4} \cdot \frac{\phi}{\nu}$$

$$t'_B - t_B = (\Delta t)_0 - \frac{2(1 + 2r^2/r_g^2)}{1 + 2r^2 r_H^2/r_g^4} \cdot \frac{\phi}{\nu}$$

valid for $|\phi| \ll 1$. The angular rate of the sensor axis of rotation (or of the rate of spin in the case of spin-stabilized vehicles with sensors fixed to the vehicle) is designated by ν.

Subtracting the two equations and solving for ϕ gives the final result

$$\phi = \frac{(1 + 2r^2 r_H^2/r_g^4)\nu}{4(1 + 2r^2/r_g^2)}(t'_A - t_A - (t'_B - t_B)) \tag{6.32}$$

Measuring the time intervals $t'_A - t_A$ and $t'_B - t_B$ and carrying out the calculations indicated in (6.28) and (6.32) therefore determines the roll angle. This supposes that the orbit radius, or more generally the instantaneous spacecraft-to-earth-center distance, r, is known, derived either from known orbital data or by a spacecraft position determination.

A particular design of a horizon scanning sensor is shown in Fig. 6.9. The (single) line of sight is through an infrared-transmitting germanium window, coated with an interference filter to define the passband. A wedge-shaped lens, rotated by an electric motor, directs the incident radiation toward a bolometer. In this way a conical scan of the earth is obtained. The angle of rotation, and from it the angular rate, is obtained from magnetic pickups. Infrared radiation from spacecraft components that may intrude into the scanning cone are electronically blanked out.

Two such instruments, working in tandem, are needed to complete the system illustrated in Fig. 6.8. Pitch and roll angle accuracies achieved are of the order of 0.1°.

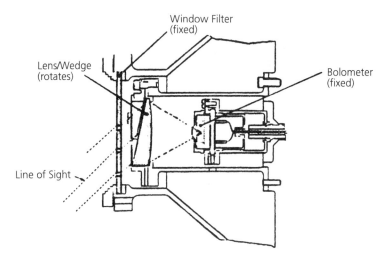

Figure 6.9 Schematic of a horizon scanning sensor. Courtesy of TELDIX GmbH, Germany.

6.4.4 Rate and Integrating Gyroscopes

Most spacecraft are equipped with gyroscopes of several different types. In particular, gyroscopes can be used for the determination of the orientation in inertial space of the chosen spacecraft reference axes and for the corresponding rates of change.

Modern gyroscopes, either of the mechanical type or laser-ring gyros, are the product of a long development and have reached unprecedented precision. A large body of technical literature exists, much of it applicable to space technology. The present discussion omits all design aspects.

Figure 6.10a illustrates schematically a *two-gimbal gyro*. The *rotor*, which is driven at a high and precisely controlled speed by an electric motor, is supported by bearings in the *inner gimbal*. This gimbal is supported by the *outer gimbal*, which in turn is supported by the **reference frame**. In some applications, the reference frame is rigidly attached to the spacecraft ("strap-down" gyro); in others it is supported by a *platform*, which is attached to the spacecraft by gimbals. Sometimes the platform orientation is controlled by feedback control from the gyros so as to be nonrotating relative to inertial space. Sometimes, in orbiting spacecraft, it is preferred to control the platform to remain at all times locally horizontal.

Gyros used for attitude control are made to be insensitive to translational accelerations by having the centers of mass of rotor and gimbals coincide with the geometric center, that is, with the common intersection of rotor and gimbal axes.

Some gyroscopes, referred to as **rate gyros**, measure one or several components of the spacecraft's angular velocity. Others, referred to as **integrating gyros**, are designed to measure the total angle through which the spacecraft has rotated about a specified axis in a given time interval. Integrating gyros therefore provide time integrals of angular velocities.

In what follows, the discussion will be limited to *single-gimbal gyros*.

Gyros of this type are schematically represented in Fig. 6.10b. The (common) principal axes of rotor and gimbal are designated by x_1, x_2, x_3. The

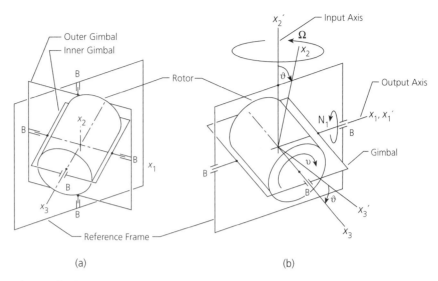

Figure 6.10 Schematics of (a) two-gimbal gyro and (b) single-gimbal rate gyro. x_1, x_2, x_3, principal axes of rotor and inner gimbal; v, rotor rate of rotation; N_1, restoring and damping torque on gimbal; Ω, spacecraft angular velocity component; B, bearings.

axes x'_1, x'_2, x'_3, also mutually orthogonal and through the common center of mass, refer to the reference frame and are parallel to spacecraft reference axes (not necessarily spacecraft principal axes). Axes x_1 and x'_1 coincide. The corresponding angular velocity components relative to inertial space are $\omega_1(t)$, $\omega_2(t)$, $v + \omega_3(t)$ for the rotor; $\omega_1(t)$, $\omega_2(t)$, $\omega_3(t)$ for the gimbal; and $\omega'_1(t)$, $\omega'_2(t)$, $\omega'_3(t)$ for the reference frame. Here v is the angular rate of the rotor relative to the gimbal.

When the spacecraft rotates, the rotor will force the gimbal to rotate about its x_1 axis, resulting in a change of the angle ϑ between gimbal and reference frame. In rate gyros a proportional restoring torque $-k\vartheta$ (k = spring constant) and damping torque $-c\dot{\vartheta}$ (c = damping constant) are applied from the reference frame to the gimbal. This torque can be visualized as being supplied by a torsion spring and mechanical damper. More precisely than by mechanical means, the torque is obtained electronically. In integrating gyros, $k = 0$.

As is evident from the figure, the shafts and bearings impose kinematic constraints. They relate ω and ω' by

$$\omega_1 = \omega'_1 + \dot{\vartheta}, \qquad \omega_2 = \omega'_2 \cos\vartheta + \omega'_3 \sin\vartheta, \qquad \omega_3 = -\omega'_2 \sin\vartheta + \omega'_3 \cos\vartheta \tag{6.33}$$

The x_1, x_2, x_3 components of the angular momentum, **L**, combined for rotor and gimbal are

$$L_1 = (I_1 + J_1)\omega_1, \; L_2 = (I_2 + J_2)\omega_2, \; L_3 = I_3(v + \omega_3) + J_3\omega_3 \tag{6.34}$$

where $I_1 = I_2$ and I_3 are the transverse and axial moments of inertia of the rotor and J_1, J_2, J_3 the moments of inertia of the gimbal. Designating time derivatives in inertial space by $d(\,)dt$, and in the space defined by the gimbal

by (˙), from (1.1) in Chap. 1,
$$dL_1/dt = \dot{L}_1 + \omega_2 L_3 - \omega_3 L_2 = N_1 \tag{6.35}$$
where $\dot{L}_1 = (I_1 + J_1)\dot{\omega}_1$ and the torque $N_1 = -k\vartheta - c\dot{\vartheta}$. From (6.33) and
$$\dot{\omega}_1 = \dot{\omega}'_1 + \ddot{\vartheta} = d\omega'_1/dt + \ddot{\vartheta}$$
[again an application of (1.1)], this becomes
$$(I_1 + J_1)\left(\frac{d\omega'_1}{dt} + \ddot{\vartheta}\right) + I_3(\nu + \omega_3)\omega_2 + J_3\omega_2\omega_3 - (I_2 + J_2)\omega_2\omega_3 = -k\vartheta - c\dot{\vartheta} \tag{6.36}$$

In the applications of interest to attitude control, the gyro rotor rates of rotation are very high in comparison with spacecraft angular velocities. The rotor may spin at 50,000 rpm or higher, whereas even a spin-stabilized spacecraft will hardly exceed 50 rpm and a three-axis stabilized vehicle much less. Similarly, the relative time rates of change of spacecraft rates of rotation will be much less than the rotor rate of rotation. Expressed in terms of strong inequalities,
$$|\omega'| \ll \nu \quad \text{and} \quad |d\omega'/dt| \ll \nu|\omega'| \tag{6.37}$$
The terms that contain $\omega_2\omega_3$ or $d(\omega'_1)/dt$ as factors are therefore negligible in comparison with the term containing ν, so that to a very high degree of approximation
$$(I_1 + J_1)\ddot{\vartheta} + c\dot{\vartheta} + k\vartheta = -I_3\gamma\omega_2 \tag{6.38}$$

The homogeneous part of this equation is seen to describe a harmonic, damped oscillation with the natural frequency
$$\omega_n = \sqrt{k/(I_1 + J_1)}$$
and the damped frequency
$$\omega_\alpha = \omega_n\sqrt{1 - \zeta^2}$$
where ζ is the nondimensional damping coefficient defined by
$$\zeta = c/2\omega_n(I_1 + J_1)$$
Up to this point, the development applies to both rate and integrating gyros.

Considering now the case of **rate gyros**, k and c are made large [with $\zeta = \mathcal{O}(1)$] so that the time for damping the gimbal becomes much shorter than the time for significant changes of the spacecraft attitude rate. Therefore, on the slow time scale of ω_2, the pseudostationary solution
$$\vartheta(t) = -\frac{I_3\nu}{k}\omega_2(t) \tag{6.39}$$
applies.

Another consequence of a large spring constant is that the gimbal deflection is small, limited only by the need for a precise electronic readout of ϑ. The difference between ω_2 and ω'_2 can then be neglected. If Ω is the component of the spacecraft angular velocity in the direction of the x'_2 axis, referred to as the gyro's *input axis*, the main result characterizing rate gyros becomes

$$\Omega(t) = -\frac{k\vartheta(t)}{I_3 \nu} \qquad (6.40)$$

The result is seen to be independent of the gimbal inertia.

In another version of the rate gyro, the reference frame, in place of being rigidly attached to the spacecraft, is rotated by an electric motor so as to keep ω_2, hence ϑ, near zero at all times. In this case ϑ serves as the error signal in a feedback control loop that controls the rotation of the reference frame. The spacecraft rate of rotation Ω can then be obtained directly as the negative of the rate of rotation of the reference frame relative to the spacecraft. The precision of this type of high-quality rate gyros is of the order of 0.01 deg/h.

The output of an **integrating gyro** is the time integral of the angular rate, hence the angle through which the spacecraft has rotated in a given time interval.

Integrating gyros are often the primary means for attitude determination. Because of the inevitable small drifts of such gyros, their output needs to be updated periodically by external sensors such as horizon sensors and sun sensors.

Distinct from rate gyros, integrating gyros do not use a restoring torque between reference frame and gimbal. As indicated by the homogeneous part of (6.38) with $k = 0$, disturbances are damped out exponentially. On the slow time scale of $\omega_2(t)$, with a sufficiently high damping coefficient, the significant solution of (6.38) is the pseudostationary one given by

$$\vartheta(t) = -\frac{I_3 \nu}{c} \int \omega_2(\tau) \, d\tau + \text{const.} \qquad (6.41)$$

In this case too, ϑ can be kept small by rotating the reference frame, controlled by a closed loop where ϑ serves as the error signal. The angle through which the spacecraft has rotated in a given time interval about the gyro input axis is then the negative of the readout of the shaft encoder on the motor that rotates the reference frame.

6.4.5 Inertial Measurement Units

Inertial measurement units consist of a platform together with gyros and accelerometers on it. The platform is usually attached to the vehicle by gimbals. Most frequently there are three gyros and three accelerometers that have mutually orthogonal sensing axes, two parallel to the reference plane of the platform and one perpendicular to it. In a frequently used arrangement, the output from the gyros is used to maintain the platform by servo controls in a *constant orientation relative to inertial space*. The platform's gimbal angles then define the Euler angles of the spacecraft reference axes with respect to inertial axes.

The accelerometers provide a measurement of the three components of the resultant of the external forces — other than the gravitational force — that act on the spacecraft. (Gravity, of course, acts on the accelerometer's sensing masses just as it does on the vehicle.) Making use of the estimated position of the vehicle, the gravitational force can be computed and added to the force derived from the accelerometers. When integrated, this will give the vehicle's velocity in inertial space. The accuracy of the type of accelerometers used in space systems is of the order of 10^{-6} times normal gravity, or better.

In orbiting spacecraft, particularly earth-pointing spacecraft on circular orbits, it is sometimes preferred to maintain the platform orientation such that it stays *locally horizontal*, instead of being constant in inertial space. In this case, gravitation does not add to the acceleration along the two axes that are parallel to the platform. The gimbals are torqued to maintain the platform at zero roll and yaw. The pitch rate is controlled to be the negative of the orbit rate (2π divided by the orbital period).

Inertial measurement units have found applications in spacecraft but are particularly suited for controlling *launch vehicles*, where they have replaced radio guidance. For a launch in the earth's atmosphere the attitude control cannot reliably use sun, star, or horizon sensors, which would be sensitive to weather and to day versus night conditions. By sensing the three components of the acceleration and using dead reckoning, inertial measurement units can control the path of launch vehicles to high precision.

On *spacecraft*, gyro drift affecting attitude control can be corrected by periodic updating, which is available from horizon, sun, or star sensors. In the long term, accelerometers introduce appreciable velocity and position errors. The primary means for correcting these is by periodic measurements of the Doppler shift of microwave signals emitted from the spacecraft.

A particular type of inertial measurement units is referred to as **strapdown IMU**. As the name implies, the platform is attached rigidly to the vehicle. The outputs from rate gyros are integrated by computation to obtain the Euler angles between spacecraft-fixed axes and inertial axes. Similarly, the output from the accelerometers serves to compute the acceleration and incremental velocity components along inertial axes.

6.5 Attitude Control Actuators

Responding to inputs from the attitude sensors, the control system commands actuators, which then cause the spacecraft to reorient itself from a perturbed attitude back to the nominal one.

A complete attitude system comprises both sensors and actuators. It is sometimes necessary to distinguish between 1) attitude changes attributable to **external torques** such as those produced by attitude control thrusters and perturbations illustrated in Fig. 6.1, and 2) **internal torques** produced by spacecraft internal actuators such as "reaction wheels" or caused by propellant sloshing or motions of spacecraft-internal masses.

External torques change the total angular momentum, that is, the combined angular momentum of spacecraft body and internal moving masses. On the other hand, *internal* torques do not affect the total angular momentum but can have an effect on the attitude control system because sensors such as horizon or sun sensors are mounted on the spacecraft body and move with it when there is an internally produced disturbance.

Discussed next are several types of actuators that are in common use.

6.5.1 Thrusters

Attitude control thrusters, sometimes also referred to as "reaction control thrusters," can be of two types. On some spacecraft relatively large thrusters are needed to produce major attitude changes in a relatively short time. An example is the reorientation of a spacecraft (and with it the thrust axis of attached rocket motors) prior to a motor firing.

Thrusters are also used to control the attitude of a spacecraft throughout its life. These thrusters can be much smaller because they only have to compensate for the long-term effects of external disturbance torques, which are typically very small. Most often, thrust levels of only a fraction of a newton are sufficient for this purpose.

The physical characteristics of thrusters have been discussed in Section 4.6.9.

The placement of thrusters on a spacecraft is strongly influenced by the vehicle's configuration. The thrust axis should have a large lever arm about the vehicle's center of mass. Also, the rocket plume must not strike any part of the spacecraft.

The number of thrusters must be sufficient to rotate the spacecraft in either direction about three independent axes. The control is simplified if these axes are the principal axes of the spacecraft. To rotate the spacecraft without at the same time imparting to it a change in velocity requires that two thrusters be fired at the same time in opposite directions to produce a pure couple. The same thrusters can also be used for imparting a velocity increment to the spacecraft, for example, for station keeping. To reduce the number of propellant lines leading from the tanks to the thrusters it is advantageous to arrange the thrusters in clusters, such as the triplets shown in the diagram, Fig. 6.11.

Because it is difficult to build small thrusters that may be required to fire a very large number of times, it is often necessary to provide redundancy by increasing their number beyond the minimum required.

Figure 6.11 shows a hypothetical, symmetric arrangement of 24 thrusters that satisfy the several requirements just listed. Because of spacecraft configurational constraints, this particular arrangement often cannot be realized in practice. It is useful, however, for indicating how by alternatively combining the firing of selected thrusters redundancy can be achieved with a minimum of thrusters and that these thrusters can be used for both attitude control and for station keeping. Referring to the figure, a positive torque about the x_1 principal axis can be produced by firing any one of the four pairs (A + 2) and (H − 2), (B + 2) and (G − 2), (E + 3) and (D − 3), or (F + 3) and (C − 3). Acceleration along the positive direction of x_1 for station keeping can be produced by firing either the pair (B + 1) and (G + 1) or the pair (C + 1) and (F + 1). Similar choices also hold for the other axes.

The propellant valves and gas pressurization valves usually also require redundancy, because otherwise a valve permanently stuck open or stuck

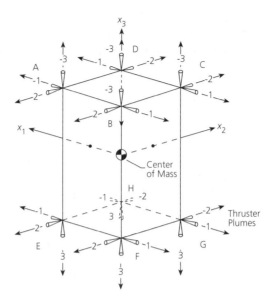

Figure 6.11 Hypothetical arrangement, with redundancy, of 24 attitude control and station-keeping thrusters. x_1, x_2, x_3; principal axes of spacecraft.

closed would probably cause catastrophic impairments. A typical redundant propellant feed system is illustrated in Fig. 4.32.

An attitude control feed system that would command the propellant flow to be proportional to demand would result in extremely small valves and thrusters and, as a consequence, would not be reliable. The fluid passages would be so narrow that clogging by impurities in the propellant would be a threat. Instead, **on–off controls** are used where, except for a short transition period, the propellant control valves are either fully open or fully closed.

The propellant control valves may stay open for as little as 10 ms, with transition times of several milliseconds. Ideally, after completion of the desired spacecraft attitude correction, there should be no residual rate of spacecraft rotation left. With on–off controls, however, there could be a (small) residual rotation as a consequence of the nonzero length of time needed for valve closure. Opposing thruster pairs may then act one against the other in succession. This phenomenon [7] which is referred to as **chatter**, will necessarily result in some waste of propellant.

To minimize the chatter of thrusters, a **dead band** is introduced into the control, meaning that as long as the spacecraft motion is within a specified narrow band, no control action is taken. In the absence of further disturbances, the system will settle into a **limit cycle**, oscillating between the boundaries of the dead band. In a well-designed system, the period of the limit cycle will be long, thereby minimizing the propellant loss.

A type of control, called **Schmitt trigger**, adds hysteresis to the dead band. Here, too, there is unavoidably a limit cycle, but its period is longer than in other on–off controls. Schmitt triggers are also used in a number of other applications outside space technology.

The theory of the Schmitt trigger assumes that the torque, say $\mathbf{M}_{th}$, produced by a thruster pair about a specified spacecraft-fixed axis has one of the three values M_0, or 0, or $-M_0$. The value assumed is made to depend on the quantity

$$\gamma + \tau \, d\gamma/dt \tag{6.42}$$

where γ is the angular displacement in inertial space of the axis relative to its desired, nominal orientation. (For systems that use a locally horizontal reference system, the definition of γ is analogous, but it now refers to the local horizontal rather than to inertial space. The angle γ may then be the pitch angle of the spacecraft, or, alternatively, the roll angle, i.e., angles that are directly inferred from horizon sensors.) The time constant τ is at the option of the designer and is chosen as a compromise between the twin objectives of fast response of the system and minimum propellant expenditure.

Figure 6.12 is a phase plane diagram that illustrates the operation of a Schmitt trigger. The various values assumed by M_{th} in different regions of the phase plane are indicated in the figure. The regions are bounded by the lines specified by

$$\gamma + \tau \, d\gamma/dt = \alpha_1, \alpha_2, -\alpha_1, -\alpha_2$$

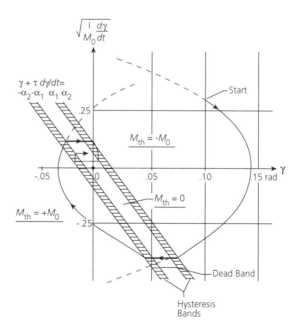

Figure 6.12 Phase plane of an on-off thruster control with deadband and hysteresis. Start of trajectory in the phase plane at $\gamma = 0.10$ rad, $\sqrt{I/M_0}\, d\gamma/dt = 0.30$ rad; $\tau = 0.15\sqrt{I/M_0}$. (Adapted from Ref. 2.)

The quantities α_1 and α_2 therefore determine the widths of the dead band and hysteresis bands. In the half-plane $\gamma + \tau \, d\gamma/dt > \alpha_2$, the thruster torque $M_{\text{th}} = -M_0$ tends to drive γ toward the origin of the phase plane. Similarly, in the half-plane $\gamma + \tau \, d\gamma/dt < -\alpha_2$, $M_{\text{th}} = +M_0$ drives γ still more toward the origin. In the deadband, $M_{\text{th}} = 0$. In the hysteresis band, the value of M_{th} that occurred immediately before the trajectory entered the band persists in it.

If I is the principal moment of inertia about the specified spacecraft axis

$$I \, d^2\gamma/dt^2 = M_{\text{th}} \tag{6.43}$$

Therefore, with constants of integration γ_0 and $(d\gamma/dt)_0$,

$$\frac{d\gamma}{dt} = \frac{M_{\text{th}}}{I} t + \left(\frac{d\gamma}{dt}\right)_0, \qquad \gamma = \frac{1}{2}\frac{M_{\text{th}}}{I} t^2 + \left(\frac{d\gamma}{dt}\right)_0 t + \gamma_0$$

Elimination in these equations of the time then results in

$$(M_{\text{th}}/I)(\gamma - \gamma_0) = \frac{1}{2}\left[(d\gamma/dt)^2 - (d\gamma/dt)_0^2\right] \tag{6.44}$$

This shows that for $M_{\text{th}} \neq 0$ the trajectories are parabolas with axes of symmetry coinciding with the γ axis. For $M_{\text{th}} > 0$ the parabolas point to the right, for $M_{\text{th}} < 0$ to the left. For $M_{\text{th}} = 0$ the trajectories are lines parallel to the γ axis.

In the example shown in the figure, the initial condition is such that γ and $d\gamma/dt$ are both positive. Following a parabolic trajectory, the error angle γ at first increases, then decreases. Between $\gamma + \tau \, d\gamma/dt = \alpha_1$, and $\gamma + \tau \, d\gamma/dt = -\alpha_2$, $d\gamma/dt = \text{const}$. The trajectory then follows a new parabola up to where $\gamma + \tau \, d\gamma/dt = -\alpha_1$.

In the absence of a new disturbance, the trajectory ends in a limit cycle bounded by the lines $\gamma + \tau \, d\gamma/dt = \pm \alpha_2$. The advantage of the Schmitt trigger is that for a given width of the deadband the period of the limit cycle is long and the propellant expenditure correspondingly low. In theory, the limit cycle would persist indefinitely; in practice, dissipative effects such as those produced by propellant sloshing will cut it short.

The choice of the constants τ, α_1 and α_2 by the designer is based in part on the expected types of disturbances induced by the space environment. A compromise must be made in the choice of α_1 and α_2: Small values reduce the residual, but only at the cost of a more rapidly oscillating limit cycle; conversely for large values.

It is of interest to note that this attitude control system is nonlinear and nonconservative. An exchange takes place from the kinetic energy of rotation of the spacecraft to the energy that is carried off by the thruster gas.

By the use of **chemical rocket thrusters**, residual errors in spacecraft attitude can be held to about 0.1 to 1.0°. This is acceptable in many applications, for instance, in broadcasting satellites. Their main antennas have half-power beam widths that are many times larger than the attitude control residual errors just mentioned.

Electric thrusters for attitude control, such as xenon ion thrusters, do not suffer from the relative imprecision of the thrust cutoff of chemical thrusters. They allow very precise control. Their specific impulse is also much higher, an important consideration in the case of long-life spacecraft.

A different and very critical problem arises when, because of a failure in the attitude control system or because of an unintentional release of pressurization gas, the spacecraft goes into an **uncontrolled tumbling** motion. The outputs from the attitude sensors, as may have been received by telemetry, are likely to be ambiguous, yet a thorough understanding of the most probable failure modes needs to be developed. Commanding a corrective torque about one axis may merely cause a still larger excursion about the other axes. Therefore, to avoid excessive propellant waste by the recovery action, a computer simulation, based on the full Euler's equations and the consequences of the proposed recovery action, is an essential task to be carried out before commanding the spacecraft.

6.5.2 Reaction Wheels and Gimbaled Momentum Wheels

Among the devices for attitude control, other than thrusters, are **reaction wheels** (Fig. 6.13). The wheels are driven by electric motors in either direction and are capable of high rotational speeds. They are supported by

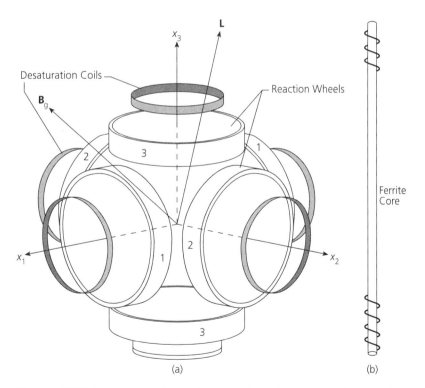

Figure 6.13 Schematic of reaction wheels and magnetic desaturation solenoids. (a) Assembly of three orthogonal wheels and their desaturation coils; (b) rodlike desaturation solenoid with ferrite core. **L**, angular momentum of spacecraft and wheels; $\mathbf{B}_g$, earth magnetic field.

bearings that are fixed to the spacecraft. In what follows, it will be assumed that the axes of the wheels are parallel to the spacecraft principal axes.

Reaction wheels serve to transfer angular momentum from the spacecraft to the wheels and vice versa. The current, and therefore the torque, of the motors is controlled with the aim of minimizing the spacecraft attitude errors. DC motors are used because they allow precise proportional control. The accuracy of the attitude control that can be achieved is superior to what can be obtained by thrusters alone.

The disturbance torques over time may change direction so that the accumulated momentum of any of the wheels may either increase or decrease. In general, however, sooner or later the rotational speed will reach the structural limit of the wheel. This then requires periodic interventions. They are referred to as **desaturations** and are intended to bring the rotational speed back to near zero.

Desaturation is accomplished by applying an external torque to the spacecraft. There are several methods available to do this: One approach is to fire an appropriate set of thrusters for the appropriate length of time. In orbiting spacecraft the burning of propellant can sometimes be avoided by making use of the gravity gradient. Still a third method, discussed in the next section, is magnetic desaturation.

Figure 6.13 shows schematically three orthogonal pairs of reaction wheels parallel to the spacecraft principal axes x_1, x_2, x_3. In some cases only a single wheel is used. This may be the case with Earth- or Mars-orbiting spacecraft at low altitudes, where aerodynamic drag becomes significant. Because of the asymmetry introduced by the antenna dishes, feeds, and other appendages, the aerodynamic torque will be predominantly along the pitch axis. A reaction wheel aligned with this axis can then provide accurate control. A similar situation occurs when descending toward a planet by aerobraking. Here, too, the aerodynamic torque will predominantly be about only one of the spacecraft axes.

Let ω be the angular velocity of the spacecraft about this axis, which is assumed to be a principal one, and let I be the corresponding moment of inertia of the spacecraft (other than the reaction wheel). The other components of the angular velocity are at first assumed to be zero. Correspondingly, let J be the wheel's moment of inertia about its axis and $\omega + \nu$ its angular velocity, where $\nu = \nu(t)$ is the rate of rotation relative to the vehicle. The centers of mass of spacecraft and wheel are assumed to coincide. The combined angular momentum about the center of mass is therefore $L = I\omega + J(\omega + \nu)$.

The equation of motion for the spacecraft is

$$I\dot{\omega} = -N_{\text{el}} + c\nu + M_{\text{d}} \tag{6.45a}$$

and for the reaction wheel

$$J(\dot{\omega} + \dot{\nu}) = +N_{\text{el}} - c\nu \tag{6.45b}$$

where N_{el} is the torque exerted on the wheel by the electric motor, M_{d} the external disturbance torque acting on the spacecraft, and c the combined friction coefficient of the bearings.

6.5 Attitude Control Actuators

The armature voltage, V, that is applied to the motor is controlled so as to bring the spacecraft attitude back to the desired nominal value. There will be an induced voltage, or "back-EMF," say E, and a resistive drop iR (i = armature current, R = resistance). The induced voltage is directly related to the motor torque by

$$iE = N_{el}\nu \quad (6.46)$$

(The equation follows immediately from conservation of energy of an ideal, lossless motor.) The motor torque per unit current, $q_i = N_{el}/i$ is a constant for a given motor. The armature voltage can therefore be expressed by

$$V = iR + E = N_{el}R/q_i + q_i\nu \quad (6.47)$$

Even in the absence of an external torque on the spacecraft, there will be a small motor torque and current caused by bearing friction. The resulting armature voltage, say, V_0, obtained from setting $M_d = \dot{\omega} = \dot{\nu} = 0$, in (6.45) and making use of (6.47), is

$$V_0 = (cR/q_i + q_i)\nu \quad (6.48)$$

so that

$$V - V_0 = (R/q_i)(N_{el} - c\nu) \quad (6.49)$$

Solving for N_{el} and substituting into (6.45a) results in the basic equation that characterizes the system:

$$I\frac{d^2\vartheta}{dt^2} = -\frac{q_i}{R}(V - V_0) + M_d \quad \textbf{(6.50)}$$

Here, $\dot{\omega}$ has been replaced by the second derivative of the spacecraft attitude angle relative to the inertial reference [noting that by virtue of (1.1), $\dot{\omega} = d\omega/dt$]. The armature voltage V is measured. Also measured, by a tachometer, is the angular velocity difference ν. From (6.48) and for given motor properties c, R, and q_i, one can obtain V_0 as a small correction to V. The wheel's moment of inertia enters (6.50) only through the armature voltage, for from (6.47) and (6.45b),

$$V = \frac{RJ}{q_i}(\dot{\omega} + \dot{\nu}) + \left(q_i + \frac{Rc}{q_i}\right)\nu$$

The inputs to the control are the deviation, ϑ, from the nominal vehicle attitude, and its derivative. The output is the armature voltage, V, applied to the motor. We assume the control law

$$V = k(\vartheta + \tau \, d\vartheta/dt) \quad (6.51)$$

where the factor k and the time constant τ are chosen to ensure stability yet rapid response. The control diagram, Fig. 6.14, shows the system with its interacting components.

The performance can be illustrated by assuming that after some instant, say t_1, external disturbances are absent. The bearing friction will be neglected, hence $V_0 = 0$. The attitude deviation and its derivative in inertial

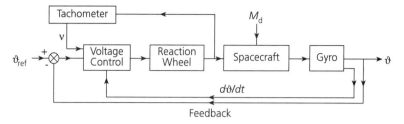

Figure 6.14 Diagram of a reaction wheel control.

space at time $t = t_1$ are designated by ϑ_1 and ϑ'_1, constituting the initial condition.

The trial solution $\vartheta = a \exp(st)$, where s is the (complex) frequency and a the amplitude, yields as the characteristic equation for s the two solutions

$$s = -\frac{q_i k \tau}{2IR} \left[1 \pm \sqrt{1 - \frac{4IR}{q_i k \tau^2}} \right] \qquad (6.52)$$

If and only if $k > 0$, the two solutions s_1 and s_2 are both in the negative real half-plane, indicating stability. For critical damping,

$$k = 4IR/(q_i \tau^2)$$

The system is seen to follow in the phase plane the path given by

$$\left. \begin{array}{l} \vartheta(t) = a_1 \exp(s_1(t - t_1)) + a_2 \exp(s_2(t - t_1)) \\ \vartheta'(t) = a_1 s_1 \exp(s_1(t - t_1)) + a_2 s_2 \exp(s_2(t - t_1)) \end{array} \right\} \qquad (6.53)$$

where

$$a_1 = -\frac{\vartheta'_1 - s_2 \vartheta_1}{s_2 - s_1}, \qquad a_2 = \frac{\vartheta'_1 - s_1 \vartheta_1}{s_2 - s_1}$$

Concerning the technical implementation of reaction wheels, critical elements in their design are the ball or roller bearings that support them in the vehicle. The wheels must operate at high speed for long periods of time, which sometimes must exceed 10 years. Even though the bearing loads in the near-weightless environment of the spacecraft are low, wear will inevitably affect the bearing friction. Nevertheless, precision bearings with space-qualified lubricant, when thoroughly tested for qualification and flight acceptance, have proved to be a satisfactory solution.

Bearing wear can be avoided if the wheels are levitated magnetically, as illustrated in Fig. 6.15. The levitation of a rotor requires the control of five degrees of freedom, three for the displacement of the center of mass and two for the tilting of the axis. In this design, a samarium–cobalt permanent magnet is used for simultaneously stabilizing against a displacement along the axis of rotation and against tilting. The remaining two degrees of freedom are the lateral displacements, which are stabilized by solenoids driven by electronic controls in response to signals from magnetic pickups. Unavoidably, there will be some eddy current losses, resulting in a small drag. The coefficient c, introduced earlier, is therefore not strictly zero.

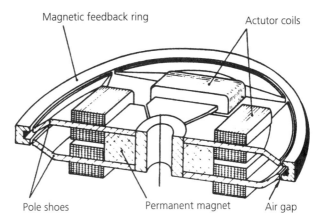

Figure 6.15 Schematic of reaction wheel with magnetic suspension. Courtesy of TELDIX GmbH, Germany.

Magnetically suspended reaction wheels in addition still require mechanical bearings. In normal operation there is no physical contact of the rotating parts with these bearings. They are needed, however, to avoid damage in case of a temporary power failure. They also serve to cage the wheels during launch, because the loads then are too high to be contained by the magnetic suspension.

Three separate, mutually orthogonal reaction wheels (Fig. 6.13) can provide complete attitude control. The analysis is virtually the same as sketched before for a single disturbance torque component about a principal axis. The only new features are the gyroscopic torques present as the wheel axes are tilted. This gives rise to nonlinear terms in the Euler rigid body equations. These terms are small and can be neglected because the wheels are operated at rates of rotation much larger than any expected angular velocity of the spacecraft.

Whereas the orientation of reaction wheels is fixed relative to the spacecraft, **gimbaled momentum wheels** are supported by a gimbaled platform. Torques from electric motors ("torquers") are applied to the gimbals so as to tilt the wheel axes relative to the spacecraft. Figure 6.16 illustrates a type.

The reactions from these torques act back on the spacecraft, returning it to its nominal attitude. The vehicle's angular momentum is therefore transferred to the angular momenta of the wheels.

In principle, a single gimbaled momentum wheel will suffice for complete three-axis control. More common are combinations of three wheels with orthogonal axes. An advantage is that they can provide redundancy in case of wheel failure. Like reaction wheels, gimbaled momentum wheels may require periodic desaturation.

The accuracy that can be achieved with either reaction wheels or gimbaled momentum wheels is of the order of 1 microradian or better, much higher than is possible with thrusters. An important application is to spacecraft-to-spacecraft and ground–spacecraft–ground laser communication. The laser pointing, acquisition, and tracking requirements are of this order. It is sufficient that the pedestal that carries the lasers, rather than the entire spacecraft, be controlled to this accuracy by reaction or momentum wheels.

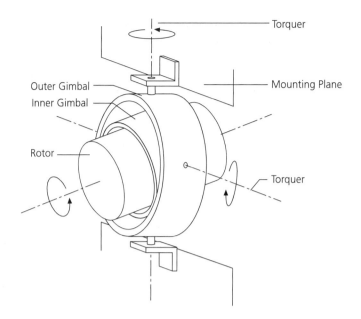

Figure 6.16 Schematic of a gimbaled momentum wheel.

6.5.3 Magnetic Desaturation

Desaturation consists of applying to the spacecraft an external torque opposite in direction to the wheel spin. By conservation of angular momentum, the wheel's rate of rotation will decrease, forced by the electric motor or torquer. By firing thrusters, or by a torque derived from a planetary magnetic field or gravity gradient, the wheel's rate of rotation can be returned to near zero or even reversed. This procedure is initiated autonomously by the spacecraft computer or can also be commanded by ground control. The discussion here is limited to **magnetic desaturation**.

Magnetic desaturation is feasible for low-earth-orbiting satellites and spacecraft that orbit other planets with an appreciable magnetic field.

Near the earth, the magnetic field can be represented approximately by a dipole field. The axis presently intersects the earth surface at about 78°N, 69°W, near Thule in Greenland, and at 78°S, −111°E, near the Vostok Station in Antarctica.

Based on the dipole approximation, the two magnetic poles and a magnetic equator can be defined. In terms of the spherical coordinates r, θ, ϕ (r = radius; θ = geomagnetic colatitude, i.e., latitude angle measured from the magnetic pole in the northern hemisphere; ϕ = geomagnetic longitude) the components of the dipole field are (e.g., Ref. 8)

$$B_r = -\frac{2D_m}{r^3}\cos\theta, \qquad B_\theta = -\frac{D_m}{r^3}\sin\theta, \qquad B_\phi = 0 \qquad (6.54)$$

where D_m is the magnetic dipole moment. At present D_m, in SI units, is approximately $8 \cdot 10^{15}$ (tesla·m³). It follows that on the earth's surface, on the geomagnetic equator, $B_\theta = -D_m/r_g^3 = -3.1 \cdot 10^{-5}$ tesla (or in cgs units -0.31 gauss). At the magnetic poles, the field strength is twice this value. The field lines (outside the earth's iron–nickel core) run from south to north.

The dipole approximation fails at distances near and beyond the nominal position of the magnetopause and even at shorter distances during magnetic storms. Magnetic desaturation is not practical at the altitude of geostationary satellites, because the magnetic field, which falls off with the third power of the distance, is too weak and too irregular there.

Solenoids on the spacecraft are often arranged in pairs for each axis, as shown in Fig. 6.13. If each solenoid is approximated as a planar current ring of circular area A, the torque, $\mathbf{M}_m$, exerted on the pair of rings by the planetary magnetic field is

$$\mathbf{M}_m = ni(\mathbf{A} \times \mathbf{B}) \tag{6.55}$$

where i is the current and n the total number of turns for the pair. The vector $\mathbf{A}$, with magnitude equal to the area, is directed along the solenoid axis in the direction given by the right-hand rule. Carrying out the vector product results in

$$\left.\begin{aligned} M_{m,r} &= ni D_m r^{-3} A_\phi \sin\theta, \qquad M_{m,\theta} = -2ni D_m r^{-3} A_\phi \cos\theta \\ M_{m,\phi} &= -ni D_m r^{-3}(2A_\theta \cos\theta - A_r \sin\theta) \end{aligned}\right\} \tag{6.56}$$

For instance, for a 400 km altitude orbit over the magnetic poles, going from south to north, and a pair of solenoids with $i = 1$ amp, $n = 2000$, $A = 0.50$ m^2, the torque exerted on the spacecraft as it passes the magnetic equator is

$$M_{m,\phi} = ni D_m r^{-3} A = 0.025 \text{ (Nm)}$$

If the current flows in the direction such that $\mathbf{A}$ is pointing vertically downward, there will be a pitch-up torque on the spacecraft. This magnitude of 0.025 (N m) is about 1000 times the combined torques of solar radiation, magnetic, and gravity gradient at this altitude.

Spacecraft with magnetic desaturation are equipped with three magnetometers having mutually orthogonal sensing axes. The magnetometers serve to measure the strength and direction of the planetary magnetic field prior to the initiation of the desaturation.

In contrast to gravity gradient desaturation (which cannot desaturate the yaw control), magnetic desaturation can be used to desaturate not only pitch and roll but also yaw. Depending on the orbit, suitable spacecraft locations must be chosen. For instance, for a polar orbit, this may be a location not too far from a magnetic pole (which allows pitch and roll desaturation) or from the magnetic equator (which allows pitch and yaw desaturation).

6.6 Spin-Stabilized Vehicles

A useful distinction among spin-stabilized vehicles can be made between **spin-stabilized** vehicles in the more narrow sense and **dual-spin** vehicles. In the first case, the entire spacecraft is spinning about some axis; in the second case only a portion (usually the more massive one) does so. In a still wider sense of the term, **momentum-biased** spacecraft, that is, spacecraft that incorporate a fast spinning momentum wheel (Sect. 6.5.2), can also be classified as spin stabilized. In all cases, the orientation in inertial space of the spin axis is maintained by the gyroscopic effect that is produced by the spinning part.

The attitude control of spin-stabilized spacecraft (Fig. 6.2a) is simpler than it is for three-axis stabilized vehicles. This advantage can be offset, however, by the dimensional restrictions that are imposed by the available launch vehicle payload space (i.e., the size of the "shroud" or "fairing"). Because the maximum possible solar cell area of spinning vehicles is proportional to their diameter and length, high-power communications and broadcasting satellites with their high demand for electric power, when spin-stabilized, can become too large for available launch vehicles.

Spinning spacecraft can be either prolate (Fig. 6.4a) or oblate (Fig. 6.4b). As discussed in Sect. 6.3.2, the precession of prolate bodies is unstable in the sense that the nutation angle tends to increase. Oblate bodies are stable in this sense. However, the attitude control of spinning vehicles requires more than merely a stable precession at a constant nutation angle: a control is needed to reduce to near zero the nutation angle that may have been caused by a disturbance.

In spite of their basic instability (which can be corrected by nutation dampers; see Sect. 6.6.3), prolate spacecraft are often preferred because of their better geometrical fit to the launch vehicle payload space.

Horizon sensors can take advantage of the spin. Mounted on the side of the vehicle, the line of sight — by virtue of the spin — can be made to sweep over the earth or planet (Fig. 6.17). If a stepping motor is used in addition to change the line of sight in the fore–aft direction, a single sensor can serve for the sweeps A to A' and B to B' indicated in Fig. 6.8.

6.6.1 Jet Damping

Rocket motor gas, as it streams through the motor case and the nozzle, generally has a damping effect on spacecraft motions. The phenomenon is referred to as **jet damping** [9]. As in aircraft, it can dampen pitch and yaw

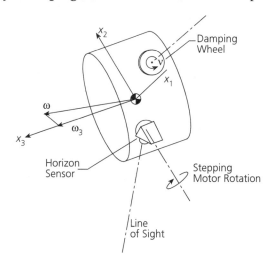

Figure 6.17 Schematic of a spinning spacecraft with nutation damping wheel and horizon sensor (wheel axis parallel to x_1 principal axis).

motions. More significant, however, is the jet damping of *spinning vehicles*, because it counteracts the precessional motion by decreasing the nutation angle.

The beneficial effect of jet damping obtained during rocket motor firings is important because it mitigates the relatively large disturbance torques that are caused by unavoidable motor performance fluctuations.

Nevertheless, by itself, jet damping may not be sufficient to keep the nutation angle from growing during motor firings. There have been some upper stage vehicles with solid-propellant motors that exhibited unacceptably large nutation angles. To keep the angle within one or two degrees then required the addition of special thrusters [10–12].

Jet damping can be explained by the torque that is produced by the Coriolis forces resulting from the angular velocity of the spacecraft and the rocket gas velocity relative to it. The jet damping torque is the integrated moment of these Coriolis forces, starting at the combustion chamber and ending at the nozzle exit. An explanation that is both simpler and also makes it clear that the combustion chamber and the nozzle are not separately involved, is as follows.

A simple but sufficiently accurate model of jet damping is illustrated in Fig. 6.18. The model assumes that the mass-averaged gas exit velocity, u_{ex}, at the nozzle exit and relative to the spacecraft is in the direction of the nominal thrust. (This is not quite accurate because the Coriolis forces cause the gas stream at the nozzle exit to deviate slightly from this direction. The gas already enters the nozzle with some angle of attack. But because

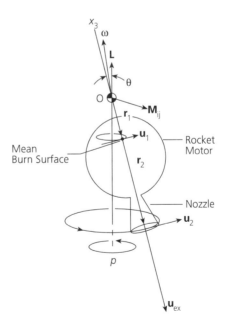

Figure 6.18 Jet damping of a spinning spacecraft: O, instantaneous center of mass; θ, nutation angle; p, precession rate; $\mathbf{u}_{ex}$, gas exit velocity relative to spacecraft; $\mathbf{M}_{jd}$, jet damping moment. Shown for case of prolate spacecraft.

of the subsequent large acceleration in the nozzle, the effect of the angle of attack is largely canceled and the gas stream is redirected to leave the nozzle close to the nominal thrust direction. The most thorough investigation, both theoretically and experimentally, of nozzle flows with initial angle of attack has been published by Pirumov and Roslyakov [13].)

The figure refers to a spinning, inertially symmetric, prolate vehicle with a solid-propellant motor, but the results derived in the following are equally valid in the oblate case. Let x_3 designate the spacecraft principal axis that coincides with the nominal thrust axis. Also, let $\mathbf{r}_1$ be the position vector in the aft direction of x_3 from the (instantaneous) center of mass, O, to the average location of the burn surface at a specified time. Similarly, $\mathbf{r}_2$ is the position vector from O to the nozzle exit midpoint. The spacecraft angular velocity is designated by ω, the momentum vector by $\mathbf{L}$, the nutation angle by θ, the precession rate by p, and the gas mass flow rate by $\tilde{m}$. A one-dimensional, steady-state description is used. Because the damping rate is slow compared with ω, it suffices to compute the jet damping moment separately at each instance, assuming that all quantities are constant at that time.

As a consequence of the precession, there is at the nozzle exit, relative to inertial space, an additional velocity component of the gas

$$\mathbf{u}_2 = p(\mathbf{L}/L) \times \mathbf{r}_2$$

transverse to the thrust axis. The reactive force resulting from the momentum per unit time, $\tilde{m}\tilde{\mathbf{u}}_2$, is opposite to the direction of the precession and tends to decrease the nutation angle. It accounts for the major part of the jet damping effect.

A closed control surface can be introduced that contains the spacecraft and crosses the nozzle exit plane. The angular momentum per unit time (in inertial space and relative to the center of mass) of the gas that leaves the control surface at the nozzle exit is

$$\tilde{m}p\mathbf{r}_2 \times (\mathbf{L}/L \times \mathbf{r}_2) \tag{6.57a}$$

The angular momentum loss in the control volume, per unit time, caused by the propellant diminution at the burn surface is

$$-\tilde{m}p\mathbf{r}_1 \times (\mathbf{L}/L \times \mathbf{r}_1) \tag{6.57b}$$

The torque on the spacecraft, balancing these two angular momentum changes, is the **jet damping moment**

$$\mathbf{M}_{jd} = \tilde{m}p[\mathbf{r}_2 \times (\mathbf{L}/L \times \mathbf{r}_2) - \mathbf{r}_1 \times (\mathbf{L}/L \times \mathbf{r}_1)] \tag{6.58}$$

The vector $\mathbf{M}_{jd}$ is seen to be perpendicular to the thrust axis and to rotate around it. The second term in the square bracket is generally much smaller than the first, since the center of mass of most spacecraft is closer to the burn surface than to the nozzle exit plane. Therefore, the averaging of the location of the burn surface, as was done in defining $\mathbf{r}_1$, will not generally introduce appreciable errors.

It follows that the magnitude M_{jd} of the jet damping moment is

$$M_{jd} = \tilde{m}p(r_2^2 - r_1^2)\sin\theta = \tilde{m}\frac{I_3}{I_1}\omega_3(r_2^2 - r_1^2)\tan\theta$$

Therefore, for small nutation angles

$$M_{jd} = \tilde{m}\frac{I_3}{I_1}\omega(r_2^2 - r_1^2)\theta \qquad (6.58')$$

6.6.2 Passive Nutation Damping of Oblate Spacecraft

Spinning, inertially symmetric spacecraft that are **oblate**, after disturbances have subsided, will continue to precess at a constant nutation angle (Sect. 6.3.1). In this sense, they are stable. The need, however, is to reduce the nutation angle to near zero after a disturbance.

A number of different passive nutation dampers are in use. The principle upon which they are based is the same: The precessional motion of the spacecraft induces a motion of the damping element relative to the spacecraft, thereby dissipating energy. The arrangement shown in Fig. 6.17 can serve as an example:

A small wheel is mounted with its axis of rotation perpendicular to the vehicle's nominal spin axis x_3. The wheel is passive in the sense that it is not driven by a motor but merely responds to the spacecraft motion. Provision is made for damping the wheel's rotation (which can be in either direction), typically by eddy currents induced by a magnetic field.

Let $\omega(t)$ be the angular velocity of the spacecraft body relative to inertial space, $v(t)$ the rate of rotation of the wheel, and J the axial moment of inertia of the wheel. The x_1 principal axis is taken parallel to the wheel's axis. With $\mathbf{i}_1, \mathbf{i}_2, \mathbf{i}_3$ designating the unit base vectors of the principal axes coordinate system, the combined angular momentum, $\mathbf{L}$, of spacecraft body and wheel is

$$\mathbf{L} = (I_1\omega_1 + Jv)\mathbf{i}_1 + I_2\omega_2\mathbf{i}_2 + I_3\omega_3\mathbf{i}_3$$

Let $\mathbf{M}_d$ be the disturbance torque that acts on the spacecraft. Then

$$\mathbf{M}_d = d\mathbf{L}/dt = \dot{\mathbf{L}} + \boldsymbol{\omega} \times \mathbf{L}$$

where, as before, ($\dot{}$) indicates time derivatives in the spacecraft-fixed space and $d(\,)/dt$ in inertial space. Substituting for $\mathbf{L}$, and noting that the basis vectors and the moments of inertia are constant in the dot differentiation, it follows that

$$I_1\dot{\omega}_1 + J\dot{v} + (I_3 - I_1)\omega_2\omega_3 = cv + M_{d,1} \qquad (6.59a)$$

$$I_1\dot{\omega}_2 - (I_3 - I_1)\omega_1\omega_3 + J\omega_3 v = M_{d,2} \qquad (6.59b)$$

$$I_3\dot{\omega}_3 - J\omega_2 v = M_{d,3} \qquad (6.59c)$$

where c is the damping coefficient, hence the torque cv the reaction from the damped wheel on the spacecraft. The wheel satisfies the condition

$$J(\dot{\omega}_1 + \dot{v}) = -cv \qquad (6.60)$$

The transverse angular velocity ω_2 will be much smaller than the vehicle's rate of spin. It follows from the last of equations (6.59) that ω_3, substituted into the first two, can be approximated by a constant, say ω_0. These two equations, together with (6.60), then result in the set of linear equations represented by

$$\begin{bmatrix} I_1 & 0 & J \\ 0 & I_1 & 0 \\ J & 0 & J \end{bmatrix} \begin{bmatrix} \dot{\omega}_1 \\ \dot{\omega}_2 \\ \dot{v} \end{bmatrix} + \begin{bmatrix} 0 & (I_3 - I_1)\omega_0 & -c \\ -(I_3 - I_1)\omega_0 & 0 & J\omega_0 \\ 0 & 0 & c \end{bmatrix} \begin{bmatrix} \omega_1 \\ \omega_2 \\ v \end{bmatrix} = \begin{bmatrix} M_{d,1} \\ M_{d,2} \\ M_{d,3} \end{bmatrix}$$

(6.61)

To examine the stability of the system and its rate of approach to equilibrium after an external disturbance has subsided, it suffices to examine the homogeneous part of (6.61). Making the substitutions

$$\omega_1 = a_1 e^{st}, \qquad \omega_2 = a_2 e^{st}, \qquad v = a_v e^{st}$$

where s is the (complex) frequency and a_1, a_2, a_v are the (complex) amplitudes, the existence of nontrivial solutions therefore requires that

$$\begin{vmatrix} I_1 s & (I_3 - I_1)\omega_0 & Js - c \\ -(I_3 - I_1)\omega_0 & I_1 s & J\omega_0 \\ Js & 0 & Js + c \end{vmatrix} = 0 \qquad \textbf{(6.62)}$$

If the root locus method is applied for examining the stability of the motion, it becomes convenient to cast this equation into Evan's form, which makes evident the zeros and poles in the complex plane of s. In this form, (6.62) becomes

$$\frac{s(s^2 + \lambda^2 \omega_0^2)}{s^2 + \frac{(I_3/I_1 - 1)^2 \omega_0^2}{1 + J/I_1}} = -\frac{(1 + J/I_1)c}{(1 - J/I_1)J} \qquad (6.63)$$

where

$$\lambda^2 = \frac{(I_3/I_1 - 1)(I_3/I_1 + J/I_1 - 1)}{1 - J/I_1}$$

It is found that, with suitable choices of J and c, this passive means of stabilizing is adequate to reduce the nutation of *oblate* vehicles to zero after a disturbance. For *prolate* vehicles, either an active system (e.g., with wheels that are driven by electric motors in response to gyro inputs) or the passive system described in the next section is usually required.

An example is shown in Fig. 6.19 for moment of inertia ratios $I_3/I_1 = 1.50$ (an oblate vehicle) and $J/I_1 = 0.025$. Shown as functions of the nondimensional damping coefficient (defined in the figure) are the loci of the three roots in the complex plane of s. There is one pair of conjugate complex roots and one purely real root. The latter is highly damped. The critical ones are the complex ones, because they result in the longest time for the exponential

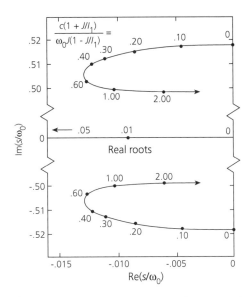

Figure 6.19 Passive nutation damping of oblate spacecraft: root loci for $I_3/I_1 = 1.50$ and $J/I_1 = 0.025$. From Eq. (6.63).

decay of a disturbance. The optimum damping coefficient is approximately 0.53 (where the tangent to the curves is vertical).

Figure 6.20 shows as a function of $\omega_0 t$ the damped oscillations of a vehicle with the same ratio $I_3/I_1 = 1.50$. The angular velocities ω_1, ω_2 and the nutation angle θ are proportional to the ordinate shown. Represented are the least damped modes for each of three values of J/I_1, each with its optimal damping coefficient. As is to be expected, the larger the wheel's

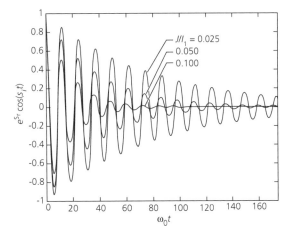

Figure 6.20 Passive nutation damping of oblate spacecraft: least damped modes with optimal damping. For $I_3/I_1 = 1.50$ and three values of J/I_1. From Eq. (6.63).

moment of inertia, the more rapid is the decay from an external disturbance.

6.6.3 Dual-Spin Spacecraft

Orbiting, spin-stabilized vehicles usually require a despun platform for the support of antennas, sensors, imagers, or scientific instruments that are intended to point toward the earth or a planet. The rate of rotation of the platform is therefore comparable to the orbital rate, which differs from and is much smaller than the vehicle spin. These vehicles are often called **dual-spin**.

A typical configuration of spin-stabilized vehicles for space communications or earth observations is shown in Fig. 6.2a. The *despun platform* and *rotor* are connected by a shaft supported by bearings in the rotor. The torque from a small electric motor compensates for the bearing friction. This motor is controlled by an error signal, derived, for instance, from horizon sensors that detect deviations from the intended orientation of the platform.

Integral parts of the bearing assembly are slip rings that provide the needed electrical connections between platform and rotor for power and signal circuits. Usually, only low-frequency signals are transmitted in this way. Microwave frequencies are usually generated directly on the platform, to avoid rotating waveguide joints.

The larger part of the spacecraft mass resides in the rotor, which carries the solar cells, power conditioning equipment, batteries, spacecraft propellant, and frequently also a solid-propellant motor as the last stage in the flight trajectory. Concentrating the mass in the rotor is desirable for it enhances the short-term attitude stability. The rate of rotor spin may be quite low (e.g., one revolution per second, which would still make it about 5000 times larger than the platform rate of a low-altitude earth-orbiting satellite).

Active damping, for instance by wheels similar to the wheel shown in Fig. 6.17, but driven, can bring the nutation angle back to zero, even in the case of prolate vehicles. However, it was discovered [14], that the same effect can also be obtained by purely passive means, provided that a **nutation damper** is located on the *despun platform*. An example of such a damper is shown schematically in Fig. 6.21. Other arrangements, such as a damped pendulum, can be equally effective.

For the purpose of analyzing the stabilizing effect of the nutation damper that is illustrated in the figure, it will be assumed that both the rotor and the despun platform are inertially symmetric about their axis of relative rotation, x_3. The combined center of mass, O, of rotor and despun section is located on this axis and is taken as the origin of the spacecraft's principal axes x_1, x_2, x_3. These are taken as corotating with the *despun platform*. The vectors of the coordinate system formed by the principal axes are designated by $\mathbf{i}_1, \mathbf{i}_2, \mathbf{i}_3$.

The *rotor's* angular velocity relative to inertial space is $\boldsymbol{\omega} = \omega_1 \mathbf{i}_1 + \omega_2 \mathbf{i}_2 + \omega_3 \mathbf{i}_3$ and its moments of inertia about O are $I_1 = I_2$ and I_3. The corresponding data for the *despun section* are $\boldsymbol{\Omega} = \Omega_1 \mathbf{i}_1 + \Omega_2 \mathbf{i}_2 + \Omega_3 \mathbf{i}_3$, $J_1 = J_2$

6.6 *Spin-Stabilized Vehicles* 259

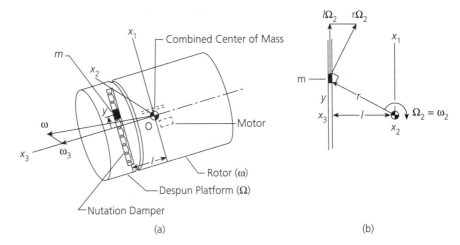

Figure 6.21 Nutation damper on dual-spin spacecraft: (a) schematic, (b) derivation of Eq. (6.67). (x_1, x_2, x_3) = plateform fixed principal axes; ω = rotor angular velocity; Ω = plateform angular velocity. For clarity, the length of the nutation damper relative to the spacecraft is greatly exaggerated.

and J_3. By virtue of the constraint imposed by the common axis of rotation,

$$\Omega_1 = \omega_1, \qquad \Omega_2 = \omega_2 \tag{6.64}$$

As shown in the figure, a mass, m, is free to move inside a tube located on the despun section, through x_3 and parallel to x_1. The mass is acted on by a pair of springs with combined spring constant k and by a damping force proportional to the relative velocity of the mass and the tube, with damping coefficient c. The distance of the tube from the center of mass, O, is designated by l and the distance of m from the x_3 axis by $y(t)$, where, by assumption, $|y| \ll l$.

To be effective as a damper, k and c are chosen such that the spring–mass system is approximately in resonance with the vehicle's precession frequency [given by (6.21)]. The latter, to second-order accuracy for small nutation angles, is a constant for a given spacecraft with fixed moments of inertia and rate of spin of the rotor.

That the nutation angle is assumed to be small implies that $|\Omega_1|, |\Omega_2|, |\Omega_3| \ll \omega_0$, where ω_0 is the rotor's nominal rate of spin, a constant. Time derivatives in the space fixed to the despun section will be designated by ($\dot{\ }$), those in inertial space by $d(\ )/dt$. To estimate the orders of magnitude of the various terms, it is convenient to write

$$\Omega_1 = \omega_0 \varepsilon_1, \qquad \Omega_2 = \omega_0 \varepsilon_2, \qquad \Omega_3 = \omega_0 \varepsilon_3, \qquad y = l\eta$$

where $|\varepsilon_1|, |\varepsilon_2|, |\varepsilon_3|, |\eta| \ll 1$. Since the rate of precession of the vehicle is of the same order as w_0, each ($\dot{\ }$) derivative introduces a factor of ω_0 into the estimated order of magnitude of the terms. Thus, for instance, $\dot{\Omega}_1 = \mathcal{O}(\omega_0^2 \varepsilon_1)$ and $y\dot{\Omega}_2 = \mathcal{O}(l\omega_0^2 \eta \varepsilon_2)$.

Considering next the motion of the mass m, its position vector from the vehicle center of mass, O, will be designated by $\mathbf{r}(t)$. Therefore

$$\mathbf{r} = y\mathbf{i}_1 + l\mathbf{i}_3, \qquad \dot{\mathbf{r}} = \dot{y}\mathbf{i}_1, \qquad \ddot{\mathbf{r}} = \ddot{y}\mathbf{i}_1$$

If $\mathbf{F}$ is the resultant force exerted on m by the spring, the viscous tube friction, and the lateral force from the tube, then

$$m\ddot{\mathbf{r}} = \mathbf{F} - m[\mathbf{\Omega} \times (\mathbf{\Omega} \times \mathbf{r}) + 2\mathbf{\Omega} \times \dot{\mathbf{r}} + \dot{\mathbf{\Omega}} \times \mathbf{r}] \qquad (6.65)$$

where the term with the square bracket represents the inertia force, as in (1.4). (The term $d^2\mathbf{R}_0/dt^2$ is absent here because the vehicle is in free fall and gravity acts both on the vehicle and on the mass m.) Evaluating the vector products in terms of their x_1, x_2, x_3 components and estimating the orders of magnitudes, one finds that all terms in the square bracket are of second or third order in ε and η, except for two first-order terms that arise from the fourth vector product,

$$\dot{\mathbf{\Omega}} \times \mathbf{r} = l\dot{\Omega}_2\mathbf{i}_1 - l\dot{\Omega}_1\mathbf{i}_2$$

Dropping all terms of higher order than the first, (6.65) becomes

$$\mathbf{F} = (m\ddot{y} + ml\dot{\Omega}_2)\mathbf{i}_1 - ml\dot{\Omega}_1\mathbf{i}_2 \qquad (6.66)$$

The force exerted by m on the vehicle is $-\mathbf{F}$, which results in a torque, say $\mathbf{N}_m$, on the vehicle about O, given therefore by

$$\mathbf{N}_m = -ml^2\dot{\Omega}_1\mathbf{i}_1 - l(m\ddot{y} + ml\dot{\Omega}_2)\mathbf{i}_2 \qquad (6.67)$$

The combined angular momentum, $\mathbf{L}$, of rotor, despun platform, and damper mass is

$$\mathbf{L} = (I_1 + J_1)\Omega_1\mathbf{i}_1 + [(I_1 + J_1)\Omega_2 + ml\dot{y}]\mathbf{i}_2 + (I_3\omega_3 + J_3\Omega_3)\mathbf{i}_3 \qquad (6.68)$$

From conservation of momentum

$$d\mathbf{L}/dt = \mathbf{N}_m + \mathbf{M}_d$$
$$= (-ml^2\dot{\Omega}_1 + M_{d,1})\mathbf{i}_1 + (-ml\ddot{y} - ml^2\dot{\Omega}_2 + M_{d,2})\mathbf{i}_2 + M_{d,3}\mathbf{i}_3 \qquad (6.69)$$

where $\mathbf{M}_d(t)$ is the moment of the external disturbance that acts on the vehicle and where, from (1.1)

$$d\mathbf{L}/dt = \dot{\mathbf{L}} + \mathbf{\Omega} \times \mathbf{L}$$

Differentiating $\mathbf{L}$ and evaluating the components of the vector product yields for the first and second components

$$(I_1 + J_1 + ml^2)\dot{\Omega}_1 + I_3\omega_0\Omega_2 = M_{d,1} \qquad (6.70a)$$

$$(I_1 + J_1 + ml^2)\dot{\Omega}_2 - I_3\omega_0\Omega_1 + 2ml\ddot{y} = M_{d,2} \qquad (6.70b)$$

The third component is not needed; in its place one has the equation of motion of the damper mass, which from the first component of (6.65) is

$$m\ddot{y} + c\dot{y} + ky + ml\dot{\Omega}_2 = 0 \qquad (6.70c)$$

6.6 Spin-Stabilized Vehicles

For the study of the stability of the system, it suffices to consider the homogeneous parts. Setting

$$\Omega_1 = a_1 e^{st}, \qquad \Omega_2 = a_2 e^{st}, \qquad y = a_y e^{st}$$

where s is the (generally complex) frequency and a_1, a_2, a_y are the (generally complex) amplitudes, a system of three linear, homogeneous equations for the amplitudes is obtained. The existence of a nontrivial solution requires that

$$\begin{vmatrix} (I_1 + J_1 + ml^2)s & I_3\omega_0 & 0 \\ -I_3\omega_0 & (I_1 + J_1 + ml^2)s & 2mls^2 \\ 0 & mls & ms^2 + cs + k \end{vmatrix} = 0 \qquad (6.71)$$

Evaluation of the determinant results in the fourth-degree characteristic polynomial,

$$[(I_1 + J_1 + ml^2)^2 m - 2(I_1 + J_1 + ml^2)m^2 l^2]s^4 + (I_1 + J_1 + ml^2)^2 cs^3$$
$$+ [(I_1 + J_1 + ml^2)^2 k + I_3^2\omega_0^2 m]s^2 + I_3^2\omega_0^2 cs + I_3^2\omega_0^2 k = 0 \qquad (6.72)$$

The system is stable if all four roots are in the left half of the complex plane, unstable if one or more roots are in the right half-plane. If stable, the angular velocity components and the nutation angle will, after the external disturbance has subsided, decay toward zero, in the most general case as a superposition of exponentially damped oscillations and of pure exponentials.

An example is shown in Fig. 6.22 for a spacecraft with $I_1/I_3 = 1.00$, $J_1/I_3 = 0.50$. For the nutation damper it is assumed that $ml^2/I_3 = 0.05$ and $k/(m\omega_0^2) = 1$. The nondimensional damping coefficient $c/(m\omega_0)$ is taken as the parameter. Depending on its range, there are two pairs of conjugate complex roots or else one such pair and two real roots. An optimum damping coefficient is one that results in the most rapid damping of the least damped mode. In the present case, its value is 1.85, as is indicated in the figure.

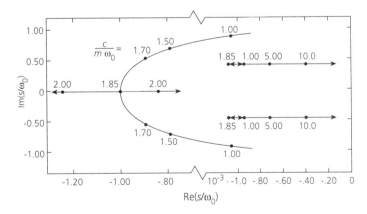

Figure 6.22 Dual-spin spacecraft: root loci for a spacecraft with $I_1/I_3 = 1.00$, $J_1/I_3 = 0.50$, $ml^2/I_3 = 0.05$, $k/(m\omega_0^2) = 1$. From Eq. (6.72).

6.7 Gravity Gradient Stabilization

The parts of a spacecraft that are closer to the center of gravitational attraction than those at a larger distance are acted upon by a slightly larger gravitational force. Hence, in general there will be a gravitational torque. Although very small, it can be used to stabilize the spacecraft attitude by means that are purely passive. Because of the smallness of the torque, **gravity gradient stabilization** is useful only when the disturbance torques (Fig. 6.1) are merely a fraction of the available gravity torque.

Figure 6.23 shows, as an example, an orbiting spacecraft with a long mast that extends toward the earth. The spacecraft is assumed to be in a circular orbit with radius R_0 to the spacecraft center of mass. It is convenient in this case to specify the attitude by means of a locally horizontal, orthogonal system of reference axes through the center of mass. This system is familiar from aircraft practice. Thus the x_1' axis is in the direction of motion, the x_3' axis downward toward the center of attraction, and the x_2' axis such as to form with the others a right-handed system.

The principal axes of the spacecraft are x_1 (the "roll axis"), x_2 (the "pitch axis"), and x_3 (the "yaw axis"). The angles from the primed to the unprimed axes are the pitch angle θ, roll angle ϕ, and yaw angle ψ. All are assumed to be very small. The moments of inertia about the principal axes are I_1, I_2, I_3.

The expression for the gravitational torque has been derived in Chap. 2, Eq. (2.9). Carrying out component by component the matrix multiplication needed to obtain the tensor product $\mathbf{J} \cdot \mathbf{R}_0$ and then the vector product $\mathbf{R}_0 \times (\mathbf{J} \cdot \mathbf{R}_0)$, one obtains for the first component M_{g_1} of the gravitational torque

$$M_{g_1} = -3\mu R_0^{-5}(I_2 - I_3)x_2' x_3' \tag{6.73}$$

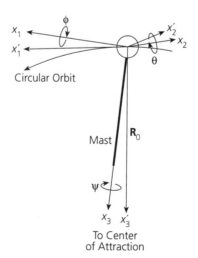

Figure 6.23 Gravity gradient stabilization: spacecraft with mast in circular orbit. θ, pitch; ϕ, roll; ψ, yaw.

where μ is the gravitational parameter ($\mu = 3.986\ 10^5$ km^3 s^{-2} for the earth). The corresponding results for the other components are obtained by cyclic interchange of the subscripts 1, 2, 3.

With ω designating the angular velocity of the spacecraft relative to inertial space and $\mathbf{M}_d$ an external (nongravitational) disturbance torque, Euler's equations (6.16′) for constant moments of inertia become

$$I_1\dot{\omega}_1 + (I_3 - I_2)\omega_2\omega_3 = -3\mu R_0^{-5}(I_2 - I_3)x_2'x_3' + M_{d,1} \quad (6.74)$$

for the first component and by cyclic permutation of the subscripts for the others. For small angles θ, ϕ, ψ

$$x_1' = R_0\theta, \qquad x_2' = -R_0\phi, \qquad x_3' = -R_0$$

Therefore the full set of Euler's equations becomes

$$\left.\begin{aligned}
I_1\dot{\omega}_1 + (I_3 - I_2)\omega_2\omega_3 &= -3\mu R_0^{-3}(I_2 - I_3)\phi + M_{d1} \\
I_2\dot{\omega}_2 + (I_1 - I_3)\omega_3\omega_1 &= +3\mu R_0^{-3}(I_3 - I_1)\theta + M_{d2} \\
I_3\dot{\omega}_3 + (I_2 - I_1)\omega_1\omega_2 &= +M_{d,3}
\end{aligned}\right\} \quad (6.75)$$

where a second-order term with the factor $\theta\phi$ has been dropped.

One can introduce the orbital angular velocity, n, which is the rate at which the attitude relative to inertial space changes by virtue of the motion along the circular orbit. It follows from (3.40) that $n^2 = \mu/R_0^3$. The angular rates $\dot{\theta}$, $\dot{\phi}$, and $\dot{\psi}$ are assumed to be small compared with n. The orbital angular velocity causes gyroscopic effects, which are essential to the gravity gradient stabilization.

For small angles θ, ϕ, ψ the angular velocity components become

$$\omega_1 = \dot{\phi} - n\psi, \qquad \omega_2 = \dot{\theta} - n, \qquad \omega_3 = \dot{\psi} + n\phi \quad (6.76)$$

The equations governing the *pitch* are seen to be independent of those for roll and yaw. Neglecting a second-order term with the factor $\omega_1\omega_3$, one obtains from the second of equations (6.75)

$$\left.\begin{aligned}
I_2\dot{\omega}_2 + 3n^2(I_1 - I_3)\theta &= M_{d2} \\
\omega_2 &= \dot{\theta} - n
\end{aligned}\right\} \quad (6.77)$$

for ω_2 and θ.

The *roll* and *yaw* are coupled [2]. It follows from the first and the third of Eqs. (6.75), neglecting $\dot{\theta}$ in comparison with n, that

$$\left.\begin{aligned}
I_1\dot{\omega}_1 + n(I_2 - I_3)\omega_3 + 3n^2(I_2 - I_3)\phi &= M_{d,1} \\
I_3\dot{\omega}_3 + n(I_1 - I_2)\omega_1 &= M_{d3} \\
\omega_1 &= \dot{\phi} - n\psi \\
\omega_3 &= \dot{\psi} + n\phi
\end{aligned}\right\} \quad (6.78)$$

for ω_1, ω_3, ϕ and ψ.

The stability of the motion can be studied by limiting oneself to the homogeneous parts. Letting

$$\omega_2 = a_2 e^{st}, \qquad \theta = a_\theta e^{st}$$

for the **pitch** equations leads to the characteristic equation

$$\begin{vmatrix} I_2 s & 3n^2(I_1 - I_3) \\ 1 & -s \end{vmatrix} = 0 \quad (6.79)$$

for the complex frequency s, hence to the second-degree polynomial

$$s^2 + 3n^2(I_1 - I_3)/I_2 = 0$$

If $I_1 > I_3$, there will be, following an external disturbance, an undamped harmonic oscillation, referred to as "pitch libration." Its angular frequency is

$$n\sqrt{\frac{3(I_1 - I_3)}{I_2}}$$

If $I_1 < I_3$, the motion is unstable with an exponential increase of the pitch angle at the rate

$$n\sqrt{\frac{3(I_3 - I_1)}{I_2}}$$

Letting

$$\omega_1 = a_1 e^{st}, \quad \omega_3 = a_3 e^{st}, \quad \phi = a_\phi e^{st}, \quad \psi = a_\psi e^{st}$$

for the **roll/yaw**, the characteristic equation becomes

$$\begin{vmatrix} I_1 s & n(I_2 - I_3) & 3n^2(I_2 - I_3) & 0 \\ n(I_1 - I_2) & I_3 s & 0 & 0 \\ -1 & 0 & s & -n \\ 0 & -1 & n & s \end{vmatrix} = 0 \quad (6.80)$$

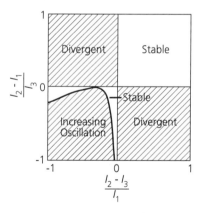

Figure 6.24 Stability of the roll–yaw motions of gravity-gradient stabilized spacecraft. (Ref. 2.) Adapted from Bryson, A. E., "Control of Spacecraft and Aircraft," Copyright © 1994 Princeton University Press. By permission.

leading to a fourth-degree polynomial for s. The regions for stable and unstable motions are shown in Fig. 6.24 (after Bryson [2]).

Nomenclature

A	area
B	magnetic induction
c	damping/friction coefficient
$\mathbf{D}_m$	magnetic dipole moment
e	eigenvector
E	back EMF
i	current
$\mathbf{i}_i$	base vectors
I, **J**	Cartesian inertia tensors
I_i, J_i	principal moments of inertia
k	spring constant [Eq. (6.36)]; also control gain [Eq. (6.51)]
l_{ij}	direction cosines
L	angular momentum
m	spacecraft mass
m_c	damper mass
M	spacecraft external torque
N	spacecraft internal torque
n	orbital rate of rotation [Eq. (6.76)]; also number of turns [Eq. (6.55)]
p	precession rate
q_i	motor torque per unit current
r	position vector; radius
R	resistance
s	complex frequency
T	kinetic energy of rotation; **T**: Cartesian tensor
u	gas velocity
V	armature voltage
x_i	principal axes
δ	declination of sensor axis
δ_{ij}	Kronecker delta
θ	nutation angle (Fig. 6.4); also pitch angle (Figs. 6.7, 6.23)
ϑ	rotor axis angle (Fig. 6.10)
λ	eigenvalue
ρ	density
τ	time constant [Eq. (6.42)]
ω	angular velocity relative to inertial space
$d\mathbf{v}/dt$	time derivative of a vector **v** in inertial space
$\dot{\mathbf{v}}$	time derivative of a vector **v** in body-fixed space
$()_d$	disturbance
$()_{el}$	electric
$()_{ex}$	exit plane
$()_{jd}$	jet damping

()$_m$ magnetic
()$_{th}$ thruster

Problems

(1) In a Cartesian coordinate system with unit base vectors $\mathbf{i}_1, \mathbf{i}_2, \mathbf{i}_3$, the inertia tensor of a rigid body is represented by the matrix (in arbitrary units)

$$\begin{bmatrix} 1 & 0 & 0 \\ 0 & 1 & 2 \\ 0 & 2 & 1 \end{bmatrix}$$

(a) Find the eigenvalues. Also find the directions of the principal axes of inertia. (Choose the algebraic signs of the eigenvectors such that they form a right-handed system.) Express the eigenvectors in terms of $\mathbf{i}_1, \mathbf{i}_2, \mathbf{i}_3$.
(b) As a check on (a), verify that the principal axes of inertia, as calculated, are mutually orthogonal.
(c) Determine the numerical values of the principal moments of inertia.

(2) Consider a spinning, rigid body with principal axes x_1, x_2, x_3 and corresponding moments of inertia I_1, I_2, I_3. The x_3 axis is an axis of symmetry, so that $I_1 = I_2$. The components of the body's angular velocity about its principal axes are designated by $\omega_1, \omega_2, \omega_3$. An external torque is acting on the body, with components M_1, M_2, M_3 about the principal axes.

Assume that $M_1 = M_0 =$ constant and $M_2 = M_3 = 0$. Initially, the angular velocity components are assumed to be equal, say ω_0.

Find $\omega_2(t)$ for $t \geq 0$. Express the result in terms of ω_0, M_0, and the moments of inertia.

(3) The attitude of a spacecraft is assumed to have been disturbed impulsively about one of its principal axes. Following the disturbance, an on–off control, provided by thrusters, is attempting to return the attitude of the spacecraft back to its nominal (zero) orientation. The torque by the thrusters has constant magnitude but can be in either direction.

The moment of inertia pertaining to this axis is 1000 kg m^2. The magnitude of the thruster torque is 0.30 Nm. Let γ the angular deviation of the attitude from its nominal value. Immediately following the disturbance $\gamma = 0$ and $d\gamma/dt = 0.005$ rad/s. A control law is assumed of the form

$$\gamma + \tau \, d\gamma/dt$$

with time constant $\tau = 10$ s. To improve the stability, a deadband between $\gamma = \pm 0.015$ rad is provided. There are two hysteresis zones, one between $\gamma = 0.015$ rad and 0.008 rad, the other between -0.008 rad and -0.015 rad (see Fig. 6.12).

(a) Draw a phase diagram with coordinates γ and $d\gamma/dt$ that shows the path followed by the spacecraft attitude.
(b) Compute the period of the limit cycle.

(4) A spacecraft is in a circular orbit that takes it over the earth's magnetic poles. It is equipped with a reaction wheel and a magnetic desaturation solenoid.

The wheel's axis is perpendicular to the orbital plane. The axis of the solenoid is at all times aligned with the local vertical.

The orbit is at 400 km altitude (orbit radius approximately 6760 km). At this altitude the earth's magnetic field can be approximated by a dipole field with magnetic lines running from the magnetic south pole to the magnetic north pole. The dipole moment is 8 10^{15} tesla m^3 (= volt s m). The only component that needs to be taken into account is the one along the orbit.

The reaction wheel has a moment of inertia of 0.02 kg m^2. At saturation it spins at 12,000 rpm.

Magnetic desaturation is accomplished with a rodlike solenoid with a ferrite core (Fig. 6.13b). The number of turns is 2000, the current is 1.0 amp. The permeability (relative to the permeability of vacuum) of the ferrite core is 200. The effective cross section of the rod is 10 cm^2. The approximations valid for long, slender solenoids can be used.

Compute the reaction wheel's angular momentum at saturation. If desaturation is applied over one-half of the orbit, from magnetic pole to magnetic pole, what fraction of the saturation angular momentum can be eliminated?

References

1. Kane, T. R., Likins, P. W., Levinson, D. A., "Spacecraft Dynamics," McGraw-Hill, New York, 1983.
2. Bryson, A. E., "Control of Spacecraft and Aircraft," Princeton University Press, Princeton, NJ, 1994.
3. "Spacecraft Pointing and Control," North Atlantic Treaty Organization Advisory Group for Aerospace Research and Development, AGARDograph No. 260, 1982.
4. Hughes, P. C., "Spacecraft Attitude Dynamics," John Wiley & Sons, New York, 1986.
5. Kaplan, M. H., "Modern Spacecraft Dynamics and Control," John Wiley & Sons, New York, 1976.
6. Wertz, J. R., ed., "Spacecraft Attitude Determination and Control," Kluver Academic Publishers, Dordrecht, The Netherlands, 1978.
7. Zelikin, M. I. and Borisov, V. F., "Theory of Chattering Control," translated from the Russian, Birkhäuser, Basel, 1994.
8. Stratton, J. A., "Electromagnetic Theory," McGraw-Hill, New York, 1941.
9. Thomson, W. T. and Reiter, G. S., "Jet Damping of a Solid Rocket: Theory and Flight Results," *AIAA Journal*, Vol. 3, No. 3, pp. 413–416, 1965.
10. McIntyre, J. E. and Tanner, T. M., "Fuel Slosh in a Spinning On-Axis Propellant Tank: An Eigenmode Approach," *Space Communication and Broadcasting*, Vol. 5, No. 4, pp. 229–251, 1987.
11. Mingori, D. L., Halsmer, D. M., and Yam, Y., "Stability of Spinning Rockets with Internal Mass Motion," *American Astronautical Society Proceedings*, pp. 93–135, February 1993.

12. Meyer, R. X., "Coning Instability of Spacecraft During Periods of Thrust," *Journal of Spacecraft and Rockets*, Vol. 33, No. 6, pp. 781–788, 1996.
13. Pirumov, U. G. and Roslyakov, G. S., "Gas Flow in Nozzles," Springer Series in Chemical Physics, Vol. 29, Springer Verlag, Berlin, translated from the Russian, 1986.
14. Iorillo, A. J., "Analysis Related to the Hughes Gyrostat Systems," Hughes Aircraft Company Report 70438 B, 1967.

7
Spacecraft Thermal Design

Spacecraft absorb and emit radiation. Depending on the physical characteristics of their external surfaces, they absorb more or less of the incident solar radiation. Spacecraft operating in the near-earth environment also absorb to a significant degree thermal radiation (in the infrared) that is emitted by the earth and solar radiation that is reflected by the earth (albedo effect). Analogous effects also play a role when operating near the inner planets and the earth's moon.

In turn, spacecraft emit thermal radiation into space. When in thermal equilibrium, their emission equals the sum of (1) their absorption of the incident radiation of all types (in the case of solar panels their net thermal absorption plus the electric power generated by them) and (2) the heat produced by spacecraft internal sources such as the heat dissipated by electrical components, by heat transfer from the combustion of propellants, and sometimes by radioisotope sources.

The reader will find much additional information related to spacecraft radiative heat transfer in Refs. 1 to 3.

Spacecraft temperatures must be controlled to within close lower and upper limits. These are imposed by the different characteristics of the various components on the spacecraft. Particularly sensitive to temperature extremes are storage batteries, spacecraft propellants, many electronic components, and certain scientific payloads. The allowable limits for electronic components are not necessarily the same for operating and nonoperating conditions (when operating, the additional electric stress often imposes a lower upper temperature limit).

The temperature assumed by these components will be determined by their own heat production and, on the other hand, by the transfer of heat by *thermal conduction* and *radiation* to and from other components and, directly or indirectly, to and from the spacecraft surface or space. Because of the vacuum environment, *convective cooling* normally used in electronic components is absent. Radiative transfer therefore plays a more important role than is the case in earth-bound electronics.

7.1 Fundamentals of Thermal Radiation

The reader may already be familiar with much of the material on the theory of radiative heat transfer that is presented in this section. Much more can be found in standard texts, such as in the book by Siegel and Howell [2]. We present some basic material here primarily to define the concepts and notation that will be needed later. Also, this section will provide the opportunity to discuss early some applications specific to space technology before describing the more elaborate methods that are needed to calculate spacecraft temperatures.

7.1.1 Blackbody Thermal Absorption and Emission

A *blackbody* is defined as an idealized material that absorbs all radiation incident on it, irrespective of the wavelength and of the angle of incidence. Reflection and transmission are therefore absent. As will be discussed here, the *absorption* and *emission* properties of a blackbody can be derived from one another, so that no new definition of the blackbody will be needed to describe its emission.

The expression "blackbody"—which has historical roots—is somewhat inaccurate, because, as is clear from the definition, almost always only the *surface* of the body is important. Thus, if there is heat flowing by conduction to the surface, there will be a distribution of temperature in the body different from the surface temperature. Yet, conventionally, one still refers to the latter in this case as "blackbody temperature."

There are no materials that match exactly the definition of a blackbody, although some—such as black paint in the visible range of wavelengths—come very close. Ideal blackbodies, however, are an important standard with which real surfaces can be compared.

The following thought experiment, illustrated in Fig. 7.1, allows one to draw some simple conclusions from purely thermodynamic arguments: One considers an evacuated enclosure at a uniform temperature and of arbitrary shape. The exterior is thermally insulated. The interior surface emits and absorbs as an ideal blackbody. Also, there is a test body (assumed for simplicity to be convex so that radiation from it strikes only the enclosure) which is also a black body.

Initially, the temperatures of enclosure and test body are arranged to be the same. It then follows from the first and second laws of thermodynamics

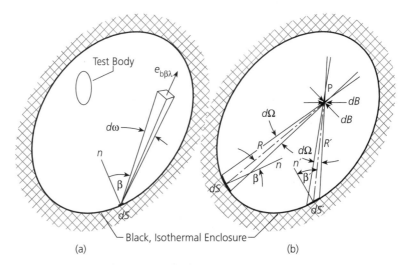

Figure 7.1 Schematic illustrating derivation of Lambert's law.

that neither temperature can either increase or decrease and that the test body's emission equals its absorption. Because the argument is independent of the test body's position or orientation, the radiation in the space between enclosure and test body must be **homogeneous** (i.e., has constant radiation density) and **isotropic**.

In Fig. 7.1a let dS be a surface element of the enclosure and β the angle from the normal n of dS to some chosen direction of emitted radiation. Also let $d\omega$ be the solid angle bounding this direction.

Whereas in Fig. 7.1a all rays emitted from a point of dS that fall within the solid angle $d\omega$ are shown, in Fig. 7.1b the rays from all points of dS are shown that converge to an aperture with area dB (located at point P at distance R.) Similarly, a second surface element dS' is shown, with rays converging to an aperture with the same area dB at P and at distance R'. The area dB can be chosen arbitrarily small compared with dS and dS'. Also, for convenience, dS and dS' are chosen such that the solid angles $d\Omega$ extended from the apertures to dS and dS' are the same.

The radiation from a point of dS or dS' striking the respective aperture therefore has directions bounded by the solid angles

$$d\omega = dB/R^2 \quad \text{and} \quad d\omega' = dB/R'^2$$

(see Fig. 7.1.a).

Another such geometrical relation, evident from Fig. 7.1.b, is

$$d\Omega = dS \cos\beta / R^2 = dS' \cos\beta' / R'^2$$

Let $e_{b\beta\lambda}\, dS\, d\omega\, d\lambda$ be the black body radiant power from dS, emitted within a cone with solid angle $d\omega$ centered about β and within a wavelength interval $d\lambda$ centered about λ. The $e_{b\beta\lambda}$ is referred to as the *directional, monochromatic emitted power* (per unit area and unit wavelength interval; in the present case for a blackbody). It is a function of β, λ, and T, and its magnitude is usually expressed in units of $W/(m^2\,\mu m)$. Analogous statements hold for $e_{b\beta'\lambda}$.

Isotropy of the radiation at point P then requires that the two radiant powers be equal, hence

$$e_{b\beta\lambda}\, dS\, d\omega\, d\lambda = e_{b\beta'\lambda}\, dS'\, d\omega'\, d\lambda$$

Substitution of the geometrical relations found for $d\omega$, $d\omega'$, and $d\Omega$ yields

$$e_{b\beta\lambda} \cos\beta' = e_{b\beta'\lambda} \cos\beta \tag{7.1}$$

In particular, when expressing the power emitted in a direction given by β with the power emitted along the normal, one obtains by setting $\beta' = 0$ the important relation

$$e_{b\beta\lambda} = e_{bn\lambda} \cos\beta \tag{7.1'}$$

The subscript notation used here and in what follows is more or less standard in the theory of thermal radiation. Thus n refers to the normal direction. Omitting among the subscripts the letter b means that a surface more general than a blackbody is considered. Omitting the angle β or n means that the radiant power has been integrated over all pertinent angles (usually a half-space). Omitting the wavelength λ means that the radiant power has been integrated over all wavelengths.

Equation (7.1′) is known as **Lambert's cosine law** (Lambert, astronomer and physicist, 1728–1777). The dependence on β only through the cosine has a simple representation: As is easily shown, if the radiant power of a surface element is drawn as the lengths of vectors pointing along the directions of propagation of the radiation, the end points are located on a *sphere* that is tangent to the radiating surface.

The radiant power, $e_{b\lambda}$, per unit area and unit wavelength interval, emitted into the *half-space* above dS, is known as the *hemispherical, monochromatic emitted power* (per unit area and unit wavelength interval) and is readily obtained by integrating $e_{b\beta\lambda}$ over the half-space:

$$e_{b\lambda} = 2\pi e_{bn\lambda} \int_{\beta=0}^{\pi/2} \cos\beta \sin\beta \, d\beta$$

so that

$$e_{b\lambda} = \pi \, e_{bn\lambda} \tag{7.2}$$

It was first shown by Planck (1858–1947), based on an electromagnetic calculation of the emission and treating the photons in the enclosure as a perfect gas, that

$$e_{b\lambda} = \frac{2\pi C_1}{\lambda^5 (\exp(C_2/\lambda T) - 1)} \tag{7.3}$$

where $C_1 = hc^2 = 0.59544 \, 10^{-16}$ W m^2 and $C_2 = hc/k = 1.4388 \, 10^4$ μm K (h = Planck's constant = $6.6252 \, 10^{-34}$ J s; k = Boltzmann's constant = $1.3806 \, 10^{-23}$ kg m^2/(s^{2}K); c = speed of light in vacuum = $2.99792 \, 10^8$ m/s). Common units for $e_{b\lambda}$ are W/(m^2 μm), therefore the same as for $e_{b\beta\lambda}$.

Equation (7.3) is known as **Planck's radiation law**. Its derivation became the basis for the later development of quantum mechanics.

In Fig. 7.2, the hemispherical monochromatic emitted power $e_{b\lambda}$ is plotted for several temperatures that are important in space technology. The curve labeled 5760 K corresponds to the effective blackbody temperature of the solar photosphere. Its radiation, to a large extent, determines the skin temperature of spacecraft. The curve labeled 300 K is representative of the majority of spacecraft components. They are designed for, and work best, in a fairly narrow temperature range around room temperature (e.g., storage batteries in the range of 5 to 30°C). The third curve applies to the boiling temperature (77 K) of nitrogen. This temperature is typical of many scientific instruments that work at infrared wavelengths. Liquid nitrogen is also used in spacecraft thermal testing, where it cools the test chamber walls, thereby suppressing unwanted thermal radiation from the chamber.

With increasing temperature, the maximum emitted power shifts toward shorter wavelengths. If $\lambda_{\max}$ designates the wavelength for which at a chosen temperature the hemispherical monochromatic emitted power has a maximum, then

$$\lambda_{\max} T = C_3 \tag{7.4}$$

where the constant $C_3 = 2.8978 \, 10^3$ μm K. First established experimentally,

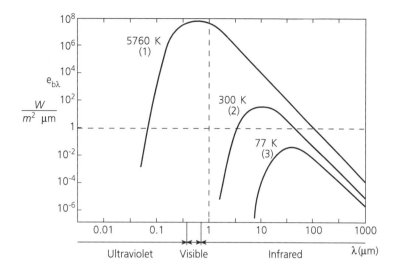

Figure 7.2 Planck's radiation law (Eq. 7.3) for three temperatures significant in space technology: (1) effective temperature of sun's photosphere, (2) ambient temperature, (3) liquid nitrogen temperature.

but also an easily derived consequence of (7.3), the equation is known as **Wien's displacement law** (Wien, 1864–1928).

The factor $e_{b\beta\lambda}$ of the directional, monochromatic emitted power, expressed by $e_{b\lambda}$ from Planck's radiation law, becomes, by using (7.1) and (7.2),

$$e_{b\beta\lambda} = \pi^{-1} e_{b\lambda} \cos\beta \tag{7.5}$$

The *hemispherical emitted power* (per unit area), e_b, is defined as the integral of $e_{b\lambda}$ over all wavelengths. The integration over Planck's relation can be carried out in closed terms (although this is not immediately obvious). The result is the **Stefan–Boltzmann law**, originally found by experiment,

$$e_b = \sigma T^4 \tag{7.6}$$

where

$$\sigma = 2C_1\pi^5/(15C_2^4) = 5.6693\ 10^{-8}\ \text{W}/(\text{m}^2\ \text{K}^4)$$

From this, the factor $e_{b\beta}$ of the *directional emitted power*, that is, the power emitted per unit solid angle in the direction defined by the angle β, when integrated over all wavelengths is

$$e_{b\beta} = \pi^{-1} \cos\beta\, \sigma T^4 \tag{7.7}$$

7.1.2 Solar Thermal Emission

The sun's thermal emission originates in the photosphere. Its temperature can be obtained from spectral measurements and is found to be about 5950 K. Absorption by the layers above the photosphere modifies the radiation. Nevertheless, as illustrated in Fig. 7.3, the sun's thermal radiation into space is well approximated by blackbody radiation, albeit at the somewhat

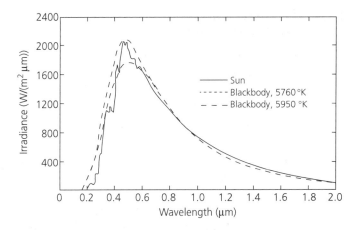

Figure 7.3 Solar spectral irradiance at 1.0 AU compared with the blackbody effective temperature (5760 K) and with the blackbody temperature of the photosphere (5950 K).

lower effective temperature of 5760 K. (Not shown in the figure are the Fraunhofer absorption lines. Although some of them, such as the prominent calcium K II line, are quite deep, they are far too narrow to show on the scale of the figure.)

At the distance of 1 AU (the mean sun–earth distance), the sun's radius has an apparent angle of 0.267°. Even at the position of Mercury, this angle is still only 0.69°. In applications to the thermal control of spacecraft, it can therefore be assumed that all rays of solar radiation that intercept a spacecraft come from a point source and are in effect *parallel* to each other.

At these distances, the sun appears as a disk of uniform luminosity. This is consistent with Lambert's law: The geometric factor resulting from the projection of an element of the sun's surface along the line of sight, is just canceled by the cosine resulting from (7.1) or (7.7).

It follows from conservation of energy that at distances large compared with the solar radius, where the sun can be approximated by a point source, the intensity (i.e., the radiant power per unit area perpendicular to the ray), j_h, falls off as the inverse square of the distance. At earth vicinity, the intensity has been measured by satellite instrumentation. It fluctuates somewhat during the course of a year, depending on the earth's distance from the sun. Thus

$$j_h = 1353 \text{ W/m}^2 \text{ (the so-called } solar\ constant) \text{ at 1 AU}$$
$$= 1309 \text{ W/m}^2 \text{ at aphelion (about July 4)}$$
$$= 1399 \text{ W/m}^2 \text{ at perihelion (about January 3)}$$

Some of the solar radiation in the thermal range is absorbed in the earth's atmosphere, primarily by water vapor and carbon dioxide. However, this effect is negligible for thermal calculations pertaining to satellites, even when the orbits are low (e.g., 300 km).

7.1.3 Thermal Emission, Absorption, and Reflection of Technically Important Surfaces

The thermal radiation properties of technically important surfaces can differ greatly from those of the ideal blackbody surface. In principle, they could be obtained by quantum mechanical calculations. In practice, however, such calculations would be impossibly difficult to carry out, or else, if approximated, would be very inaccurate. Therefore such radiation properties can be obtained only by direct measurements. Whereas Eqs. (7.1) to (7.7) are thermodynamically precise, no such claims can be made for the equations that follow (with the exception of Kirchhoff's law in its most general form).

The thermal radiation properties of a surface are characterized by its emission, absorption, reflection, and transmission. *Transmission* hardly plays any role for surfaces that are employed in spacecraft and will therefore not be considered here. The transparent layers of fused silica or quartz that cover the surfaces of optical solar reflectors, or the covers of solar cells, are sufficiently thin that their temperature is virtually the same as that of their opaque substrate; for the purposes of thermal calculations they can therefore be considered as single, opaque units.

Absorption and *reflection* depend not only on the properties of the surface but also on the direction and spectral characteristics of the *incident* radiation. For this reason, they are more difficult to take into account than emission. Even if a surface is isotropic, hence its emission independent of the azimuthal angle, reflection does not need to be so because it will in general depend on the direction of the incident radiation. Opaque surfaces reflect what is not absorbed. The reflected component can often be approximated by the sum of a specular reflection and of a fully diffuse one.

Let $i_{\phi\beta\lambda}\, dS\, d\omega\, d\lambda$ be the radiant power incident on dS, bounded within the solid angle $d\omega$ of a cone that is centered on the incident direction, and for a wavelength interval $d\lambda$ centered on the wave length λ. The angles ϕ and β are the azimuthal angle and the angle with the surface normal, respectively, of the incident radiation.

For some purposes it is convenient to introduce in place of the *incident power* $i_{\phi\beta\lambda}$ the *intensity* $j_{\phi\beta\lambda}$, which is defined relative to a surface element with the same area, but perpendicular to the incident radiation, so that

$$i_{\phi\beta\lambda} = j_{\phi\beta\lambda} \cos\beta \tag{7.8}$$

Also, let correspondingly $a_{\phi\beta\lambda}$ be the power per unit area, unit solid angle, and unit wavelength interval that is absorbed from radiation incident at the angles ϕ and β and wavelength λ. In all applications considered in this text, $a_{\phi\beta\lambda}$ is proportional to $i_{\phi\beta\lambda}$. (Nonlinear effects are important in the case of radiation from high-power lasers but are insignificant in the analysis of radiation heat transfer between surfaces.) One then defines as the coefficient of proportionality the *directional monochromatic absorptivity* $\alpha_{\phi\beta\lambda}$ by the relation

$$a_{\phi\beta\lambda} = \alpha_{\phi\beta\lambda} i_{\phi\beta\lambda} \tag{7.9}$$

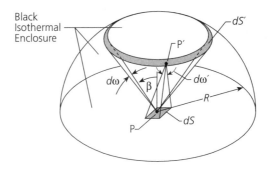

Figure 7.4 Derivation of Kirchhoff's law.

One also defines, relative to the blackbody emission $e_{b\beta\lambda}$, the *directional monochromatic emissivity* $\varepsilon_{\phi\beta\lambda}$ by the relation

$$e_{\phi\beta\lambda} = \varepsilon_{\phi\beta\lambda}\, e_{b\beta\lambda} \tag{7.10}$$

For the remainder of this chapter it will be assumed that all surfaces, in addition to being *opaque*, emit and absorb *isotropically* and *diffusely*. (Fluorescent materials, i.e., materials that absorb at one wavelength and emit at a different, longer wavelength, hardly ever occur in engineering applications and are excluded in what follows.)

The absorptivity and emissivity are related by **Kirchhoff's law** (Kirchhoff, 1824–1887), which can be derived by the following thought experiment:

As illustrated in Fig. 7.4, a surface element dS of the sample material is placed at the center of a hemispherical enclosure of radius R. With the exception of dS, all interior surfaces of the enclosure are ideal blackbody surfaces. The entire configuration is assumed to be initially at a common temperature. The enclosure therefore is filled with a homogeneous, isotropic blackbody radiation. As is illustrated for the point P, but equally applies to all points of dS, the solid angle extended to a surface element dS' (an annulus on the hemisphere) is $d\omega = dS'/R^2$. Similarly, the solid angle extended from a point P' of dS', or from any other point of dS', is $d\omega' = dS\cos\beta/R^2$, where β is the angle of incidence of the radiation on dS from dS'. Therefore

$$dS'\, d\omega' = dS \cos\beta\, d\omega \tag{7.11}$$

The only radiation reaching dS within the solid angle $d\omega$ is that coming from dS'. Hence the radiant power incident on dS within the intervals $d\omega$ and $d\lambda$ is

$$i_{\beta\lambda}\, dS\, d\omega\, d\lambda = e_{bn\lambda}\, dS'\, d\omega'\, d\lambda = e_{bn\lambda}\, dS \cos\beta\, d\omega\, d\lambda$$

(keeping in mind the convention adopted for the notation of subscripts as introduced in Sect. 7.1.1).

As before, it follows from the first and second laws of thermodynamics that the temperatures of dS and of the blackbody enclosure cannot change and that the absorption of radiation by dS must equal its emission. Therefore $a_{\beta\lambda} = e_{\beta\lambda}$. Hence, with the definitions (7.9) and (7.10),

$$\alpha_{\beta\lambda} = \varepsilon_{\beta\lambda} \quad \text{(opaque, isotropic surfaces)} \tag{7.12}$$

It follows from the definitions of the absorptivity and emissivity that for the ideal blackbody $\alpha_{\beta\lambda} = \varepsilon_{\beta\lambda} = 1$.

It should be noted that in the most general case $\alpha_{\beta\lambda}$ and $\varepsilon_{\beta\lambda}$ are functions of the angle β, the wavelength λ, and the temperature of the absorbing surface. There are, however, important instances in which some or all of these functional dependences can be neglected. In particular, the dependence on β can be neglected without incurring a major error when the surface is rough on the scale of the wavelength, as occurs with a majority of technical materials. Such surfaces reflect, absorb, and emit diffusively. It then follows from (7.8) and (7.9) that

$$a_{\beta\lambda} = \alpha_{\beta\lambda} j_{\beta\lambda} \cos\beta \tag{7.13}$$

and from (7.10) and (7.1') that

$$e_{\beta\lambda} = \varepsilon_{\beta\lambda} e_{\text{bn}\lambda} \cos\beta \tag{7.14}$$

showing that the absorption and emission of an ideal diffuse surface have the same $\cos\beta$ dependence as is the case in Lambert's law (7.1') for blackbody emission. (Actual diffuse surfaces show deviations from the ideal cosine dependence near the glancing angle, i.e., near $\beta = 90°$. For metals $a_{\beta\lambda}$ and $e_{\beta\lambda}$ are somewhat larger there, for insulators somewhat smaller than what would be indicated by the simple cosine dependence.) In what follows, diffuse surfaces will be assumed throughout this chapter.

Another form of Kirchhoff's law is obtained by integrating the angle β over the half-space above the diffuse surface. In a notation that is self-explanatory, one defines α_λ and ε_λ by

$$a_\lambda = \alpha_\lambda i^\lambda \tag{7.15}$$

$$e_\lambda = \varepsilon_\lambda e_{\text{b}\lambda} \tag{7.16}$$

The coefficients α_λ and ε_λ that are defined by these equations are referred to as the *monochromatic absorptivity* and *emissivity*, respectively. Hence, because $\alpha_{\beta\lambda}$ and $\varepsilon_{\beta\lambda}$ are now independent of β,

$$a_\lambda = \alpha_{\beta\lambda} \int_{\text{h.s.}} i_{\beta\lambda}\, d\omega = \alpha_{\beta\lambda} i_\lambda, \qquad e_\lambda = \varepsilon_{\beta\lambda} \int_{\text{h.s.}} e_{\text{b}\beta\lambda}\, d\omega = \varepsilon_{\beta\lambda} e_{\text{b}\lambda}$$

(h.s. = half-space) so that $\alpha_{\beta\lambda} = \alpha_\lambda$ and $\varepsilon_{\beta\lambda} = \varepsilon_\lambda$. Therefore, from (7.12),

$$\alpha_\lambda = \varepsilon_\lambda \qquad \text{(opaque, isotropic, diffuse surfaces)} \tag{7.17}$$

A third form of Kirchhoff's law is obtained by integrating (7.15) and (7.16) over the wavelength. One then defines the absorptivity α and emissivity ε by

$$a = \alpha i, \qquad e = \varepsilon e_{\text{b}} \tag{7.18}$$

so that from

$$a = \int_{\lambda_1}^{\lambda_2} \alpha_\lambda i_\lambda\, d\lambda, \qquad e = \int_{\lambda_1}^{\lambda_2} \varepsilon_\lambda e_{\text{b}\lambda}\, d\lambda$$

follows

$$\alpha = \frac{\int_{\lambda_1}^{\lambda_2} \alpha_\lambda i_\lambda \, d\lambda}{\int_{\lambda_1}^{\lambda_2} i_\lambda \, d\lambda} \qquad \varepsilon = \frac{\int_{\Lambda_1}^{\Lambda_2} \varepsilon_\lambda e_{b\lambda} \, d\lambda}{\int_{\Lambda_1}^{\Lambda_2} e_{b\lambda} \, d\lambda} \qquad (7.19)$$

It should be noticed that in spite of Kirchhoff's law in the form (7.17), it is in general *not* true that $\alpha = \varepsilon$. The reason is that the weighing factors used in defining α and ε depend on the spectra of the incident, respectively emitted, radiation.

As is evident from the examples shown in Figs. 7.2 and 7.3, the weighting factors i_λ and $e_{b\lambda}$ approach zero as $\lambda \to 0$ and ∞. To compute α and ε from (7.19), it is therefore sufficient to know α_λ in a wavelength interval λ_1 to λ_2 and ε_λ in the interval Λ_1 to Λ_2, that is, in an interval where most of the radiant power is concentrated.

For instance, for the blackbody emission, it is easily established by Planck's radiation law (7.3) that the emitted power below $\lambda T = 1450 \; \mu$m K and above 23 200 μm K is only 1% of the total. In the case of solar radiation, with $T = 5760$ K, the corresponding limits of integration are 0.25 μm for the lower and 4.0 μm for the upper limit. Assuming 300 K for the temperature of a spacecraft component, these limits are 4.8 and 77 μm. It is important to note that for the absorptivity the limits depend on the temperatures of the *sources* of the radiation.

Surfaces for which there is a *common interval* of wavelength in which most of the radiant power is found and in which α_λ and ε_λ are approximately *independent of the wavelength* are called (somewhat misleadingly) **gray surfaces**. In that case, it follows from (7.17) that for a given material at a given temperature

$$\alpha = \varepsilon \quad \text{(opaque, isotropic, diffuse, gray surfaces)} \qquad \textbf{(7.20)}$$

In the analysis of radiant heat transfer between spacecraft components, the temperatures are most often in a relatively narrow range, say from 270 to 350 K. The limits of integration for α and ε in (7.19) then can be taken to be the same. All other conditions for the validity of (7.20) are also assumed satisfied. In what follows, this temperature range will be referred to as *ambient* and designated by the subscript $(\,)_a$.

Of course, **for solar radiation**, designated by $(\,)_h$, the wavelength interval in which most of the power occurs is quite different from that for absorption by surfaces that are at ambient temperature. To summarize:

$$\alpha_a \approx \varepsilon_a \quad \text{but} \quad \alpha_h \not\approx \alpha_a \qquad \textbf{(7.21)}$$

[Of course, ε_h would be only of astrophysical interest; for spacecraft applications, instead, it is only the solar constant (Sect. 7.1.2) that matters.]

7.2 Spacecraft Surface Materials

In Fig. 7.5, adapted from Agrawal [4], $\alpha_a = \varepsilon_a$ (the absorption and emission coefficients at room temperature) as functions of the wavelength, are shown for three different surfaces that are frequently used in applications to

7.2 Spacecraft Surface Materials

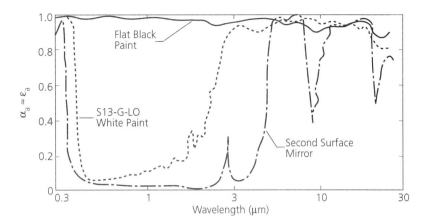

Figure 7.5 Monochromatic absorptivity/emissivity at ambient temperature at beginning of life for three typical spacecraft surfaces. From Ref. 4, Agrawal, B. N., "Design of Geosynchronous Spacecraft," Prentice-Hall. Copyright © 1986 Prentice Hall, Inc. By permission.

space technology. White and black paints are used on external surfaces of spacecraft. Electronic boxes are usually painted black to promote radiative transfer among the various components interior to the spacecraft, thereby equalizing to the extent possible their temperatures.

Optical solar reflectors (OSRs) are second surface mirrors, often made from a thin sheet of fused silica, silvered on the back. They strongly reflect solar radiation, yet emit well in the infrared. They are therefore used to cool by radiation into space components that have high rates of electric–thermal dissipation, such as is the case for traveling wave tubes.

Notable are the strong absorption and emission bands in the infrared, resulting from the vibrational bands of surface atoms. Paint that is normally white is essentially black at wavelengths above 3 μm.

That the difference between $\alpha_a = \varepsilon_a$ and α_h can be very large is seen for white paints. When they are first exposed to the space environment, the absorptivity α_h may vary from 0.2 to 0.3, whereas α_a may vary from 0.8 to 0.9.

Figure 7.6, adapted from Wingate in Pisacane and Moore [5], and from Schmidt in Hallmann and Ley [6], indicates the range of $\alpha_a = \varepsilon_a$ and α_h of various materials and coatings that are used in spacecraft design. (As before, the data apply to surfaces when first exposed to the space environment.) It is noteworthy that all four corners of the diagram, that is, all extreme combinations of the coefficients, can be approximately realized by the proper choice of materials. Intermediate values of the coefficients can be obtained with other materials or, as is frequently done, by applying to the spacecraft surface aluminized or other tapes in patterns of alternating absorptivity that then produce the desired mean absorptivity. The application of tapes to surfaces of spacecraft is also a convenient method when there is a need to correct discrepancies between design temperatures and the true temperatures measured in thermal vacuum tests.

Although α_a and ε_a change relatively little with time of exposure to the space environment, this is not the case with solar absorptivity. Particularly

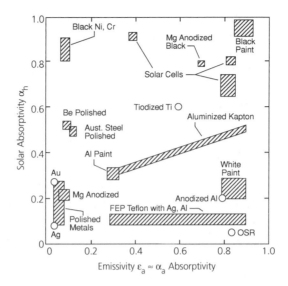

Figure 7.6 Solar absorptivity versus ambient temperature absorptivity/emissivity at beginning of life. OSR = optical solar reflectors. After Wingate [5] and Schmidt [6]. Adapted from and added to from Ref. 5, C. A. Wingate in "Fundamentals of Space Systems," Pisacane, V. L. and Moore R. C. Copyright © 1994 by Oxford University Press, Inc. Used by permission of Oxford University Press.

surfaces that initially have a low value of α_h, such as white paints and to a lesser extent anodized aluminum, suffer major increases in solar absorptivity over time (e.g., over the 5 to 10 years of the useful life of many spacecraft). The most important contributor to the deterioration of the surface is usually the sun's **ultraviolet radiation**.

Micrometeoroids, in the long term, also affect the surface properties by causing dings in the surface. In the near-earth environment, typically even more important is small **space debris**, particularly aluminum oxide particles from solid-propellant motors that had been fired in the past at the vehicle's altitude. Although effective micrometeoroid shields have been developed for use on space stations, particularly on the sides receiving the largest number of impacts, the shields themselves are also subject to the long-term increase in solar absorptivity. Therefore there may still be a requirement for active thermal control.

For optical solar reflectors, Teflon-based second-surface mirrors have been found to be particularly useful. But even they show some significant increases of solar absorptivity when the dose of 5 keV to 1 MeV protons and electrons exceeds about 10^{15} particles per cm^2.

In the case of low-altitude earth-orbiting spacecraft, it has been found that coated Kapton and Mylar deteriorate quickly as a consequence of the chemical interaction of the material with the atomic oxygen in the upper atmosphere.

Still another cause of increased solar absorptivity over time is the effect produced by **outgassing** of organic spacecraft materials, such as electrical insulators. When vented to the outside, the vapors can be adsorbed on exterior surfaces of the spacecraft, thereby increasing the absorption of solar radiation. A frequently employed remedy is intentional venting of the vapor at an opening in the spacecraft skin where vapor adsorption next to the vent is of less concern.

The result of these effects is a tendency toward an increased temperature of the spacecraft. The designer of long-life spacecraft must take account of this by planning for a relatively low temperature at the beginning of the mission or by providing active thermal control (discussed at the end of this chapter). Toward the end of life of the spacecraft, its operator may also have to reduce the operating level to minimize the thermal dissipation of one or more of the spacecraft's electronic and communications components.

7.3 Model of a Spacecraft as an Isothermal Sphere

We consider a spherical shell of radius R, uniform temperature T, uniform solar radiation absorptivity α_h, and uniform emissivity ε_a. Electric–thermal dissipation in the interior of the sphere is (at first) neglected.

Except for several spherical shells that have been orbited for the purpose of radar calibration, spacecraft of course are not spherical but come in a great variety of shapes. Considerations of isothermal spheres, however, provide a simple means for a comparison of the effect that different surface coatings can have on the overall temperature of vehicles.

By integrating (7.13) over a hemisphere, or more simply by noting that the incident radiant power is proportional to the circular disk formed by the (parallel) incident solar rays, the absorbed power is $\pi R^2 \alpha_h j_h$. In the case of thermal equilibrium, the absorbed power is equal to the emitted power $4\pi R^2 \varepsilon_a \sigma T^4$. Solving for T gives

$$T = \left(\frac{\alpha_h j_h}{4 \varepsilon_a \sigma} \right)^{1/4} \qquad (7.22)$$

The equilibrium temperature therefore depends only on the ratio of α_h and ε_a. (It follows from equating absorbed and emitted radiant power for an isothermal body of *general shape* in a fixed orientation to the sun that also in this case the equilibrium temperature depends only on the same ratio.) Table 7.1 shows some equilibrium temperatures for spheres at average distances from the sun for Earth (1 AU), Venus (0.723 AU), and Mars (1.523 AU).

Next we consider the temperature change of the sphere as it becomes eclipsed and as it exits again from the eclipse. Such is the case with geostationary satellites. These spacecraft are periodically eclipsed each spring and fall at and around the equinox. The eclipse lasts a maximum of 72 minutes [taken from the center of the first (entering) penumbra to the center of the second (exiting) penumbra]. This maximum duration occurs at equinox.

Table 7.1 Equilibrium Temperatures of Spheres

	α_h	ε_a	α_h/ε_a	Equilibrium temperature (K) of spheres at		
				Earth	Venus	Mars
White paint, BOL[a]	0.2	0.9	0.22	191	365	82
EOL	0.6	0.9	0.67	251	480	108
Black paint	0.9	0.9	1.00	278	532	120
Gold	0.08	0.03	2.67	355	679	153

[a] BOL, Beginning of life; EOL, end of life.

As before, we consider an isothermal, thin shell of uniform emissivity and absorptivity. Electric power P_{el} is assumed to be generated with conversion efficiency η_{el} by a solar panel equal in size to the projected area of the sphere. Therefore $P_{el} = \pi R^2 \eta_{el} j_h$. The generated power is assumed to be dissipated in the interior of the vehicle, both out of and during the eclipse, in the latter case supplied by storage batteries.

If m is the mass of the shell and c the specific heat per unit mass of the material, steady-state energy balance requires that

$$mc \, dT/dt = \pi R^2 \alpha_h j_h - 4\pi R^2 \varepsilon_a \sigma T^4 + P_{el}$$
$$= \pi R^2 [(\alpha_h + \eta_e) j_h - 4\varepsilon_a \sigma T^4] \quad \text{(out of eclipse)} \quad (7.23a)$$

$$mc \, dT/dt = -4\pi R^2 \varepsilon_a \sigma T^4 + P_{el}$$
$$= \pi R^2 [\eta_e j_h - 4\varepsilon_a \sigma T^4] \quad \text{(in eclipse)} \quad (7.23b)$$

If T_h and T_{-h} designate the temperatures out of and in the eclipse, as they would be if steady-state equilibrium prevailed, then

$$T_h = \left[\frac{(\alpha_h + \eta_{el}) j_h}{4\varepsilon_a \sigma}\right]^{1/4}, \quad T_{-h} = \left[\frac{\eta_{el} j_h}{4\varepsilon_a \sigma}\right]^{1/4} \quad (7.24)$$

Also, let τ_h and τ_{-h} designate time constants defined by

$$\tau_h = \frac{mc}{4\pi R^2 \varepsilon_a \sigma T_h^3}, \quad \tau_{-h} = \frac{mc}{4\pi R^2 \varepsilon_a \sigma T_{-h}^3} \quad (7.25)$$

Using these quantities, the energy balance equations become

$$\frac{dT}{dt} = \frac{T_h}{\tau_h}\left[1 - \left(\frac{T}{T_h}\right)^4\right] \quad \text{(out of eclipse)} \quad (7.26a)$$

$$\frac{dT}{dt} = \frac{T_{-h}}{\tau_{-h}}\left[1 - \left(\frac{T}{T_{-h}}\right)^4\right] \quad \text{(in eclipse)} \quad (7.26b)$$

therefore two nonlinear equations. Their solutions are coupled by the requirements of continuity of the temperature at the entrance [designated by the subscript ( )$_1$] and exit [designated by ( )$_2$] from the eclipse.

7.3 Model of a Spacecraft as an Isothermal Sphere

Then from

$$\int \frac{dT}{1-(T/T_h)^4} = \frac{T_h}{2}\left(\tanh^{-1} T/T_h + \tan^{-1} T/T_h\right) + \text{const.}, \qquad T < T_h$$

$$\int \frac{dT}{1-(T/T_{-h})^4} = \frac{T_{-h}}{2}\left(\coth^{-1} T/T_{-h} - \cot^{-1} T/T_{-h}\right) + \text{const.}, \qquad T > T_{-h}$$

follows, observing that (7.26a) and (7.26b) are each separable, that

$$\tanh^{-1} T_1/T_h + \tan^{-1} T_1/T_h - \tanh^{-1} T_2/T_h - \tan^{-1} T_2/T_h$$
$$= 2[p-(t_2-t_1)]/\tau_h \tag{7.27a}$$
$$\coth^{-1} T_2/T_{-h} - \cot^{-1} T_2/T_{-h} - \coth^{-1} T_1/T_{-h} + \cot^{-1} T_1/T_{-h}$$
$$= 2[t_2-t_1]/\tau_{-h} \tag{7.27b}$$

where a periodic dipping, with period p, in and out of the eclipse has been assumed.

T_1 and T_2 represent the extremes in the cyclic temperature variation. They can be found by solving the two coupled, transcendental equations (7.27) by Newton's or one of the related numerical methods.

To provide a general estimate of the magnitude of the temperature fluctuation, the following example is useful: The sphere is assumed to be in geostationary orbit at the time of equinox, so that $t_2 - t_1 = 72$ minutes, $p = 24$ hours. The shell is assumed to be 3.4 mm thick, with a specific heat per unit mass of 960 J/(kg K) and density 2800 kg/m^3 (for aluminum 2014-T6), solar radiation absorptivity 0.55, ambient temperature emissivity 0.30, and an electric conversion efficiency of 0.15.

The extremes of the temperatures, as calculated from (7.27), are $T_1 = 343$ K (virtually the same in this case as the steady-state equilibrium temperature T_h) and $T_2 = 290$ K, resulting in a maximum temperature swing of 53 K. The result is independent of the radius of the sphere.

In this calculation, instantaneous transitions of the spacecraft from being in the sun to being totally eclipsed, and back again, were assumed. In fact, as shown schematically in Fig. 7.7, before entering the umbra (Latin, shadow), where it is completely eclipsed, the spacecraft passes briefly through the penumbra, where the earth blocks the solar disk only partially. The crossing by a geostationary satellite of the penumbra takes 2.1 minutes at equinox. After being completely eclipsed, the spacecraft passes the penumbra once more.

For the purpose of calculating the electrical power available from solar panels and for calculating the spacecraft temperature, the modifications to the assumed instantaneous transition in and out of an eclipse are negligible.

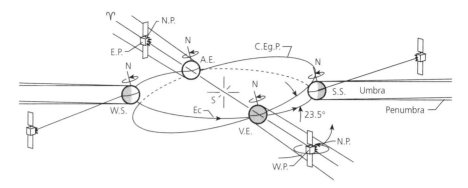

Figure 7.7 Schematic of diurnal and annual rotations of the earth and of a three-axis stabilized, geostationary spacecraft, shown at local midnight. S, sun; Ec, ecliptic; C.Eq.P., celestial equatorial plane, V.E. (A.E.), vernal (autumnal) equinox; S.S. (W.S.), summer (winter) solstice; E.P., W.P., N.P., east, west, north panel.

7.4 Earth Thermal Radiation and Albedo

The earth's surface, atmosphere, and clouds radiate into space at infrared wavelength (as corresponds to their relatively low temperature). At its surface, the earth is approximately in thermal equilibrium between the power absorbed from the incident solar radiation and its thermal emission. By comparison, the flow from the earth's interior of heat, produced by the radioactive decay of uranium and its daughter products, is only a minor contributor.

The **earth's thermal radiation** follows roughly Lambert's law. The emitted power per unit area (outside the atmosphere and averaged over the earth's surface and time) will be designated by e_g. It has been measured by spacecraft instrumentation at 237 ± 7 W/m^2.

For *low-altitude satellites*, the earth's thermal radiation can be an important contributor to the spacecraft's temperature. Taking as an example a satellite at an altitude $h = 300$ km, it is easily calculated from the geometry of the satellite position that the spacecraft has a field of view of the earth that extends in arc length from the nadir to about 6.4 times the altitude. For the purpose of heat transfer calculations, the earth's surface can therefore be replaced approximately by an infinite plane.

To obtain a rough estimate of the effect, we represent the satellite as a sphere of radius R and absorptivity α_a for ambient temperature incident radiation. If $dS = r\, d\phi\, dr$ (r = distance from the nadir, ϕ = azimuthal angle) is a surface element of this plane, then, since the dimensions of the spacecraft are negligible compared with its altitude, the solid angle $d\omega$ extended from dS to the spacecraft is $\pi R^2/(h^2 + r^2)$. The power emitted by dS, incident on the half of the sphere that faces dS, is $e_g\, dS \cos\beta\, \pi R^2/(h^2 + r^2)$ where β is the angle from the local vertical to the line from dS to the satellite. Therefore, with $\varrho = r/h$, the total power, A_g, absorbed by the satellite from earth thermal radiation is

$$A_g = \pi R^2 \alpha_a e_g h \int_{r=0}^{\infty} \int_{\phi=0}^{2\pi} \frac{r\, dr\, d\phi}{(h^2 + r^2)^{3/2}} = \pi R^2 \alpha_a e_g \int_{\varrho=0}^{\infty} \int_{\phi=0}^{2\pi} \frac{\varrho\, d\varrho\, d\phi}{(1 + \varrho^2)^{3/2}}$$

$$= 2\pi^2 R^2 \alpha_a e_g \qquad (7.28)$$

Comparing this result with the absorption, A_h, from direct solar radiation (Sect. 7.1.2), the ratio is seen to be

$$\frac{A_g}{A_h} \approx \frac{2\pi \cdot 237 (\text{W/m}^2)}{1353 (\text{W/m}^2)} \frac{\alpha_a}{\alpha_h} \approx 1.10 \frac{\alpha_a}{\alpha_h} \tag{7.29}$$

For low-altitude satellites, depending on the ratio α_a/α_h, the earth thermal and direct solar radiation therefore can be of comparable magnitude.

At *higher altitudes*, earth thermal radiation becomes increasingly unimportant. For instance, in the case of a geostationary satellite (orbit radius = 42,164 km) the earth can be approximated roughly as a point source. A simple calculation then shows that the factor of 1.10 in (7.29) is replaced by 0.004. For geostationary spacecraft, earth thermal radiation therefore is negligible.

The method for calculating the effect of earth thermal radiation on spacecraft of *arbitrary* shape is virtually the same as the method outlined in Sect. 7.8 for solar radiation. Calculations analogous to those discussed for the earth apply to other planets and to the moon as well. In the case of the outer planets, their surface temperatures are so low that the effect of their thermal emission on a spacecraft becomes negligible.

The solar radiation incident on the earth is in part absorbed and in part reflected. The latter is referred to as **earth albedo** (Latin *albus*: white). The maximum effect on the spacecraft occurs when sun, spacecraft, and earth are approximately aligned.

The albedo is a highly variable quantity because it largely depends on the presence or absence of clouds. At optical wavelengths, clouds, snow, the polar ice, and the oceans all have high reflectivity compared with land masses. Reflected radiation is frequently approximated by the sum of a specular and a diffuse component, but predictions of the albedo are difficult because, in addition to depending on meteorological data, they depend on the position of the spacecraft relative to the sun.

The reflected radiant power per unit area can be expressed by the product $a_g j_h$ where $j_h = 1353$ W/m² is the solar constant (Sect. 7.1.2) and a_g is known as the **earth albedo coefficient**. As an average over the satellite orbit, the annual mean value of a_g is about 0.3, somewhat lower for equatorial orbits and somewhat higher for polar orbits.

Like the earth thermal radiation, and for the same reasons, albedo radiation is important only in the case of low-altitude spacecraft. If the satellite always faces the earth with the same side, as is the case with three-axis stabilized vehicles, the effect on the spacecraft of the variability of the albedo can be reduced by selecting materials for the earth-facing side that have low solar absorption coefficients.

7.5 Diurnal and Annual Variations of Solar Heating

The thermal input to a spacecraft by solar radiation over the course of 24 hours and over a year is illustrated in Fig. 7.7. The figure applies to a geostationary, three-axis stabilized vehicle.

Seen from the north, the earth's daily rotation about its axis is anticlockwise. So is the direction of the earth's annual orbit about the sun in the ecliptic plane. Geostationary satellites, as the name implies, orbit the earth in the equatorial plane synchronously with the earth. The equatorial and ecliptic planes include an angle of 23.5°, referred to as the obliquity of the ecliptic. The four positions of the satellite shown in the figure are at local midnight, that is, when the spacecraft's projection onto the earth is at 24.00 hours local solar time.

At the time of the vernal or autumnal **equinox**, that is, when the earth crosses the line of intersection of the ecliptic and celestial equatorial planes, the sun, earth, and spacecraft in its midnight position are collinear, and the spacecraft is eclipsed. At summer or winter **solstice**, there will be no eclipse because the distance of a geostationary spacecraft from the earth is sufficiently large for the spacecraft to pass the earth's umbra and penumbra below or above the ecliptic plane. However, near the equinox positions there will be about 90 days per year, centered about the equinox and the spacecraft midnight positions, where the spacecraft will be eclipsed. The longest duration of the eclipse is about 72 minutes.

Three-axis stabilized satellites rotate about one of their axes so as to point their high-gain antennas at all times toward the earth. For the solar panels to point toward the sun, they must rotate relative to the body of the spacecraft through 360° every 24 h. In the usual arrangement, the axis of rotation of the solar array is normal to the equatorial plane, so that the angle of incidence of the solar radiation stays roughly constant at 90°.

By a nomenclature that has become conventional for three-axis stabilized satellites, the sides of the main body are called east, west, north, south, earth, and antiearth (or space-facing) panels. The north and south panels, because solar radiation is incident on them at most at a glancing angle of nominally 23.5°, have the advantage over the other panels of having a nearly constant, cool temperature. They are therefore favored for the placement of electronic components that have large heat rejection.

7.6 Thermal Blankets

Thermal blankets, also referred to as **multilayer insulation**, are used on spacecraft to insulate thermally against radiative heat transfer. They are used in such applications as insulation against solar heating and against heat transfer from hot motors and exhaust plumes. They can also serve as a thermal barrier during the launch phase when the spacecraft is no longer protected by the launch vehicle's payload shroud.

Thermal blankets are used not only to reduce heat gain but also to reduce heat loss. Particularly in deep-space missions, electric heating of key components may be needed. The application of multilayer insulation then reduces the need for electric power.

As shown in Fig. 7.8, thermal blankets consist of a number of layers, most often aluminized Mylar or Kapton, typically 20 to 30 μm thick. For higher temperatures, up to 340°C, Kapton is preferred over the less costly Mylar. For the outermost layer of blankets that must insulate against radiation from hot rocket motors, titanium or stainless steel thin sheets can be used. Such sheets can withstand for short periods temperatures of up to 1400°C.

7.6 Thermal Blankets

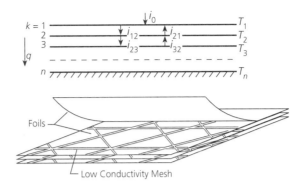

Figure 7.8 Thermal blanket for insulation against solar radiation.

Ideally, the outer surface of a blanket that insulates against radiant heat input should have an absorptivity as low and an emissivity as high as possible. A compromise, however, is needed where high-temperature-resistant materials must be used.

To minimize thermal conduction from layer to layer, a spacer is used to prevent contact of the sheets. As shown in the figure, this may take the form of a mesh of low-conductivity material, such as Dacron or glass fibers. Crimping or embossing of the sheets can serve the same purpose by reducing the contact areas. Because of the difficulty in defining the conduction path with any accuracy, the part played by conduction through the mesh must be determined experimentally.

An additional consideration in designing thermal blankets is that perforations of the sheets are needed to vent, during launch, the air otherwise trapped between layers.

Because the thermal capacity of the layers is very low, a quasi-steady calculation that assumes a steady-state temperature distribution at each instant of time gives sufficiently accurate results. Also, because the separation between layers is very small compared with their lateral dimensions, edge and curvature effects of the blanket can be neglected.

Except for the outermost surface, which usually has a different coating, we assume the same absorptivity α_a ($= \varepsilon_a$) for all *inward*-facing surfaces and the absorptivity α'_a for all *outward*-facing surfaces of each layer. The temperature of the kth layer is designated by T_k.

Let i_0 be the radiant power incident on the thermal blanket. The solar absorptivity of the outermost surface will be designated by α_{h0}, its emissivity by ε_{a0}.

Also let $i_{k,k+1}$ be the radiant power, summed over all directions and wavelengths, that crosses in the inward direction a control surface between layers k and $k+1$. It therefore includes the emission from the inner surface of layer k, as well as all reflections from it. Similarly, $i_{k+1,k}$ is defined as the radiant power crossing the same control surface in the outward direction. Therefore

$$i_{k,k+1} = \varepsilon_a \sigma T_k^4 + (1 - \alpha_a) i_{k+1,k}$$
$$= \alpha_a \sigma T_k^4 + (1 - \alpha_a) i_{k+1,k} \qquad (k = 1, 2, \ldots, n-1)$$

where the last term is the radiant power reflected from the outward-facing surface of the $(k+1)$th layer. Similarly,

$$i_{k+1,k} = \varepsilon'_a \sigma T^4_{k+1} + (1-\alpha'_a) i_{k,k+1}$$
$$= \alpha'_a \sigma T^4_{k+1} + (1-\alpha'_a) i_{k,k+1} \qquad (k=1,2,\ldots,n-1)$$

Eliminating $i_{k,k+1}$ and $i_{k+1,k}$ and taking the difference $q = i_{k,k+1} - i_{k+1,k}$ gives for the net inward thermal flux

$$q = \frac{\sigma(T^4_k - T^4_{k+1})}{1/\alpha_a + 1/\alpha'_a - 1} \qquad (k=1,2,\ldots,n-1) \tag{7.30}$$

which, by conservation of energy, is the same for all spaces between the layers and equals the net thermal power per unit area that reaches the component wrapped in the blanket.

Adding these equations results in

$$(n-1)q = \frac{\sigma(T^4_1 - T^4_n)}{1/\alpha_a + 1/\alpha'_a - 1} \tag{7.31}$$

Using the thermal balance for the outermost layer

$$q = \alpha_{h0} i_0 - \varepsilon_{a0} \sigma T^4_1 \tag{7.32}$$

to eliminate T_1 results in

$$[1 + (n-1)\varepsilon_{a0}(1/\alpha_a + 1/\alpha'_a - 1)]q = \alpha_{h0} i_0 - \varepsilon_{a0}\sigma T^4_n \tag{7.33}$$

where the right side

$$q_0 = \alpha_{h0} i_0 - \varepsilon_{a0} \sigma T^4_n \tag{7.34}$$

is recognized as the heat transferred to the component if no blanket was used and if the component surface had the absorptivity and emissivity of the outermost surface.

A commonly encountered problem is one in which T_n is prescribed and one wishes to calculate q and T_1 to T_{n-1}. Thus, T_n might be the boiling temperature of a cryogen or the specified maximum allowed temperature of the component. The final result is therefore conveniently written as

$$\frac{q}{q_0} = \frac{1}{1+(n-1)\varepsilon_{a0}(1/\alpha_a + 1/\alpha'_a - 1)} \tag{7.35}$$

It follows that for an effective insulation the solar absorptivity of the outermost layer should be as low as possible, and the emissivity at ambient temperature as high as possible, compatible with material limitations. It also follows that there is no theoretical advantage in choosing α_a and α'_a differently; both should be as small as possible.

The temperatures of the individual layers are readily obtained from (7.30) by finding T_{n-1} from the last equation, then T_{n-2} from the second to the last, and so on.

7.7 Thermal Conduction

Heat transfer among spacecraft components is in part by radiation and in part by conduction. (Convective heat transfer, as it applies to rocket motors, is discussed in Sect. 4.14.)

The reader is likely to be already familiar with the fundamentals of thermal conduction. Much material is readily available in several standard texts, such as the one by Carslaw and Jaeger [7]. The present section therefore merely lists some basic relations and adds some comments that apply to spacecraft.

For an isotropic material, the rate of heat, q, conducted per unit area is

$$\mathbf{q} = -k \operatorname{grad} T \qquad (7.36)$$

where k is the **thermal conductivity**. In some applications, particularly when cryogens are part of a system, it is necessary to take into account the dependence of k on the temperature. Thus, elemental aluminum has a thermal conductivity of 210 W/(m K) at 300 K, but 420 W/(m K) at the normal boiling temperature (78 K) of nitrogen and 5700 W/(m K) at 20 K. This last temperature is typical for many spaceborne scientific instruments operating in the far infrared.

Including time dependence and heat production, conservation of energy requires that

$$\operatorname{div} \mathbf{q} = -\varrho c \, \partial T / \partial t + \varrho w \qquad (7.37)$$

where ϱ is the density of the material, c its specific heat per unit mass, and w the rate of internal heat produced per unit mass. Examples of the latter are the heat production that occurs in electrical conductors and in radioisotope sources used in deep-space probes for energy generation.

It follows from (7.36) and (7.37) that

$$\nabla^2 T + \frac{1}{k} \operatorname{grad} k \cdot \operatorname{grad} T - \frac{1}{a^2} \frac{\partial T}{\partial t} + \frac{\varrho}{k} w = 0 \qquad \textbf{(7.38)}$$

where $a^2 = k/(\varrho c)$, called the **thermal diffusivity**. It is a measure of the rapidity with which temperature changes propagate through the material.

Fourier's heat conduction equation

$$\Delta^2 T - \frac{1}{a^2} \frac{\partial T}{\partial t} = 0 \qquad (7.38')$$

is an important special case, obtained when there are no internal sources of heat generation and when k can be assumed to be constant.

A substantial number of exact solutions of (7.38′), satisfying various boundary conditions, are known. In spacecraft design, they have been largely replaced by numerical methods that are based on finite-difference or finite-element methods. Software is available that, once the geometry and boundary conditions have been specified, allows one to obtain rapidly numerical solutions that would be unattainable by classical analysis. The same software also provides for presenting the results graphically for inspection by the spacecraft designer.

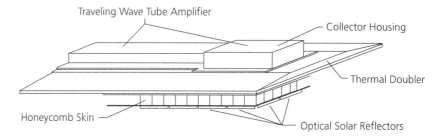

Figure 7.9 Heat rejection from a traveling wave tube amplifier by means of a thermal doubler and optical solar reflectors.

The functioning of integrated circuits and of their assembly into circuit boards depends critically on the removal of heat by conduction. Because on spacecraft the electronic enclosures are usually vented, convection is absent. Typically, the enclosures are attached to aluminum shelves whose function is not only to support the units during launch but also to remove by conduction the heat generated by electric–thermal dissipation.

To remove the heat from devices with high power consumption, for instance traveling wave tube amplifiers, **thermal doublers** are used to conduct the heat from the relatively small device to a larger area that then radiates the heat—directly or indirectly—into space. An example is shown in Fig. 7.9. In this case the principal heat source is the anode or collector of a traveling wave tube. The conduction path leads from the collector and its housing to an aluminum alloy thermal doubler and from there through the spacecraft skin to optical solar reflectors. These are typically small, about 5 by 5 cm silica glass squares, at most a few millimeters thick, silvered on the back side, and attached to the spacecraft skin by adhesive.

Joints between two metal surfaces that are bolted together sometimes present an unwanted barrier to heat transfer because actual metal-to-metal contact may be confined to small areas only, often just around the bolt holes. In such cases, to improve the heat transfer, special greases with high thermal conductivity and low vapor pressure can be used.

7.8 Lumped Parameter Model of a Spacecraft

The design of a new spacecraft requires extensive thermal calculations. Their purpose is to make certain that the specified temperature limitations of the various components are not exceeded at any time during the life of the spacecraft. Such calculations must be performed for the different orientations of the spacecraft relative to the sun as they may occur during normal operating conditions. The same types of calculations are also carried out for times when the spacecraft is eclipsed.

During the launch phase, after the launch vehicle's shroud is removed and the spacecraft released, solar array panels, antennas, and booms are often still folded against the main body. The thermal configuration then may be quite different from the normal operating condition with its deployed panels and antennas. Separate calculations are also needed for this case.

7.8 Lumped Parameter Model of a Spacecraft

The thermal calculations go hand in hand with the design process. Their results will indicate required design changes. Conversely, design changes will induce modifications of the thermal calculations.

Later in the buildup of the vehicle, components, subsystems, and the entire spacecraft (in the case of large craft often without solar panels, for reasons of size limitations of existing test facilities) will undergo thermal tests. For thermally sensitive components, discrepancies between test and calculation of no more than about 10°C are usually acceptable. Larger discrepancies require an investigation of their cause and corrective action. The latter may be in form of a modification of the design or, if there is separate justification for it, by changes in the thermal model.

7.8.1 Basic Relations

Thermal models, such as the simplified version illustrated in Fig. 7.10, divide the spacecraft into **nodes**, each assumed to have uniform temperature, absorptivity, and emissivity. The nodes interact with each other thermally by radiation and conduction.

To obtain the required accuracy, depending on the complexity of the spacecraft, one hundred or more such nodes are often needed. To some extent the choice of nodes is arbitrary; comparison with test results and

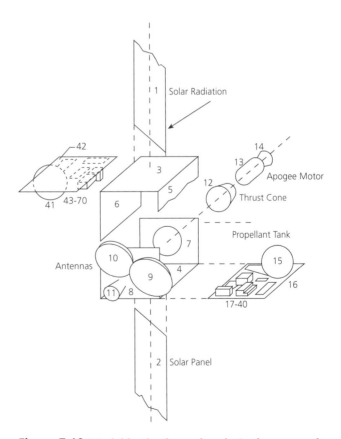

Figure 7.10 Model for the thermal analysis of a spacecraft.

resulting modifications of the model will improve the choice. Thermally sensitive subsystems will require a more extensive breakdown into nodes than is needed for those less sensitive. Typically selected as nodes are the main structural elements of the spacecraft, electronic boxes, solar array panels, antennas, rocket motor components, temperature-sensitive components such as storage batteries, propellant tanks, cryogenic subsystems, and scientific instruments. Still others are thermal control devices such as louvers and those using phase-changing materials.

As formulated in the following, the nodes are defined such that the radiative exchanges are between *one-sided* surfaces, such as the outer surface of an electronic enclosure. Panels, such as shelves or solar arrays, may have different absorptivities and emissivities on the two sides but typically are thin enough so that the temperatures of either side are virtually the same. They are represented by two different nodes, although at the same temperature.

In what follows, all surfaces are assumed to be *diffuse* and their emission, absorption, and reflection to obey the simple cosine dependence on the angle between the direction of the radiation and the normal, in accordance with (7.13) and (7.14). (The active sides of solar panels reflect the incident solar radiation at least in part specularly. But because of their orientation toward the sun, this radiation does not reach other parts of the spacecraft, hence has no effect on the temperature of the latter.)

Solar radiation is included in the following formalism. For simplicity, earth thermal and albedo radiations are omitted. However, if needed, earth thermal radiation can be dealt with by introducing the earth as just one more node in addition to the spacecraft nodes. Albedo radiation, because its spectrum is similar to the solar spectrum, can be treated in the same way, as will be carried out here for the incident solar radiation.

Some of the nodes will be connected by radiative and/or conductive heat transfer. They can be viewed as forming a **double directed graph**, one for radiation and one for conduction (Fig. 7.11).

The radiant power *incident* on the ith node is composed of radiation directly emitted from and also reflected from other nodes. In the most

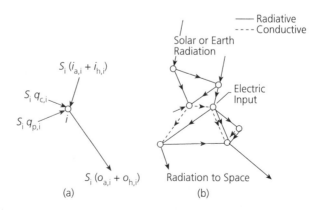

Figure 7.11 (a) Thermal inputs to ith node. (b) Directed, double graph for spacecraft thermal calculations.

general case, it includes radiant power, $i_{a,i}$, per unit area, from ambient temperature sources and similarly, $i_{h,i}$, from solar radiation, either direct or reflected. The total power per unit area that is directed *outward* from the ith node is (e = emitted power per unit area, i = incident power per unit area)

$$o_{a,i} = e_{a,i} + (1 - \alpha_{a,i})i_{a,i} \qquad (i = 1, 2, \ldots, n) \qquad (7.39a)$$

for ambient temperature radiation and

$$o_{h,i} = j_h F_{h,i}(1 - \alpha_{h,i}) + (1 - \alpha_{h,i})i_{h,i} \qquad (i = 1, 2, \ldots, n) \qquad (7.39b)$$

for solar radiation, where

$$e_{a,i} = \varepsilon_{a,i} \sigma T_i^4 \qquad (7.40)$$

The quantities j_h and $F_{h,i}$ both refer to the solar radiation. The former is defined as in (7.8). The latter is a purely geometrical factor, the "view factor" as defined below.

In addition, as illustrated in the figure, there may be added to the node thermal power by conduction, $q_{c,i}$, defined per unit area of the node. In addition, there may be external thermal power, $q_{p,i}$, per unit area, that may be added either by electric dissipation, from combustion of propellant, or from a radioisotope source. (In the case of solar arrays, $q_{p,i}$ is the electrical power delivered by a unit area of the panel and is negative.)

The thermal conduction power, $q_{c,i}$, can be expressed by

$$q_{c,i} = \sum_{j=1}^{n} \Gamma_{ij}(T_j - T_i) \qquad (i = 1, 2, \ldots, n) \qquad (7.41)$$

where Γ_{ij} is called the **conductance** between nodes i and j.

In what follows, steady-state conditions are assumed. Conservation of energy at each node therefore requires that

$$i_{a,i} - o_{a,i} + i_{h,i} - o_{h,i} + q_{c,i} + q_{p,i} = 0 \qquad (7.42)$$

7.8.2 View Factors

As a preliminary to the radiative part of the calculation, **view factors** F_{ij} (also called "configuration factors") of pairs (i, j) of nodes are introduced. They are purely geometrical quantities, depending only on the node surfaces S_i and S_j and their relative configuration.

The nodes, in addition to each having constant temperature, absorptivity, and emissivity, are defined such that each node either has a *direct*, uninterrupted ray path or none to and from each other node, including itself.

The several types of geometric configurations that may arise are illustrated in Figs. 7.12 and 7.13. The simple case of two convex surfaces S_i and S_j with direct ray paths between them is illustrated in the first figure. Also shown there is a schematic of a Cassegrain antenna as an example of the radiative exchange between a concave and a convex surface. The nodes S_i and S_j are seen to have the required property that all points of S_i have direct ray paths with all points of S_i itself and with all points of S_j.

Another geometrical feature that may occur is illustrated by the three plates A, B, C shown in Fig. 7.13a. (B is a thin, two-sided plate; only single

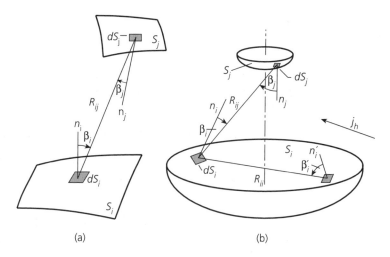

Figure 7.12 (a) For derivation of Eq. (7.43). (b) Schematic of Cassegrain antenna as an example of a concave (S_i) and of a convex (S_j) surface; j_h = solar radiation intensity.

surfaces of A and C participate in the radiative exchange.) To satisfy the definition of nodes, it is necessary here to divide A and C each into two nodes. Hence a total of six nodes with surfaces S_1 to S_6 are required for a complete specification. For instance, S_6 has a direct ray path (with $0 \le \beta_i, \beta_j < \pi/2$) to and from S_1, S_2, and S_4 but not to and from S_3, S_5, and S_6.

One then defines the view factor F_{ij} of S_i relative to S_j as the ratio of the radiant power emitted and reflected by S_i and incident on S_j to the total radiant power emitted and reflected by S_i.

For diffuse surfaces, as assumed, the outward-directed radiant power (including emission and all reflections), per unit area, at the angle β is $o_\beta = o_n \cos \beta$ within the solid angle $d\omega = 2\pi \sin \beta \, d\beta$. Therefore

$$o = 2\pi o_n \int_{\beta=0}^{\pi/2} \cos \beta \sin \beta \, d\beta = \pi o_n$$

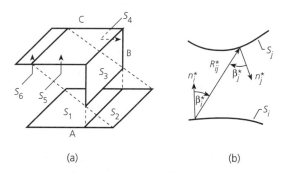

Figure 7.13 (a) Definition of nodes: panels A and C each consist of two nodes. (b) Illustration of a two-dimensional node pair.

and
$$o_\beta = \pi^{-1} o \cos \beta$$

From this follows, as illustrated in Fig. 7.12a, that the outward directed radiant power from an element of surface dS_i of S_i that is incident on the surface element dS_j of S_j is

$$dS_i \pi^{-1} o_i \cos \beta_i \, d\omega_i = \pi^{-1} o_i \cos \beta_i \cos \beta_j \, dS_i \, dS_j / R_{ij}^2$$

where β_i and β_j are the angles made by the ray with the normals to the surfaces and where R_{ij} is the distance between points on S_i and S_j, respectively. Therefore

$$F_{ij} = \begin{cases} \dfrac{1}{\pi S_i} \displaystyle\int_{S_i} \int_{S_j} \dfrac{\cos \beta_i \cos \beta_j}{R_{ij}^2} dS_j \, dS_i & \text{(for direct paths between } S_i \text{ and } S_j, 0 \le \beta_i, \beta_j < \pi/2) \\ 0 & \text{(otherwise)} \end{cases} \quad (7.43)$$

If $i = j$, (7.43) can be simplified to

$$F_{ii} = \begin{cases} \dfrac{1}{\pi} \displaystyle\int_{S_i} \dfrac{\cos \beta_i \cos \beta_i'}{R_{ii}^2} dS_i & \text{(for direct paths from } S_i \text{ to } S_i, 0 \le \beta_i, \beta_i' < \pi/2) \\ 0 & \text{(otherwise)} \end{cases} \quad (7.43')$$

The integration, particularly when involving the fourfold integral in (7.43), tends to be exceedingly cumbersome or else trivial (such as for the parallel surfaces considered in Sect. 7.6 where $F_{ij} = 1$ and $F_{ii} = 0$). In spacecraft design practice, the view factors are therefore evaluated numerically by special software.

A reciprocity relation

$$S_i F_{ij} = S_j F_{ji} \qquad (i, j = 1, 2, \dots, n) \tag{7.44}$$

holds. It follows immediately from interchanging in (7.43) the indexes i and j.

For **two-dimensional** configurations, such as is the case, for instance, for parallel tubes that are very long compared with their diameters, evaluation by analysis, rather than numerically, is often feasible. Figure 7.13b is representative of such a two-dimensional configuration. The unit vectors $\mathbf{n}_i^*$ and $\mathbf{n}_j^*$ normal to the surfaces, the angles of incidence β_i^* and β_j^*, and the distance vectors $\mathbf{R}_{ij}^*$ are all in the transverse plane. It follows easily (with z the lengthwise coordinate) that between dS_i at $z = 0$ and dS_j at z,

$$\cos \beta_i = \frac{\mathbf{n}_i \cdot \mathbf{R}_{ij}}{R_{ij}} = \frac{R_{ij}^* \cos \beta_i^*}{\sqrt{R_{ij}^{*2} + z^2}}, \qquad \cos \beta_j = -\frac{\mathbf{n}_j \cdot \mathbf{R}_{ij}}{R_{ij}} = \frac{R_{ij}^* \cos \beta_j^*}{\sqrt{R_{ij}^{*2} + z^2}}$$

and $R_{ij} = \sqrt{R_{ij}^{*2} + z^2}$.

The outward-directed radiant power from the surface element dS_i of S_i incident on the surface element dS_j of S_j is therefore

$$\frac{1}{\pi} o_i \frac{R_{ij}^{*2} \cos \beta_i^* \cos \beta_j^*}{\left(R_{ij}^{*2} + z^2\right)^2} dS_i \, dS_j$$

and the radiant power from dS_i incident on a strip of width db of S_j is

$$\frac{1}{\pi} o_i \frac{\cos\beta_i^* \cos\beta_j^*}{R_{ij}^{*2}} dS_i\, db \int_{z=-\infty}^{+\infty} \frac{dz}{(1+z^2/R_{ij}^{*2})^2}$$

The value of the definite integral is $(\pi/2) R_{ij}^*$. Therefore, the radiant power from dS_i incident on all of S_j is

$$\frac{1}{2} o_i\, dS_i \int_{S_j} \frac{\cos\beta_i^* \cos\beta_j^*}{R_{ij}^*}\, db$$

The ratio, therefore, of the outward-directed power from a strip of width dz of S_i incident on all of S_j to the total outward-directed power from this strip is

$$\frac{1}{2 S_i} \int_{S_i} \int_{S_j} \frac{\cos\beta_i^* \cos\beta_j^*}{R_{ij}^*}\, dS_j\, dS_i$$

Therefore the view factor for two-dimensional configurations becomes

$$F_{ij} = \begin{cases} \dfrac{1}{2 S_i} \displaystyle\int_{S_i} \int_{S_j} \dfrac{\cos\beta_i^* \cos\beta_j^*}{R_{ij}^*}\, dS_j\, dS_i & \text{(for direct paths between } S_i \text{ and } S_j, 0 \le \beta_i^*, \beta_j^* < \pi/2) \\ 0 & \text{(otherwise)} \end{cases} \quad (7.45)$$

If $i = j$, (7.44) can be simplified to

$$F_{ii} = \begin{cases} \dfrac{1}{2} \displaystyle\int_{S_i} \dfrac{\cos\beta_i^* \cos\beta_i^{*\prime}}{R_{ii}^*}\, dS_i & \text{(for direct paths from } S_i \text{ to } S_i, 0 \le \beta_i^*, \beta_i^{*\prime} < \pi/2) \\ 0 & \text{(otherwise)} \end{cases} \quad (7.45')$$

Finally, the view factor $F_{h,i}$ for **solar radiation** incident on the ith node is

$$F_{h,i} = \begin{cases} \dfrac{1}{S_i} \displaystyle\int_{S_i} \cos\beta_i\, dS_i & \text{(for direct paths from the sun)} \\ 0 & \text{(otherwise)} \end{cases} \quad (7.46)$$

7.8.3 Computational Model

Next, the radiative transfer by emission and reflection among the nodes is considered, first for solar radiation and then for ambient temperature (infrared) radiation.

The **solar** radiant power, $S_i i_{h,i}$, incident on the ith node is the sum of directly incident solar radiation and of solar radiation reflected from other spacecraft nodes including itself (for $F_{ii} \neq 0$). Therefore, making use of the reciprocity relation (7.44),

$$S_i i_{h,i} = S_i F_{h,i} j_h + \sum_{j=1}^{n} F_{ji} S_j o_{h,j} = S_i F_{h,i} j_h + S_i \sum_{j=1}^{n} F_{ij} o_{h,j}$$

7.8 Lumped Parameter Model of a Spacecraft

and substituting for $o_{h,j}$ from (7.39b)

$$i_{h,i} = F_{h,i} j_h + \sum_{j=1}^{n} F_{ij}[(1-\alpha_{h,j})F_{h,j}j_h + (1-\alpha_{h,j})i_{h,j}] \qquad (7.47)$$

Here the first term on the right represents the effect of direct solar radiation incidence, whereas the sum accounts for reflections of solar radiation, either direct or indirect via still other nodes, that reach the ith node from the various nodes (including the ith).

Let

$$\delta_{ij} = \begin{cases} 1 & \text{when } i = j \\ 0 & \text{when } i \neq j \end{cases}$$

Therefore

$$\sum_{j=1}^{n} [\delta_{ij} - F_{ij}(1-\alpha_{h,j})] i_{h,j} = f_{h,i} \qquad (7.48)$$

where

$$f_{h,i} = j_h \left[F_{h,i} + \sum_{j=1}^{n} F_{ij} F_{h,j}(1-\alpha_{h,j}) \right] \qquad (7.49)$$

We define the square matrix $[\mathbf{M_h}]$ and the column matrices $[\mathbf{i_h}]$ and $[\mathbf{f_h}]$ by

$$[\mathbf{M_h}] = \begin{bmatrix} 1 - F_{11}(1-\alpha_{h,1}) & -F_{12}(1-\alpha_{h,2}) & \cdots & -F_{1n}(1-\alpha_{h,n}) \\ -F_{21}(1-\alpha_{h,1}) & 1 - F_{22}(1-\alpha_{h,2}) & \cdots & -F_{2n}(1-\alpha_{h,n}) \\ \vdots & & & \\ -F_{n1}(1-\alpha_{h,1}) & \cdots & \cdots & 1 - F_{nn}(1-\alpha_{h,n}) \end{bmatrix}$$

(7.50a)

$$[\mathbf{i_h}] = [i_{h,1}, i_{h,2}, \cdots i_{h,n}]^T \qquad (7.50b)$$

$$[\mathbf{f_h}] = [f_{h,1}, f_{h,2}, \cdots f_{h,n}]^T \qquad (7.50c)$$

so that

$$[\mathbf{M_h}][\mathbf{i_h}] = [\mathbf{f_h}]$$

Premultiplication with $[\mathbf{M_h^{-1}}]$ results in the final result

$$[\mathbf{i_h}] = [\mathbf{M_h}]^{-1}[\mathbf{f_h}] \qquad \mathbf{(7.51)}$$

Having found the incident radiant power, $i_{h,i}$, per unit area, the corresponding outward-directed power, $o_{h,i}$, is found from (7.39b).

The development for the **ambient** temperature, infrared radiative exchange among the nodes is similar. Using the reciprocity relation, the radiant power incident on the ith node is

$$S_i i_{a,i} = \sum_{j=1}^{n} F_{ji} S_j o_{a,j} = S_i \sum_{j=1}^{n} F_{ij} o_{a,j}$$

CHAPTER 7 Spacecraft Thermal Design

from which, by substituting for $o_{a,j}$ from (7.39a)

$$i_{a,i} = \sum_{j=1}^{n} F_{ij}[e_{a,j} + (1 - \alpha_{a,j})i_{a,j}] \tag{7.52}$$

so that

$$\sum_{j=1}^{n} [\delta_{ij} - F_{ij}(1 - \alpha_{a,j})]i_{a,j} = \sum_{j=1}^{n} F_{ij}e_{a,j} \tag{7.53}$$

Here, $e_{a,j}$ can be replaced by means of the equation of conservation of energy for the jth node, because it follows from combining (7.39a), (7.39b), and (7.42) that

$$e_{a,j} = \alpha_{a,j}i_{a,j} + \alpha_{h,j}i_{h,j} - j_h F_{h,j}(1 - \alpha_{h,j}) + q_{c,j} + q_{p,j} \tag{7.54}$$

Therefore (7.53) becomes

$$\sum_{j=1}^{n} [\delta_{ij} - F_{ij}]i_{a,j} = f_{a,i} \tag{7.55}$$

where

$$f_{a,i} = \sum_{j=1}^{n} F_{ij}[\alpha_{h,j}i_{h,j} - j_h F_{h,j}(1 - \alpha_{h,j}) + q_{c,j} + q_{p,j}] \tag{7.56}$$

Defining the matrix $[\mathbf{M}_a]$ and the column matrices $[\mathbf{i}_a]$ and $[\mathbf{f}_a]$ by

$$[\mathbf{M}_a] = \begin{bmatrix} 1 - F_{11} & -F_{12} & \cdots & -F_{1n} \\ -F_{21} & 1 - F_{22} & \cdots & -F_{2n} \\ \vdots & & & \\ -F_{n1} & \cdots & \cdots & 1 - F_{nn} \end{bmatrix} \tag{7.57a}$$

$$[\mathbf{i}_a] = [i_{a,1}, i_{a,2}, \cdots i_{a,n}]^T \tag{7.57b}$$

$$[\mathbf{f}_a] = [f_{a,1}, f_{a,2}, \cdots f_{a,n}]^T \tag{7.57c}$$

(7.56) can be expressed by

$$[\mathbf{M}_a][\mathbf{i}_a] = [\mathbf{f}_a]$$

resulting in the final result for the ambient temperature radiative exchange

$$[\mathbf{i}_a] = [\mathbf{M}_a]^{-1}[\mathbf{f}_a] \tag{7.58}$$

If conduction among the nodes can be neglected, that is, $q_{c,i} = 0$, the algorithm is now complete, because from (7.51) and (7.58), together with the auxiliary equation (7.54), the temperature of each node can be calculated from (7.40).

When both radiative and conductive heat transfers are present, an iterative procedure is needed. One such procedure, which in spacecraft thermal calculations is in most cases sufficiently rapidly convergent, is to initially assume $q_{c,i} = 0$, then to calculate an initial set of temperatures that can be used to compute a new value for $q_{c,i}$, and then to repeat the step to obtain increasingly accurate results. (For two-sided surfaces, such as thin sheets or panels, rather than setting the conductance initially to zero, the temperatures of the two surfaces can usually be taken to be the same.) The inversions of the matrices need to be calculated only once.

In spacecraft, the majority of node pairs will have no direct radiation path connecting the two nodes of the pair. Hence a majority of view factors will be zero, resulting in *sparse matrices* for [M_h] and [M_a]. This greatly facilitates the numerical inversion and makes it practical to include in the calculation, if needed, several hundred nodes. Software that is both fast and robust is available for this purpose.

To summarize this section: The steady-state thermal transport by near-infrared and solar radiation and by conduction has been considered. For spacecraft in the vicinity of planets, where planetary thermal and albedo radiation can play a role, these effects can also be included in the analysis, the former by adding the planet as an additional node, the latter by adding terms to the solar radiation. The distinction between the absorptivity α_a for ambient temperature radiation and α_h for solar radiation must almost always be made.

7.9 Thermal Control Devices

As mentioned in Sect. 7.2, **passive means** of spacecraft temperature control can be obtained by selecting for the exterior surfaces of the spacecraft coatings or tapes that have the appropriate absorptivity and emissivity.

Surfaces that are interior to the spacecraft are also treated in this manner. Black paint is applied to electronic boxes and other devices when electric-to-thermal dissipation is significant. An example is the digital module shown in Fig. 7.14. Although some of the generated heat flows from the components to

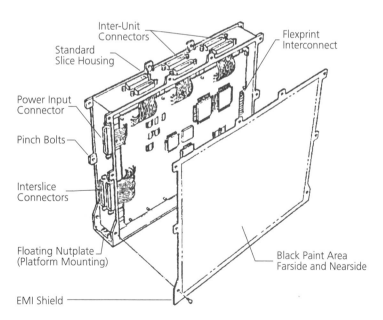

Figure 7.14 Digital module. The EMI (electromagnetic interference) shields are painted black to increase the radiative heat transfer. (From Ref. 8.)

the module covers and then to the mounting platform by conduction, additional heat transfer is obtained by radiation from the black paint (applied in this case to the covers that serve to suppress electromagnetic interference).

Ideally, thermal control of the spacecraft and of its components would be based on purely passive means, because they are reliable and require no electric power. However, variations in thermal dissipation in on versus off operation, variations in the thermal environment, degradation of the spacecraft exterior surfaces, and a very precisely maintained temperature need for all require **active** thermal control.

Electric heaters, controlled by solid-state devices, are commonly applied for this purpose. Frequently used heating elements are in the form of flexible patches consisting of a conductor in the shape of a pattern that fills the area of the patch, where the conductor is sandwiched between two sheets of Kapton or similar material. Conservative design of the heating elements is very important because otherwise they may fail over long periods of time. Redundancy is obtained by providing two independent circuits, often on the same patch.

Spacecraft storage batteries require particularly close thermal control. The requirements can vary greatly, depending on whether the battery is used at a low charge or discharge rate or at high ones. Often, both heating and cooling are needed to meet the varying conditions. Cooling can be obtained by mounting the battery such that one of its surfaces can radiate into space. As indicated in the example shown in Fig. 7.15, this surface is designed as a radiator with low solar absorptivity and high ambient temperature emissivity. In the case of satellites in equatorial or near-equatorial orbits, preferred mounting positions are the north and south panels of the spacecraft, where the solar radiation incidence and its variation are minimal. External heating is provided in this example by patch heaters.

Other electric heating elements have the form of standard cartridges that can be inserted and potted into a hole in the component. An example is the cartridge used for preheating the hydrazine catalyst bed of a thruster.

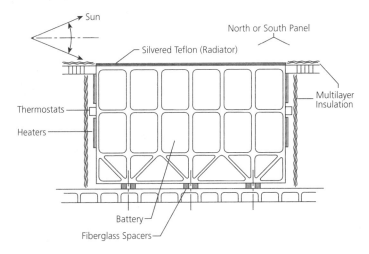

Figure 7.15 Thermal protection of a storage battery on a three-axis stabilized spacecraft.

7.9 Thermal Control Devices

Special forms of heaters have also been developed for wrapping around tubes, for example, for the purpose of preventing hydrazine from freezing and possibly rupturing the tube.

The heater current is controlled by means of a temperature-sensing element, usually a thermistor, and a solid-state controller or relay. A small dead band of the controller reduces the temperature swing of the component but also increases the number of on–off switch actions, which may reduce the reliability of the circuit. Dead bands of 5 to 10°C are common. Much more precise control, however, is needed for maintaining the alignment and surface figure of some optical systems. Thus, some components of the Hubble Space Telescope are controlled to better than 0.1°C.

There may be as many as a hundred heating circuits on a spacecraft. They operate autonomously, that is, normally without intervention by the controlling ground station. The temperature readings and controller status are periodically telemetered.

There is an advantage to be gained by controlling all heating circuits by the spacecraft's onboard computer, rather than by individual controllers. In case of an unforeseen event or failure, the computer can often be reprogrammed by ground command to circumvent the problem. Even so, autonomous temperature control, as the basic mode of operation, is essential because diagnosing a fault and reprogramming are time consuming. Sending the command may have its own difficulties or, in the case of deep-space probes, may suffer from long communications delays.

Louvers represent another type of active control. Typically they are activated by a bimetal spiral spring. Therefore, they require no electric power. Venetian vane-type louvers (Fig. 7.16) consist of rectangular blades that can be rotated into any position between fully open and fully closed. If open, solar radiation is admitted to the surface of the regulated component

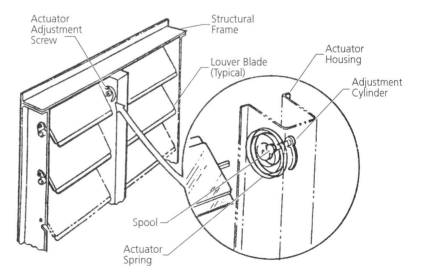

Figure 7.16 Bimetal, spring-actuated, rectangular blades ("venetian blind") louvers. Courtesy of LORAL Space & Communications Ltd. By permission.

or spacecraft skin underneath the louvers. If this surface is provided with a high ratio of solar radiation absorptivity to ambient temperature emissivity, control over a wide range of temperatures can be obtained. A ratio of about 6:1 in heat rejection from fully closed to fully open has been achieved. To augment the reliability, each vane is driven separately from the others by its own spring. Other actuators are constructed from bellows, which are sometimes filled with fluid.

Louvers with bimetal springs are accurate to better than 10°C. More precise control can be obtained by providing a temperature sensor, controller, and a small heater that is in contact with the bimetal spring. This arrangement also has desirable redundancy characteristics: If the heater should fail, the bimetal, mechanical actuator provides a backup, although with less accuracy.

Venetian blind louvers can also be used for the temperature control of a component on the side of a spacecraft that is not exposed to solar radiation. In this case, the positions of the louvers merely change the net thermal emission to space by the regulated component.

Pinwheel louvers (Fig. 7.17) are similar, but in place of rectangular blades incorporate vanes (typically four) in the form of sectors of a circle that together cover one-half of the area under the pinwheel. They are also rotated by bimetal actuators. Depending on the position of the sectors, they will cover (uncover) areas that have high ratios of solar absorptivity to emissivity and uncover (cover) areas that have a low such ratio.

Heat pipes are used on spacecraft and in terrestrial applications for thermal transport and thermal control. Figure 7.18 illustrates a space radiator based on heat pipe technology. A detailed discussion of their theory and design is beyond the scope of this text.

For design data on various types of thermal controls on spacecraft, the handbook by Gilmore [8] may be consulted.

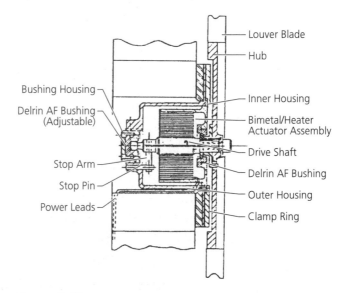

Figure 7.17 Bimetal actuated pinwheel assembly for thermal control. (From Gilmore [8].)

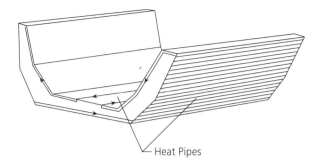

Figure 7.18 Space radiator with heat pipes.

Nomenclature

a	absorbed radiant power per unit area; a^2: thermal diffusivity [Eq. (7.38)]
a_g	earth albedo coefficient
c	specific heat
e	emitted radiant power per unit area
$f_{a,i}$, $f_{h,i}$	functions defined by Eqs. (7.56) and (7.49)
i	incident radiant power per unit area
j	intensity
k	thermal conductivity
m	mass
n	number of nodes
o	outward-directed radiant power per unit area
q	thermal flux per unit area; $q_{c,i}$: thermal flux added to ith node by conduction; $q_{p,i}$: thermal flux added to ith node by other sources (electric-to-thermal dissipation, combustion, or radioisotope sources)
r	radial coordinate
A	absorption per unit time
C_1, C_2, C_3	universal constants [Eqs. (7.3), (7.4)]
F_{ij}	view factor for nodes i, j [Eqs. (7.43) to (7.45$'$)]
$F_{h,i}$	view factor for solar radiation incident on node i
$[\mathbf{M}_a]$, $[\mathbf{M}_h]$	matrices defined by Eqs. (7.57a), (7.50a)
R	radius
S	surface
T	absolute temperature
α	absorptivity
β	angle formed by ray with surface normal
Γ_{ij}	conductance between nodes i and j [Eq. (7.41)]
δ_{ij}	Kronecker delta
ε	emissivity
λ	wavelength
σ	Stefan–Boltzmann constant $= 5.6693 \, 10^{-8}$ W/(m^{2}K^4)
τ	nondimensional characteristic time
ϕ	azimuthal angle

ω, Ω	solid angles
$(\ )_a$	ambient temperature condition
$(\ )_b$	blackbody
$(\ )_c$	conduction
$(\ )_g$	earth
$(\ )_h$	solar; $(\ )_{-h}$ eclipse condition
$(\ )_n$	normal
$(\ )_p$	power

Problems

(1) According to Greek mythology, Daedalus fashioned from feathers held by beeswax flying machines for himself and for his son Icarus. But Icarus soared too close to the sun. The wax melted and he fell to his death.

   Beeswax has a melting point of 65°C. Its absorptivity for solar radiation is about 0.3; its emissivity at the melting temperature is about 0.9. The wax may be represented by small spheres, in thermal equilibrium between incident solar radiation and the emission from the spheres.

   In terms of AUs (astronomical units), to what distance from the sun did Icarus soar before the wax melted?

(2) Consider a radioisotope source on a deep-space probe. The source is spherical, with a radius of 0.10 m. The power resulting from the radioactive decay is uniform across the volume and has a density of 0.80×10^6 W/m³. The thermal conductivity of the material is 5.00 W/(m K). The outer surface of the sphere is kept at a temperature of 1000 K. A steady-state condition can be assumed.

   Compute the temperature at the center of the sphere. (Solution: 1333 K.)

(3) A solar array (a flat panel) of a near-earth spacecraft has solar radiation incident on it at 66.5° before being eclipsed by the earth. The solar-to-electric conversion efficiency is 0.15, the solar radiation reflectivity 0.25. The emissivity of the active side of the array is 0.80, of the back side 0.70. The array is thin enough so that the temperatures of front and back can be assumed to be the same. The specific heat of the array, per unit area, is 9000 J/(m² K). Earth thermal and albedo radiation can be neglected. The solar radiation intensity is 1350 W/m². The preeclipse temperature of the array can be assumed to be equal to the equilibrium temperature in the sun.

   Compute the array's temperature at 1000 s after the spacecraft has entered the eclipse.

(4) Consider two adjoining rectangular flat panels, both of the same width and infinitely long in the third dimension.

   Express the view factor between the two panels as a function of the angle between them.

(5) Find an expression [analogous to Eq. (7.45), which is valid for *two-dimensional* configurations] to calculate view factors for *axisymmetric* configurations.

(6) Consider a spherical shell. One half of the shell is at temperature T_1 and has emissivity ε_1 and absorptivity α_1 on its inside surface. The other half

is at temperature T_2 and has emissivity ε_2 and absorptivity α_2 on the inside surface.

In terms of these quantities, express the net power transferred by radiation from the first to the second hemisphere.

(7) Consider a spacecraft consisting of a cylindrical thin shell and a payload inside the shell. The payload may be irregularly shaped, although assumed to be convex everywhere.

The shell's diameter is 1.00 m, its length 2.50 m. The thickness of the shell is negligible. The cylindrical portion of the spacecraft is covered with solar cells. Solar radiant power at 1350 W/m^2 is incident at a right angle to the cylinder axis. The spacecraft is in the so-called rotisserie mode, spinning fast enough so that the temperature of the shell (including, by conduction, the end caps) can be assumed spacially uniform and constant in time.

The solar cells' conversion efficiency is 17%. Their packing factor (which accounts for the gap between cells) is 95%. The reflectivity is 0.20, the emissivity 0.35. The electric power delivered by the solar cells is dissipated in the payload, which has an outer surface of 5.00 m^2 and is assumed to have a uniform temperature.

The shell and payload exchange thermal energy by radiation. Conduction is neglected. The inside of the shell and the outside of the payload are painted black, with emissivities and absorptivities of 0.90.

Compute the temperature of the shell and of the payload. (Solution: shell temperature = 281 K; payload temperature = 301 K).

References

1. Horton, T. E., ed., "Spacecraft Radiative Transfer and Temperature Control," *Progress in Astronautics and Aeronautics*, Vol. 83, American Institute of Aeronautics and Astronautics, Washington, DC, 1982.
2. Siegel, R. and Howell, J. R., "Thermal Radiation Heat Transfer," McGraw-Hill, New York, 1972.
3. Mills, A. F., "Heat Transfer," Irwin, Boston, 1992.
4. Agrawal, B. N., "Design of Geosynchronous Spacecraft," Prentice Hall, Englewood Cliffs, NJ, 1986.
5. Pisacane, V. L. and Moore, R. C., eds., "Fundamentals of Space Systems," Oxford University Press, New York, 1994.
6. Hallmann, W. and Ley, W., eds., "Handbuch der Raumfahrttechnik," Carl Hauser Verlag, Munich, 1988.
7. Carslaw, H. S. and Jaeger, J. C., "Conduction of Heat in Solids," 2nd. ed., Clarendon Press, Oxford, 1959.
8. Gilmore, D. G., ed., "Satellite Thermal Control Handbook," Institute of Aeronautics and Astronautics, Washington, DC, 1994.

A
Physical Constants Used in this Text

Recommended values 1986, E. R. Cohen and B. N. Taylor. The numbers in parantheses represent the dispersion of the last two digits of the mean value.

Velocity of light in vacuum	$c = 2.99792458 \times 10^8$ m s^{-1}
Universal gravitational constant	$G = 6.67259(85) \times 10^{-11}$ m^3 kg^{-1} s^{-2}
Planck's constant	$h = 6.6260755(40) \times 10^{-34}$ J s
Boltzmanns's constant	$k = 1.380658(12) \times 10^{-23}$ J K^{-1}
	$8.617385(73) \times 10^{-5}$ eV K^{-1}
Avogadro's number	$N = 6.0221367(36) \times 10^{26}$ kmol^{-1}
Universal gas constant	$R_0 = 8.314510(70) \times 10^3$ J kmol^{-1} K^{-1}
Stefan–Boltzmann constant	$\sigma = 5.67051(19) \times 10^{-8}$ W m^{-2} K^{-4}
Atomic mass constant (^{12}C/12)	$M_0 = 1.6605402(10) \times 10^{-27}$ kg
Proton rest mass	$M_p = 1.6726231(10) \times 10^{-27}$ kg
Electron rest mass	$m_e = 9.1093897(54) \times 10^{-31}$ kg
Electronic charge	$e = 1.60217733(49) \times 10^{-19}$ C

B
Astronomical Constants

Astronomical unit	$AU = 1.495979\ 10^8$ km
Light year	$ly = 9.46054\ 10^{12}$ km
Parsec	$pc = 3.08568\ 10^{13}$ km
Ephemeris day	$d_E = 86400$ s
Siderial day	$d_S = 86164.09055$ s $+$ $(0.0015\ T)$ s*
Mean solar day	$d_h = 86400$ s $+$ $(0.0015\ T)$ s*
Tropical year	$365.24219\ d_E$
Siderial year	$365.25637\ d_E$
Julian year	$365.25\ d_E$
Gregorian calendar year	$365.2425\ d_E$

Earth:
Mean equatorial radius	6378.140 km
Polar radius (based on spheroid)	6356.755 km
Mass	$5.976\ 10^{24}$ kg
Mean density	5.517 g cm^{-3}
Gravitational parameter	$3.986013\ 10^5$ km^3 s^{-2}
Obliquity of ecliptic	$23°27'8.26''-46.86''\ T$*
Gravitational acceleration at equator	9.8142 m s^{-2}
Equatorial rotational velocity	0.4651 km s^{-1}
Centrifugal acceleration at equator	0.0339 m s^{-2}
International standard gravity	9.80665 m s^{-2}
Escape velocity at equator	11.19 km s^{-1}
Sun–earth distance, minimum	$1.4710\ 10^8$ km
maximum	$1.5210\ 10^8$ km

Moon:
Mean radius	1738.2 km
Mass	$7.350\ 10^{22}$ kg
Mean density	3.341 g cm^{-3}
Gravitational parameter	$4.90265\ 10^3$ km^3 s^{-2}
Mean orbital inclination to ecliptic	$5°8'43''$
Mean equatorial inclination to ecliptic	$1°32'30''$
Rotation period	$27.32166\ d_E$
Gravitational acceleration at surface	1.62 m s^{-2}
Escape velocity at surface	2.38 km s^{-1}
Earth–moon distance, minimum	356 400 km
maximum	406 700 km

*T is in centuries from the year 1900.

Sun:

Mean radius	6.9599×10^5 km
Mass	1.990×10^{30} kg
Mean density	1.409 g cm^{-3}
Gravitational parameter	1.32712×10^{11} km^3 s^{-2}
Inclination of equator to ecliptic	7°15'
Rotation period (at 17° latitude)	25.38 days
Gravitational acceleration at surface	273.97 m s^{-2}
Centrifugal acceleration at surface	0.0057 m s^{-2}
Radiation emitted	3.826×10^{26} W
Solar constant	1360 W m^{-2}
Effective blackbody temperature	5760 K

Planets:

	Mercury	Venus	Earth	Mars	Jupiter	Saturn	Uranus	Neptune	Pluto
Max. distance from sun (AU)	0.467	0.728	1.017	1.666	5.452	10.081	19.997	30.340	48.94
Min. distance from sun (AU)	0.307	0.718	0.983	1.381	4.953	9.015	18.272	29.682	29.64
Siderial period (tropical years)	0.2408	0.6152	1.000	1.8809	11.861	29.50	83.70	164.79	246.28
Mean orbital velocity (km s^{-1})	47.87	35.02	29.78	24.13	13.06	9.64	6.81	5.44	4.75
Equatorial radius (km)	2432	6052	6378	3402	71490	60,000	25,400	24,765	3200
Mass (Earth = 1)	0.0558	0.8150	1.000	0.1074	317.89	95.179	14.629	17.222	0.111
Mean density (g cm^{-3})	5.53	5.25	5.52	3.93	1.36	0.71	1.31	1.66	4.83
Equatorial gravitational acceleration (m s^{-2})	3.76	8.87	9.81	3.73	25.7	10.8	8.95	11.0	4.3
Equatorial centrifugal acceleration (m s^{-2})	0.0	0.0	0.034	0.017	2.23	1.75	0.66	0.31	0.0
Siderial rotation period	59^d	243^d retrograde	23^{h}56^{m}4^s	24^{h}37^{m}23^s	9^{h}50^m	10^{h}14^m	10^{h}49^m retrograde	16^{h}07^m	6^{d}9^h retrograde
Equator's inclination to orbit	0°	177°	23°27'	23°59'	3°5'	26°44'	97°55'	28°48'	
Atmosphere, main components	Virtually none	Carbon dioxide	Nitrogen oxygen	Carbon dioxide	Hydrogen helium	Hydrogen helium	Helium hydrogen methane	Hydrogen helium methane	None detected

Heliocentric Osculating Elements of the Planets for 1996

The elements listed here are referred to the mean ecliptic and equinox at the Julian date 2000.0. First row under each planet is for the Julian date 2 450 120.5, the second row for 2 450 320.5. Values given for Earth are actually for the Earth–moon barycenter. Taken from The Astronomical Almanac for the Year 1996, U.S. Government Printing Office, Washington, DC and Her Majesty's Stationery Office, London.

Appendix B *Astronomical Constants*

Orbit inclination (deg)	Ecliptic longitude		Mean distance (AU)	Eccentricity	Mean angular velocity (deg/ sidereal day)
	Ascending node (deg)	Perihelion (deg)			
Mercury					
7.00519	48.3362	77.4503	0.3870985	0.2056337	4.092342
7.00513	48.3356	77.4537	0.3870975	0.2056430	4.092358
Venus					
3.39480	76.6916	131.470	0.7233257	0.0067574	1.602152
3.39477	76.6900	131.814	0.7233235	0.0067974	1.602159
Earth					
0.00051	345.4	102.8913	1.0000082	0.0167549	0.9855970
0.00047	346.4	102.8837	1.0000108	0.0167206	0.9855931
Mars					
1.84997	49.5709	336.0217	1.5236253	0.0932886	0.5240671
1.84991	49.5705	336.0365	1.5237131	0.0933046	0.5240218
Jupiter					
1.30461	100.4708	15.7513	5.202289	0.0484146	0.08310355
1.30460	100.4707	15.7225	5.202427	0.0484362	0.08310024
Saturn					
2.48534	113.6415	90.5489	9.555375	0.0523713	0.03337295
2.48525	113.6365	89.8160	9.561943	0.0525151	0.03333857
Uranus					
0.77337	74.0909	176.6862	19.29767	0.0444402	0.01162669
0.77345	74.0971	176.6634	19.30425	0.0437359	0.01162075
Neptune					
1.76969	131.7789	2.445	30.26605	0.0084404	0.005919454
1.76902	131.7851	2.167	30.27750	0.0092486	0.005916098
Pluto					
17.11942	110.3902	224.7317	39.77030	0.2538350	0.003929758
17.11829	110.3949	224.8114	39.71320	0.2526847	0.003938236

The Julian Date of an event is the number of days, and fractions thereof, elapsed after an origin of time set at 4713 B.C., January 1, Greenwich noon. In this reckoning, there is a leap year every 4 years (hence the designation "Julian"). A Julian day starts at Greenwich mean noon until the next noon. For example, the Julian date for January 1 in the year AD 2000 in the modern calendar, at 18 hours Universal Time is 2 451 545.25.

The values given in the table are for the osculating elements. These are parameters that specify the instantaneous position and velocity of a celestial body that the body would follow if perturbations were to cease instantaneously.

C

(a) Earth Atmosphere above 100 km Altitude. 1976 U.S. Standard Atmosphere

Altitude (km)	Temperature (K)	Pressure (N/m^2)	Density (kg/m^3)
100	195	3.20×10^{-2}	5.60×10^{-7}
110	240	7.10×10^{-3}	9.71×10^{-8}
120	360	2.54×10^{-3}	2.22×10^{-8}
130	469	1.25×10^{-3}	8.15×10^{-9}
140	560	7.20×10^{-4}	3.83×10^{-9}
150	634	4.54×10^{-4}	2.08×10^{-9}
160	696	3.04×10^{-4}	1.23×10^{-9}
170	748	2.12×10^{-4}	7.81×10^{-10}
180	790	1.53×10^{-4}	5.19×10^{-10}
190	825	1.13×10^{-4}	3.58×10^{-10}
200	855	8.47×10^{-5}	2.54×10^{-10}
210	879	6.48×10^{-5}	1.85×10^{-10}
220	899	5.01×10^{-5}	1.37×10^{-10}
230	916	3.93×10^{-5}	1.03×10^{-10}
240	930	3.11×10^{-5}	7.86×10^{-11}
250	941	2.48×10^{-5}	6.07×10^{-11}
260	951	1.99×10^{-5}	4.74×10^{-11}
270	959	1.61×10^{-5}	3.74×10^{-11}
280	966	1.31×10^{-5}	2.97×10^{-11}
290	971	1.07×10^{-5}	2.38×10^{-11}
300	976	8.77×10^{-6}	1.92×10^{-11}
310	980	7.23×10^{-6}	1.55×10^{-11}
320	983	5.98×10^{-6}	1.26×10^{-11}
330	986	4.96×10^{-6}	1.03×10^{-11}
340	988	4.13×10^{-6}	8.50×10^{-12}
350	990	3.45×10^{-6}	7.01×10^{-12}
360	992	2.89×10^{-6}	5.80×10^{-12}
370	993	2.43×10^{-6}	4.82×10^{-12}
380	994	2.04×10^{-6}	4.01×10^{-12}
390	995	1.72×10^{-6}	3.35×10^{-12}
400	996	1.45×10^{-6}	2.80×10^{-12}

(b) Mars Atmosphere: Nominal Models of Daily Mean at Midlatitude

		Northern summer		Southern summer	
z (km)	T (K)	p (N/m^2)	ϱ (kg/m^3)	p (N/m^2)	ϱ (kg/m^3)
0	214	636	1.56 10^{-2}	730	1.78 10^{-2}
2	214	530	1.30	608	1.49
4	213	441	1.08	507	1.24
6	212	368	9.07 10^{-3}	423	1.04
8	209	306	7.65	351	8.78 10^{-3}
10	205	254	6.47	291	7.42
12	201	210	5.45	241	6.25
14	198	173	4.57	198	5.24
16	195	142	3.81	163	4.37
18	191	116	3.17	133	3.64
20	188	94.7	2.63	109	3.02
22	185	77.0	2.18	88.4	2.50
24	183	62.5	1.79	71.7	2.05
26	180	50.6	1.47	58.1	1.69
28	178	40.8	1.20	46.8	1.38
30	175	32.8	9.84 10^{-4}	37.6	1.13
32	173	26.3	7.98	30.2	9.16 10^{-4}
34	170	21.1	6.48	24.2	7.44
36	168	16.8	5.24	19.3	6.01
38	165	13.3	4.23	15.3	4.86
40	162	10.6	3.40	12.2	3.90
42	160	8.33	2.72	9.56	3.12
44	158	6.56	2.17	7.53	2.49
46	156	5.15	1.73	5.91	1.99
48	154	4.03	1.37	4.63	1.57
50	152	3.15	1.08	3.62	1.24
60	144	8.79 10^{-1}	3.19 10^{-5}	1.01	3.66 10^{-5}
70	140	2.33	8.75 10^{-6}	2.67 10^{-1}	1.00
80	139	6.09 10^{-2}	2.29	6.99 10^{-2}	2.63 10^{-6}
90	139	1.60	6.03 10^{-7}	1.84	6.92 10^{-7}
100	139	4.24 10^{-3}	1.60	4.87 10^{-3}	1.84

Pressures are inferred from post-Viking Mission parachute descents. Altitude z is from height above reference ellipsoid. Reprinted from A. Seiff, in "Advances of Space Research," Vol. 2, No. 2. Copyright © 1982, with permission from Elsevier Science.

Mars Atmosphere: Principal Composition

Gas	Mole fraction
CO_2	0.9555 ± 0.0065
N_2	0.027 ± 0.003
Ar	0.016 ± 0.003
O_2	0.0015 ± 0.0005

Mean molecular weight $= 43.49 \pm 0.07$; gas constant $= 191.18 \pm 0.29$ J/kg K.

D

Properties of Selected Rocket Propellants

Sources: *Chemical Propulsion Information Agency (CPIA)* manuals, Johns Hopkins University, Baltimore, MD (continuing database); Sutton, G. P. and Ross, D. M., *Rocket Propulsion Elements*, 5th ed., John Wiley & Sons, New York, 1986.

(a) Liquid, Noncryogenic Propellants

Propellant	Chemical symbol	Use	Molecular weight	Freezing point (°C)	Boiling point at 1 atm (°C)	Vapor pressure (N/cm²)	Density (g/cm²)	Compatible materials
Nitrogen tetroxide	N_2O_4	Oxidizer	92.0	−12	21	77 at 70°C	1.44 at 20°C	Stainless steel; Al, Ni alloys; Teflon
IFRNA (inhibited fuming red nitric acid)	82% HNO_3 15% NO_2 2% H_2O 1% HF	Oxidizer, coolant	55.9	−49	66	11.9 at 70°C	1.57 at 20°C	Stainless steel; Al alloys; polyethylene
Bromine pentafluoride	BrF_5	Oxidizer, coolant	174.9	−62	40	28.3 at 70°C	2.48 at 20°C	Al alloys; 18–8 stainless steel; Teflon
Tetranitro methane	$C(NO_2)_4$	Oxidizer	196.0	14	126	1.64 at 75°C	1.64 at 20°C	Al alloys; mild steel; Teflon
Chlorine trifluoride	ClF_3	Oxidizer	92.5	−76	11.8	55 at 60°C	1.83 at 20°C	Al alloys; 18–8 stainless steel; Ni alloys; Teflon
RP-1 (rocket propellant)	Hydrocarbons	Fuel, coolant	165 to 195	−44 to −53	172 to 264	0.23 at 70°C	.80 to .82 at 20°C	Al, steel, Ni alloys; Teflon; neoprene
UDMH (unsym. dimethyl-hydrazine)	$(CH_3)_2NNH_2$	Fuel, coolant	60.8	−58	63	12.1 at 70°C	0.789 at 20°C	Al alloys; stainless steel; Teflon
92.5% ethyl alcohol	C_2H_5OH	Fuel, coolant	41.2	−123	78	8.95 at 70°C	0.81 at 15°C	Al, steel, Ni alloys; Teflon, polyethylene
JP-4 (jet propulsion fuel)	Hydrocarbons	Fuel, coolant	128	−60	130 to 240	4.95 at 70°C	0.75 to 0.82	Al, steel, Ni alloys; neoprene; Teflon
Pentaborane	B_5H_9	Fuel,	63.2	−47	60	13.1 at 70°C	0.61 at 20°C	Al alloys; steel; copper; Teflon; viton
Propyl nitrate	$C_5H_7NO_3$	Fuel, coolant	105.1	−91	111	2.55 at 70°C}	1.06 at 20°C	Al alloys; stainless steel; Teflon
Trimethyl amine	$(CH_3)_3N$	Fuel	59.1	−117	2.8	74.5 at 70°C	0.603 at 20°C	Al alloys; steel; copper; Teflon
95% hydrogen peroxide	H_2O_2	Monopropl. oxidizer, coolant	32.6	−5.6	146	0.035 at 25°C	1.414 at 25°C	Al alloys; stainless steel; Teflon
Hydrazine	N_2H_4	Monopropl., coolant	32.1	1.4	113	1.93 at 70°C	1.01 at 20°C	Al alloys; 304, 307 stainless steel; Teflon

Appendix D *Properties of Selected Rocket Propellants*

(b) Cryogenic Propellants

Propellant	Chemical symbol	Use	Molecular weight	Freezing point (°C)	Boiling point (°C)	Critical pressure (N/cm^2)	Critical temp. (°C)	Density at boiling point (g/cm^3)	Compatible materials
Liquid oxygen	O_2	Oxidizer	32.0	−219	−183	508	−119	1.142	Stainless steel; Al, Ni alloys; copper; Teflon
Oxygen difluoride	OF_2	Oxidizer	54.0		−184	496	−58	1.521	Al alloys; 300 series stainless steel; Ni alloys; brass
Liquid fluorine	F_2	Oxidizer	38.0	−220	−188	557	−129	1.509	Al alloys; 300 series stainless steel; Ni alloys; brass
Liquid hydrogen	H_2	Fuel, coolant	2.016	−259	−253	127	−240	0.071	Stainless steel; Ni alloys
Ammonia	NH_3	Fuel, coolant	17.03	−78	−33	1092	132	0.683	Al alloys; steel; Teflon

(c) Solid Propellant Oxidizers

Oxidizer	Chemical symbol	Molecular weight	Density (g/cm^3)	Oxygen content by mass
Ammonium perchlorate	NH_4ClO_4	117.5	1.95	34%
Potassium perchlorate	$KClO_4$	138.6	2.52	46%
Lithium perchlorate	$LiClO_4$	106.4	2.43	60%
Sodium perchlorate	$NaClO_4$	122.4	2.02	52%
Ammonium nitrate	NH_4NO_3	80.0	1.73	39%
Lithium nitrate	$LiNO_3$	68.9	2.38	58%
Sodium nitrate	$NaNO_3$	89.0	2.26	47%
Nitronium perchlorate	NO_2ClO_4	145.5	2.20	66%

(d) Solid Propellant Fuels

Fuel	Chemical symbol	Molecular weight	Melting point (°C)	Density (g/cm^3)	Typical I_{sp} for listed propellant combination, for expansion from 690 N/cm^2 to sea level pressure
Aluminum	Al	26.98	659	2.70	PU-AP-Al: 265 s
Beryllium	Be	9.01	1277	1.84	PU-AP-Be: 280 s
Boron	B	10.81	2304	2.30[1]	PU-AP-B: 255 s
Aluminum hydride	AlH_3	30.0	decomposes	1.42	PU-AP-AlH$_3$: 280 s
Beryllium hydride	BeH_2	11.03	decomposes	0.65[2]	PU-AP-BeH$_2$: 310 s

PU, polyurethane; AP, ammonium perchlorate.
[1] Crystalline; [2] compacted.

E

Thermal Properties of Selected Spacecraft Materials

Unless otherwise noted, the values are for 300 K

Material	Density (kg/m³)	Thermal conduct. W/(m K)	Specific heat J/(kg K)	Thermal expansion (10⁻⁶/K)
Aluminum 2014-T6 (4.5% Cu, 1.5% Mg, 0.6% Mn)	2770	177	875	73.0
at 600 K		65	473	
at 400 K		163	787	
at 200 K		186	925	
at 100 K		186	1042	
Beryllium alloy (extrusion)	1850	179	1862	11.5
Magnesium (extrusion, AZ31B)	1770	43.6	1046	25.2
Titanium 6Al-4V	4430	7.4	502	8.8
Constantan (55% Cu, 45% Ni)	8920	23	384	6.71
at 200 K		17	237	
at 100 K		19	362	
Iridium	22500	147	130	50.3
at 400 K		144	133	
at 800 K		132	142	
at 1500 K		111	172	
Carbon steel (Mn < 1%, Si < 0.1%)	7854	60.5	434	17.7
Carbon steel (1% < Mn < 1.65%, 0.1% < Si < 0.6%)	8131	41.0	434	11.6
Chromium–vanadium steel (0.2% C, 1.02% Cr, 0.15% V)	7836	48.9	443	14.1
Stainless steel AISI 302	8055	15.1	480	3.91
at 1000 K		17.3	512	
at 800 K		20.0	559	
at 600 K		22.8	585	
at 400 K		25.4	606	
Stainless steel AISI 304	7900	14.9	477	3.95
at 1500 K		9.2	272	
at 1200 K		12.6	402	
at 800 K		16.6	515	
at 400 K		22.6	582	

(*Continued*)

(Continued)

Material	Density (kg/m³)	Thermal conduct. W/(m K)	Specific heat J/(kg K)	Thermal expansion (10⁻⁶/K)
at 200 K		28.0	640	
at 100 K		31.7	682	
Nichrome (80% Ni, 20% Cr)	8400	12	420	3.4
at 800 K		14	480	
at 600 K		16	525	
at 400 K		21	545	
Inconel X-750	8510	11.7	439	3.1
at 1500 K		13.5	473	
at 1000 K		17.0	510	
at 600 K		24.0	626	
at 400 K		33.0		
Invar 36, annealed	8080	13.5	514	1.26
Tantalum	16600	57.5	140	24.7
at 2500 K		58.6	146	
at 1500 K		60.2	152	
at 1000 K		62.2	160	
at 600 K		65.6	189	
Tungsten	19300	174	132	68.3
at 600 K		137	142	
at 1000 K		118	148	
at 1500 K		107	157	
at 2500 K		95	176	
Boron fiber epoxy, 30% vol.	2080			
at 400 K			364	
parallel		2.10		
perpendicular		0.37		
at 300 K			757	
parallel		2.23		
perpendicular		0.49		
at 200 K			1122	
parallel		2.29		
perpendicular		0.59		
at 100 K			1431	
parallel		2.28		
perpendicular		0.60		
Graphite fiber epoxy, 25% vol.	1400			
at 400 K			337	
parallel		5.7		
perpendicular		0.46		
at 300 K			642	
parallel		8.7		
perpendicular		0.68		

(Continued)

(*Continued*)

Material	Density (kg/m³)	Thermal conduct. W/(m K)	Specific heat J/(kg K)	Thermal expansion (10^{-6}/K)
at 200 K			935	
parallel		11.1		
perpendicular		0.87		
at 100 K			1216	
parallel		13.0		
perpendicular		1.1		

F

Absorption and Emission Coefficients of Spacecraft Materials

Unless otherwise noted, the values are for surfaces not yet degraded by the space environment. Adapted from Gilmore, D. G., ed., "Satellite Thermal Control Handbook," American Institute of Aeronautics and Astronautics, Washington, DC, 1994.

Material	α_h solar	$\alpha_a \approx \varepsilon_a$ ambient temperature
Black coatings and plastics		
Carbon black paint NS-7	.96	.88
Black polyurethane paint Z306, 3 mil thick	.95	.87
after 3 years GEO	.93	.87
after 5 years GEO	.92	.87
CATALAC black paint	.96	.88
EBANOL carbon black	.97	.73
PALADIN black lacquer	.95	.75
3M Black Velvet paint	.97	.91
after 2.5 years GEO	.97	.84
TEDLAR black plastic	.94	.90
White coatings and plastics		
CATALAC white paint	.24	.90
Titanium oxide with methyl silicate	.20	.90
Titanium oxide with potassium silicate	.17	.92
Zinc oxide with sodium silicate	.15	.92
Barium sulphate with polyvinyl alcohol	.06	.88
TEDLAR white plastic	.39	.87
Anodized aluminum		
Black or blue anodized aluminum	.53–.67	.82–.87
Brown anodized aluminum	.73	.86
Chromic anodized aluminum	.44	.56
Gold anodized aluminum	.48	.82
Metals		
Buffed aluminum	.16	.03
Polished aluminum	.15	.05

(*Continued*)

(*Continued*)

Material	α_h solar	$\alpha_a \approx \epsilon_a$ ambient temperature
Heavily oxidized aluminum	.13	.30
Polished stainless steel	.42	.11
Stainless steel 1 mil foil	.40	.05
Inconel foil	.52	.10
Tantalum foil	.40	.05
Polished tungsten	.44	.03
Polished gold	.30	.05
same after 5 years GEO		
Electroplated gold	.23	.03
Composites		
Fiberglass epoxy	.72	.89
same after 5 years GEO		
Graphite epoxy	.93	.85
same after 5 years GEO		
Miscellaneous		
Indium oxide optical solar reflector	.07	.76
after 5 years GEO	.11	.76
Aluminized Kapton, first surface	.12	.03
Aluminized Kapton 0.5 mil, second surface	.35	.53
Mylar film 0.15 mil with aluminum backing	.14	.28
Aluminum tape	.04	.10

Index

absides, 193
absorption
   ambient temperature radiation, 278
   blackbody, 270
   solar radiation, 275
   technical surfaces, 275
acceleration, 7
   at equator, 2, 28
   residual, 2
   transformation equation, 8
accelerometers, 239
accommodation coefficient, 44
actuators, *see also* thrusters, 240
aerobraking, 37
aerodynamic forces and moments, 37
aerospike, 115, 121
albedo, 285
ammonium nitrate, 171
ammonium perchlorate, 171
anomaly
   eccentric, 65
   hyperbolic, 66
   mean, 65, 69
   true, 61, 70, 207
apoapsis, apogee, aphelion, 61, 190, 194, 203
ascending node, 69
ascent, 37
atmosphere
   atomic oxygen, 280
   density, 42, 52
   effects, 37, 39, 40, 73
   entry, *see* reentry
   ionosphere, 73
   Mars, 120
   rocket motor operation, 117
   scale height, 52
   skipping, 54
   temperature, 42, 52
   upper atmosphere wind, 39
atomic
   clocks, 15
   oxygen, 280
   time (TAI), 15, 17
attitude control, 215 to 265
azimuth, 4

back pressure, 117
barycenter, 6, 63
batteries, 300
bell nozzle, 115, 121
Bernoulli's equation, 152
beryllium, 171
bipropellants, 140, 167
blackbody, *see* thermal emission, thermal absorption
body cone, 227
Boltzmann, 272, 273
boosters, *see also* launch vehicles, 101
Bryson, A. E., 218
buffeting, 40
burn rate, 172

capillary feeds, 168
capture, 89
catalyst bed, 142, 300
cavitation, 151
colatitude, 4
collisions, 11
combustion
   chamber, 100
   instabilities, 156
composite propellants, 170
contamination, 123
Coriolis force, 9
cryogens, 140

D'Alembert, 66
Dalton's law, 128
debris, 280
declination, 6, 70
degree of reaction, 129, 132
De Laval, 115
desaturation, 246, 250
despun platform, 216
diaphragms, 146

docking, 24, 205, 208
Doppler shift, 74, 240
double base propellants, 170
drag, 37
drag free satellite, 46
dual spin, 258

earth
  albedo, 285
  diurnal rotation, 17, 65
  gravity, 26
  reference ellipsoid, 27
  thermal emission, 284
eccentricity, 62, 68
ecliptic, 3, 69, 284
  latitude, longitude, 6
  obliquity, 4, 284
eigenvectors, 219
Einstein, 92
electric
  propulsion, 99
  thrusters, 245
elevation, 4
emission, thermal
  blackbody, 270
  earth, 284
  solar, 273
  technical surfaces, 275
epoch, 14
equatorial plane, 69
  celestial, 3, 69, 284
equilibrium constant, 131
equinox, 3, 69, 284
Euler angles, 12, 226, 230, 239
Euler's equation for rigid bodies, 224, 263
  for time dependent inertia, 223
Evan's form, 256
exit plane, 100, 104
expansion-deflection nozzle, 115, 121
extendable nozzle, 115, 122

f and g series
film cooling, 149
first point of Aries, 3, 284
flow rate control, 162
flyby, 85
forces, *see also* gravity, thrust
  aerodynamic, 37
  buffeting, 40
  Coriolis, 9
  drag, lift, 37, 211
  gravitational, 21
  inertial, 9

Fourier, 289
free molecule flow, 40
  heat transfer, 47
  pressure, 45
  shear, 45
free radicals, 97
frozen equilibrium, 126, 131, 135
fuels, 141, 171
  hydrocarbons, 141
  hydrogen slush, 142
  liquid hydrogen, 141
  unsymmetric dimethylhydrazine, 141

Galilean invariance, 2
gas-dynamic shocks, 117, 119, 122
gas generators, 147, 149
gas turbines, 147
Gauss, 23, 27
geocentric reference, 5
geoid, 27
geostationary orbit, 77
Gibbs, 129, 131
gimbals, 160, 236
Global Positioning System, 16, 64
Goddard, 147
grain recession, 172
gravitational parameter, 22, 60
gravity, 21
  assist, 85
  gradient, 24, 216, 262
  loss, 35, 109
  turn, 209
Greenwich
  mean solar time (GMST), 16
  meridian, 4
grey surfaces, 278
ground track, 4
gyroscopes, 236
  integrating, 239
  rate, 238

heat pipes, 302
heat transfer, 47, 49, 127, 146, 149, 157, 269
heaters, 300
heliocentric reference, 6, 87
Hohmann transfers, 192, 202, 206
  bielliptic, 196
  modified, 195
  semitangential, 196
horizon, 71, 232
  sensors, 232
Huang, D. H., 99

Huzel, D. K., 99
hybrid
  motors, 103, 175
  propellants, 139
hydrazine, 140, 142, 165
hydrocarbons, 141
hydrogen, 141
  para- and ortho-, 97
  peroxide, 142
  slush, 142
hyperbolic anomaly, 66
hypergolic propellants, 140

igniter, 101, 156
impact parameter, 87
impulsive thrust, 188
inclination, 3, 68
inertia
  moments of, 218
  principal moments of, 221
  tensor, 25
inertial
  forces, 9
  measurement units, 239
  space, 2
inertially symmetric body, 221
  stability, 227
inhibited red fuming nitric acid, 138
injector plate, 99, 155
insertion errors, 199
  eccentricity, 202
  inclination, 199
  semimajor axis, 201
inverse square field, 23
ionosphere, 73

jet damping, 252

Kepler, 60
  equation, 65
  laws, 62
kinetic energy of rotation, 227
Kirchhoff's law, 276
Knudsen number, 41

Lagrange, 107
Lambert's
  law, 272, 284
  theorem, 184
Laplace, 27, 85
  invariable plane, 7
latitude, 4, 6
launch vehicles, *see also* boosters, 99, 104, 108, 117, 143, 209
  vibrations, 157
launch windows, 197
law of mass action, 130, 138
leap second, 16
liquid propellants, 99, 138, 140
longitude, 4, 6
louvers, 301
low earth orbits, 76
lumped parameter model, 290

magnetic
  levitation, 248
  torque, 216
Marman clamps, 104
mass
  action law, 130, 138
  fractions, 129
mean free path, 41
micrometeoroids, 280
microwave communications, 77
minimum energy paths, 181
mole fractions, 128
Molniya, 63
moments of inertia, 218, 221
momentum wheels, 249
monomethylhydrazine, 138
monopropellants, 140
  hydrazine, 142, 165
  hydrogen peroxide, 142
motor case, 175
multilayer insulation, 286

Newton, 2, 21, 23, 33, 62
nitric acid, 141
nitrogen tetroxide, 140
north-south correction, 205
nozzle, 100, 117, 149
  aerospike, 115, 121
  bell, 115, 121
  divergence angle, 125
  exit plane, 100, 104
  expansion-deflection, 115, 121
  extendible, 115, 122
  over-expansion, 119, 123
  throat, 100, 104, 113
  under-expansion, 119, 123
nutation, 229
  angle, 13, 226
  damping, 255, 258

oblate vehicles, 225, 255
orbits
  angular momentum, 60

Orbits (*contd.*)
   apoapsis, 61
   ascending node, 69
   circular, 75
   eccentricity, 62
   elements, 68
   energy, 60
   geostationary, 77
   inclination, 68
   injection errors, 199
   Kepler, 60
   low earth, 76
   on orbit drift, 196, 202
   periapsis, 61
   period, 64
   right ascension of the ascending node, 69
   semimajor axis, 63
   sun synchronous, 80
oxidizers, 140, 171
   liquid oxygen, 140
   nitric acid, 141
   nitrogen tetroxide, 140
   red fuming nitric acid, 141
   white fuming nitric acid, 141

paints, 279
panels, 286
partial pressure, 128
periapsis, perigee, perihelion, 61, 69, 93, 190, 194, 203
Planck's radiation law, 272
plane change, 188
planets
   flyby, 85
   missions to, 81, 187
planetocentric reference, 4, 86
pogo oscillation, 158
Prandtl-Meyer function, 124
precession, 226
   angle, 12
pressure thrust, 32
principal axes, 218
prolate vehicles, 225, 256
propellants, *see also* fuels, oxidizers
   binders, 171
   bipropellants, 140
   composite, 170
   cryogenic, 140
   double base, 170
   feed systems, 147
   flow rate control, 162
   hybrid, 139
   hypergolic, 140
   monopropellants, 142
   pumps, 148, 151
   solid, 139
   tanks, 143

range rate, 73
reaction control motors, *see* thrusters
reaction wheels, 245
   desaturation of, 246, 250
   magnetically levitated, 248
reentry, 49
reference frames
   accelerated, 7
   barycentric, 6, 63
   geocentric, 5
   heliocentric, 6, 87
   pitch, roll, yaw, 230, 235
   topocentric, 4
reflection of thermal radiation by technical surfaces, 275
relativistic effects, 91
rendez-vous maneuver, 205
right ascension, 6, 69
rocket motors, 99, 111, 135
   see also thrust, nozzle
   case, 175
   characteristic velocity, 106
   combustion instabilities, 156
   controls, 162
   gimbals, 160
   hybrid, 103, 175
   idealized model, 111
   igniter, 101, 156
   injector, 99, 155
   liquid propellant, 99
   operation in the atmosphere, 117
   regenerative cooling, 149
   shut-down, 155, 164
   slag, 32
   solid propellant, 101, 170
   start-up, 155, 164
   throttling, 155
   thrust chamber, 100, 155
   thrust vector control, 160, 166
Ross, D. M., 99

satellites, *see* spacecraft
Schmitt trigger, 242
second
   leap, 16
   Système International (SI), 15
second surface mirrors, 279
sensors, 230

horizon, 232
star, 230
sun, 231
shifting equilibrium, 126, 131, 137
sidereal time, 17
sodium perchlorate, 171
solar
  constant, 274
  photosphere, 274
  radiation pressure, 21, 47, 216
  sailing, 21, 47
  thermal radiation, 273
  time, 16
  ultraviolet, 280
solstice, 284
sounding rocket, 34
space cone, 227
spacecraft
  actuators, 240
  despun platform, 216
  dual spin, 258
  gravity gradient stabilized, 262
  main motors, 165
  observations, 73
  position as a function of time, 65
  sensors, 230
  spin stabilized, 216, 251
  three-axis stabilized, 217, 232
  thrusters, 165, 229
  tumbling, 14
  visibility above horizon, 71
Space Shuttle, 46, 161, 174
specific impulse, 33, 99, 137, 139
sphere of influence, 84, 86
spin, 227
squibs, 174
stages, 103, 191
  residual mass, 105
  separation, 104
Stefan-Boltzmann law, 273
stoichiometric coefficients, 130
strap-on motors, *see* boosters
Summerfield criterion, 123
Sutton, G. P., 99
synoptic period, 207

tesseral harmonics, 29
tether, 24
thermal
  absorption, 269
  blankets, 286
  conduction, 289
  control devices, 299
  diffusivity, 289
  emission, 269
thrust, 31, 99, 109
  control, 162
  ideal, 116
  in the atmosphere, 117
  impulsive, 188
  pressure thrust, 32, 116
  shut-down, 155, 164
  start-up, 155, 164
  vacuum thrust, 99, 119
  vector control, 160
  velocity thrust, 32, 116
thrusters, 127, 217, 241
  control, 242
  dead band, 242
  electric, 245
time
  atomic (TAI), 15, 17
  derivative in different reference systems, 7
  Greenwich mean solar time (GMST), 16
  sidereal, 17
  solar, 16
  standard, 16
  universal (UT), 16
torques
  aerodynamic, 37, 216
  gravitational, 24, 216
  magnetic, 216
  solar radiation, 216
Tsiolkovsky, 34
turbo-pumps, 147

ullage, 144
universal gravitational constant, 22
unsymmetric dimethylhydrazine, 138

Van Allen belts, 79
Vertregt, M., 107
vibration modes, 126
view factors, 293

wake flow, 121
Wien's displacement law, 273

yo-yo mechanism, 10

zonal harmonics, 29, 82